DECOMPOSITION OF SUPERPOSITIONS
OF DENSITY FUNCTIONS AND DISCRETE DISTRIBUTIONS

Decomposition
of Superpositions of Density Functions
and Discrete Distributions

PÁL MEDGYESSY

A HALSTED PRESS BOOK

JOHN WILEY & SONS
NEW YORK

© Akadémiai Kiadó, Budapest 1977

Published in the U.S.A. by Halsted Press,
a Division of John Wiley & Sons, Inc.,
New York

Library of Congress Cataloging in Publication Data

Medgyessy, Pál.
Decomposition of superpositions of density functions
and discrete distributions.

Bibliography: p.
Includes indexes.
1. Distribution (Probability theory) 2. Decomposition (Mathematics)
3. Numerical analysis. I. Title. II. Title: Superpositions of density functions
and discrete distributions.

QA273.6.M43 1976   519.2′4   75—46231

ISBN 0-470-15017-3

Published as a co-edition of
AKADÉMIAI KIADÓ, Budapest
and
ADAM HILGER LTD
and
HALSTED PRESS

Printed in Hungary

To the memory of my father
György Medgyessy (1889–1972),
my mother
Jolán Medgyessy, née Varga (1890–1946)
and my brother Gábor (1923–194?)

To my wife
ÉVA MEDGYESSY, née KORMOS,
and
ÉVA, PÁL, FERENC, LÁSZLÓ,
who for three years
contributed to me so many hours
which should have been
theirs

CONTENTS

FOREWORD

In 1922, Nikhilranjan Sen published a short article in the *Physikalische Zeitschrift*. The article stated that if one knew the graph of the intensity distribution of the H_α line of the hydrogen spectrum, which could be considered as the graph of a superposition of two Gaussian functions (two *components*), the data concerning the fine structure of the H_α line could be derived more accurately than was possible previously if, using *only* the ordinate values of the graph, one could construct the superposition of two Gaussian curves having the same modes as the two components. but being *narrower* than they. The data required would then be found from the modes and other characteristics of the components of this latter superposition.

N. Sen, relying on the physicist A. Sommerfeld's remarks, constructed the required superposition. By doing this he carried out, for the first time, the *decomposition* of a superposition of two density functions (more precisely: two normal density functions). However, he did not realize the significant fact that his procedure was *independent* of the number of components. This was pointed out only by G. Doetsch in 1928 when he made N. Sen's procedure exact. It was in this way that the field of problems of the *decomposition of superpositions* of density functions (and later, of discrete distributions) was discovered some 50 years ago. The main problem was as follows:

Given the graph of the superposition of an *unknown* number of components (of unimodal density functions or discrete distributions) — say N — of given type, how is it possible to determine N and, eventually, obtain approximate values of some of the unknown parameters of the superposition?

Our book is an attempt to put into perspective the sphere of problems on the decomposition of superpositions. The matter has in the last years become crucial from the viewpoint of practice. Problems of this nature arise in physics, e.g. in analysing spectra of different origin in spectroscopy or in nuclear physics; in biology, mathematical statistics, system identification and in several other scientific fields.

With these needs in mind and in the knowledge of the relevant investigations of the last fifty years, I felt some justification, following a short summary in the form of a dissertation, in publishing my little monograph *"Decomposition of super-positions of distribution functions"* (Akadémiai Kiadó, Budapest, 1961). This was the first systematic elaboration dealing with such problems. Although there was a need for it at that time, the date of publication was unfortunate: several problems that needed to be treated numerically were *incorrect* (ill-posed) and their treatment had not been developed at that time. Thus my previous book only dealt briefly with the field. As to the analytical tools then used, these have since been shown to be capable of improvement (e.g. the characterization of the narrowness of graphs). Moreover, the treatment in my book was excessively general. However, there was possibly the merit of its being the first survey of a new discipline as well as containing several new results interesting in themselves.

A systematic investigation of the handling of the incorrect problems was started in 1962; and in the course of time, I also found newer definitions and methods in connection with the decomposition problems. The whole discipline has taken a more coherent form, its investigation has been stabilized. Thus it seemed both justifiable and necessary to publish a newer summary.

The present book is not a revised — or enlarged — edition of my earlier work. Naturally, the fundamental problem and certain results are the same, although restricted here to a smaller area: to superpositions of density functions and of discrete distributions. As to the rest, however, this work is *thoroughly new*, and it preserves only certain important aspects.

To avoid possible misunderstandings, it should be emphasized here that the *fundamental scope of the present book belongs, essentially, to numerical analysis* (and *not* e.g. to mathematical statistics); in spite of this *the majority of tools used come from probability theory.*

The book consists of five chapters, divided and subdivided into paragraphs, sections and subsections. The numbering of the paragraphs, sections and subsections recommences from chapter to chapter. A Postscript deals with possible future research.

Remarks, unanswered questions, etc. remaining in the course of the treatment are collected at the end of each paragraph under the title *Supplements and problems*. These may also point out the future trends for research, but *they may be left out at the first reading.*

A short synopsis of the chapters follows.

Chapter I is an introduction. It provides the basic concepts, formulates the fundamental problems the solution of which will be called *decomposition*.

Chapter II summarizes the mathematical tools employed in the book; they originate mostly with the author. These are mainly in connection with the following topics which are of course of fundamental importance throughout: uni-modality of density functions and discrete distributions (§§ 1, 2); a new characterization of the shape (of the "narrowness") of the graphs of density functions and discrete distributions: it is felt that this is dealt with more obviously and better than previously, although its analytical treatment is possibly more difficult (§§ 4, 5). All this is grouped around the concept of the unimodality-preserving, narrowness-increasing transformations of the above-mentioned density functions and discrete distributions, which are the basis of the methods to be presented in §§ 6, 7. It should be mentioned that the characterization of the shape is a somewhat difficult problem, especially in the case of discrete distributions; and it is unlikely that the present results complete the investigations.

Paragraphs 1, 2, 4 and 5 are, at the same time, *systematic (though not definitive) summaries* of their subjects, too. *These as well as the rest of the chapter may have some interest also for specialists in probability theory.*

Chapter III is devoted to the decomposition of superpositions of density functions, which is solved, essentially, by means of a single but efficient idea, previously introduced in Chapter II: the application of unimodal transforms of increased narrowness. The realization of this provides two methods by which a great number of superpositions can be decomposed. Historically the first — and by now classical — decomposition problem: *the decomposition of superpositions of normal density functions,* first investigated by N. Sen in 1922 and later called "Gauss-Analysis" by G. Doetsch, appears here as one particular case within a group of problems.

Only such methods that can be realized numerically have been considered. Thus several former procedures analytically elegant in themselves but useless in practice had to be avoided. Even so, the methods of solution represent, in many instances, an *incorrect problem.* Although they can already be handled, *it seemed useful to work out and stress methods which imply the solution of a correct problem.* Therefore this chapter mainly deals with such methods and they should be regarded as the main point of the chapter. It is with them that future research should — in my opinion — be concerned.

This chapter also includes — within the framework of the investigation of a particular type of superposition — *the decomposition of a superposition of exponential density functions,* which is more difficult than that of a superposition of normal density functions. This seems, at present, to be an important problem in several fields of experimental science. Unfortunately, no definite solution can be given, although certain special methods are provided or, rather, proposed.

Chapter IV deals with the decomposition of superpositions of discrete distributions. Here newer methods have been constructed with the aid of the above-mentioned fundamental idea. These methods differ essentially from those in Chapter III only in the discrete character of the superpositions to be decomposed. In general, the investigated types have an analogue in Chapter III. From the viewpoint of numerical analysis the situation is easier here; in the most important cases incorrect problems are not touched upon and every theoretically good method can be proved in practice; however, despite the fortunate basic idea the value of the methods is invalidated as it is far more difficult to follow the course of the graphs revealing the unknown parameters in the application of the methods due to their discrete character, than in the continuous case.

Naturally other and earlier methods of decomposition exist besides those which utilize the above-mentioned fundamental idea. These are summarized briefly in a separate paragraph under the name "ad hoc" methods both in Chapters III and IV; nowadays they are mostly of historical interest only.

Chapter V surveys those *numerical methods* which have been found most useful in decomposition problems; but it should be emphasized here that *the construction of numerical methods is not the main objective of the book*. My fundamental principle, however, has been to elaborate only those methods that can be adequately used numerically.

Because of the incorrect problems in the book — e.g. *the solution of integral equations of the* Ist *kind* — the most important part of this chapter deals with the *numerical treatment of incorrect problems*. It was intended to give a short but comprehensive *survey* of this question, which may also be regarded as new. In this survey, among other methods described, the so-called *regularization method* invented by A. N. Tihonov in 1963 is given in detail. This proved to be very important in this field. *It is certain that this survey will be of some use not only for mathematicians concerned with decomposition problems, but for all dealing with the numerical solution of operator equations (among others, integral equations) of the* Ist *kind*.

In this chapter there are several methods of mine that are different from the others and which have proved useful in certain special cases.

Readers interested chiefly in the numerical side of decomposition methods may study Chapter V prior to Chapters III and IV.

The *Postscript* summarizes, essentially, the tasks of future research, with regard to the main problems to be solved, as crystallized in the former chapters.

The whole discipline — I emphasize — is not closed yet, and although the time has long been ripe for a summary, *the aim of my book is to present an objective summary of the existing situation with possible simplifications, preferring concrete cases to over-worked generalizations*.

The reading of the book requires only few preliminary notions of probability theory and numerical analysis. Less-known concepts are always defined.

The numbering of the formulas recommences in each paragraph, section, and subsection. If reference is made, in a paragraph, section etc., to a formula in another place of the same paragraph, section, etc., then only its number is given.

The numbering of Theorems, Lemmas, figures etc. together with any references is arranged similarly.

The greatest possible number of figures is provided to enable a quick understanding of the text. To the most important cases, numerical examples abundantly accompanied by figures, taken from practice have been attached.

At the end of the book, *References* list the works referred to, by means of the author's name and the year of publication in the text. They have been closed with 1970. If several works of an author appeared in the same year, they are distinguished by italic letters after the year. Results without any reference are all due to the present author.

Certain items from the *References* section were also collected to yield a *Chronological bibliography* which is, evidently, not complete; it contains only the more important literature referring exclusively to problems of decomposition in a chronological order. This arrangement is intended to give *a historical picture of the evolution of the theme* as well. The items in this bibliography are *not* referred to separately in the text.

Beside the usual *Author Index* and *Subject Index,* an *Index of unsolved problems* is also given. The *Index of notations* contains also all notations not mentioned separately in the text.

Acknowledgements

First of all I wish to express my sincere thanks to the Mathematical Committee of the Hungarian Academy of Sciences for supporting the publication of my book.

I commemorate here with gratitude Prof. A. Rényi (1921—1970), my Master, who always took great interest in my studies, until his untimely death; my early-deceased friend Prof. T. Szele (1918—1955), who encouraged me in my first investigations; Prof. T. Szentmártony (1895—1965), who first drew my attention to the early literature of the theme.

I am most indebted to Prof. E. Breitenberger (Athens, Ohio) for encouragement, suggestions and the collection of bibliographical data; to Mr. L. Varga for his advice and invaluable help in numerical work; to my critics Mr. P. Bártfai and Mr. K. Sarkadi for their work and helpful comments; to the reviewers of my earlier book for their remarks.

I thank Messrs S. Gacsályi, B. Gyíres, I. Körmendi, E. Makai, my cousin Gy. Medgyessy (1923—1969), Messrs J. Merza, J. Myhill (Canberra), Sz. Rochlitz and L. Vekerdi for valuable discussions and their interest; Messrs F. Berencz, G. Békési, O. Dobozy, P. Kosik and T. Volly for their help in the construction of certain figures.

My special thanks are due to Mrs. G. Palotai, Mrs. L. Ujhelyi and Mrs. É. Várnai for their efficient typing as well as to Mrs. R. Kosik and Mr. E. Makai, Jr. for their assistance in the preparation of the figures and the manuscript.

I acknowledge the work of the Akadémiai Kiadó (the Publishing House of the Hungarian Academy of Sciences) and that of the Szegedi Nyomda.

Pál Medgyessy

P. S. At the proof-reading certain new items from 1971 to 10 July, 1975 have been inserted in the *References*; they have been referred to in the text, too.

My heartfelt thanks are due to Mrs. Gyöngyi M.-Belea for her invaluable help in the proof-reading.

P. M.

I. INTRODUCTION

§ 1. SUPERPOSITIONS OF DENSITY FUNCTIONS AND DISCRETE DISTRIBUTIONS.
THE PROBLEM OF THE DECOMPOSITION OF SUPERPOSITIONS

In experimental science the following situation often arises:

A system, which may be physical or biological, etc., produces some "signal" as output. This signal can be represented by

(a) the graph of a continuous function;
(b) a sequence of discrete numerical data.

Frequently the signal is represented — or is suspected to be represented — by the graph of the function

$$k(x) = \sum_{k=1}^{N} p_k f(x, \alpha_k, \beta_k),$$

where $f(x, \alpha, \beta)$ is a two-parameter density function of *known* analytical form, $\alpha \in A$, $\beta \in B$ where A, B are given; α_k, β_k are fixed values of α and β respectively, and there are no identical pairs (α_k, β_k); p_k ($p_k > 0$) are positive parameters,

Without infringing the generality *we can suppose in the following that* $\beta_1 \leq \beta_2 \leq ... \leq \beta_N$.

The function $k(x)$ is called the **superposition** of the **components** $f(x, \alpha_k, \beta_k)$ formed with the **weights** p_k. $k(x)$ is a **superposition of density functions**.

In other cases the signal is — or is suspected to be — a set of numbers

$$\{k_n\} = \left\{ \sum_{k=1}^{N} p_k f_n(\gamma_k, \delta_k) \right\} \qquad (n = 0, 1, ...)$$

where $\{f_n(\gamma_k, \delta_k)\}$ denotes a two-parameter discrete distribution of *known* analytical form and γ, δ, γ_k, δ_k, p_k are defined in an analogous way to $\alpha, \beta, \alpha_k, \beta_k, p_k$.

The set $\{k_n\}$ is called the **superposition** of the **components** $\{f_n(\gamma_k, \delta_k)\}$, formed with the weights p_k. $\{k_n\}$ is a **superposition of discrete distributions**.

In practice it is found that $f(x, \alpha, \beta)$ is, in general, *unimodal* and often also symmetrical, and $\{f_n(\gamma, \delta)\}$ is *unimodal*. (See Definition 2 in Ch. II, § 1 and Definition 1 in Ch. II, § 2.)

In the investigation of the above system (physical, biological, etc.) the following problems frequently occur:

Problem 1. The *measured* ordinate values, i.e. numerical data of the graph of $k(x)$, are given for some x-values. The parameters N, α_k, β_k, p_k are *unknown*. On the basis of the *measured* ordinate values we have to determine:

N, the number of components and, eventually, *approximations* to the values of the parameters p_k, α_k, β_k (or of a part of these parameters).

The *numerical* procedure providing this is called the (numerical) **decomposition** of the superposition $k(x)$.

For the sake of uniformity let the set of the points (n, k_n) be called the *graph of* $\{k_n\}$.

Problem 2. The *measured* ordinate values, i.e. numerical data of the graph of $\{k_n\}$, are given for some n-values. The parameters N, γ_k, δ_k, p_k are *unknown*. On the basis of the *measured* ordinate values we have to determine:

N, the number of components and, eventually, *approximations* to the values of the parameters p_k, γ_k, δ_k (or of a part of these parameters).

The numerical procedure providing this is called the (numerical) **decomposition** of the superposition $\{k_n\}$.

We speak also of the **decomposition of a superposition of density functions** as well as of the **decomposition of a superposition of discrete distributions.**

The systematic investigation of Problems 1 and 2 is the subject of the present book.

As *we do not begin with statistical samples* and we do not consider the graphs as graphs of sample functions of stochastic processes (although the errors of the data, the "noise" might justify this viewpoint), *our problems essentially belong to the field of numerical analysis.* However, their treatment needs certain analytical tools from *probability theory.* Naturally a histogram or a diagram of relative frequencies constructed from a statistical sample may also be considered as the graph of some superposition $k(x)$ or $\{k_n\}$; their investigation *will not be based, however, on the sample.*

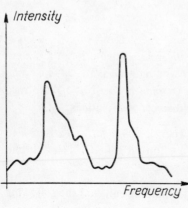

I. § 1. Figure 1

Example 1. *Investigation of absorption spectra.* Figure 1 shows the intensity distribution of the photograph of the absorption spectrum of an Fe (iron) arc in a fixed interval (Berencz 1955b). In spectroscopy the intensity distribution curve of a line of an absorption spectrum is usually considered for theoretical reasons as the graph of a normal or Cauchy density function. Then the curve of the intensity distribution will be the graph of a superposition of normal or Cauchy density functions, in which the places of the maxima of the components are the places of the different spectral lines and the areas under the components are proportional to the squares of the intensities of the lines. It often occurs that components belonging to lines situated near to each other do not appear as separated peaks in the graph (the corresponding Gaussian or Cauchy curves make each other indistinct). It is desirable to determine the *number* and the places of the lines that exist in reality on the basis of the graph. Clearly this problem is identical with that of the decomposition of a superposition of normal or Cauchy density functions.

Example 2. Separation of proteins by electrophoresis. Figure 2 shows a section of a diagram obtained in the *electrophoretic separation of the proteins of different molecular weight in human blood serum*, taken by the apparatus of A. Tiselius;

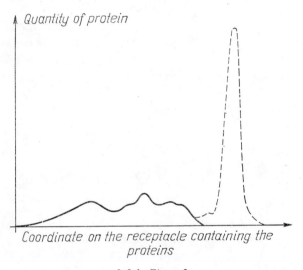

I. § 1. Figure 2

an omitted part is shown by a dashed line (Medgyessy 1953, 1954c). (Naturally proteins of other origin can also be investigated by this method.) In the relevant literature it is shown that this diagram is the superposition of Gaussian functions and each component corresponds to a protein fraction in the serum. The number, the percentage proportions and other characteristics of the "component" proteins can be read off from them; all these are of great *diagnostic and pharmaceutical* importance (production of γ-globulin). The determination of these — generally unknown — parameters evidently requires the decomposition of a superposition of normal density functions if the components blur each other.

Example 3. *Investigations by the aid of the method of counter current distribution in chemistry.* In applying the method of *"counter current distribution"* in chemical technology a mixture of chemically immiscible solutes (components) is brought into the solvents situated in cells of indices $0, 1, 2, ..., N$ of the experimental apparatus; each cell contains all solutes. The quantity of the different components in the sequence of cells varies according to the members of a binomial distribution. Thus the quantity of the material present in the solvent varies from cell to cell as do the members of a superposition of binomial distributions. This is shown in a concrete case in Fig. 3 (Medgyessy 1954a). If the number and the chemical characteristics of the components are unknown, an attempt may be made to determine them from the graph of the quantities of matter found in the different cells, i.e. from the graph of a superposition of binomial distributions. Clearly this problem is equivalent to the decomposition of this superposition of binomial distributions.

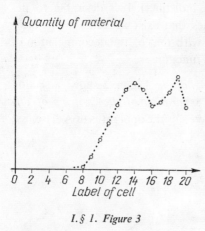

I. § 1. Figure 3

These three questions will be discussed in Chapters IV and V.

Supplements and problems to Ch. I, § 1

1. It is clear that except for a constant factor, $k(x)$ and $\{k_n\}$ are mixtures of density functions and discrete distributions respectively with respect to a discrete distribution.

2. The *superposition of distribution functions* can be defined in full analogy with the above definitions of superpositions. In Medgyessy's book (1961a) it played the central role. The present treatment, however, seems to be more advantageous.

3. Synonyms for "decomposition" are: "system identification", more precisely: "unscrambling" (Bellmann, Kagiwada, Kalaba 1965); "raffinage" (Rosenstiehl, Ghouila-Houri 1960); "analyse d'un mélange" (Thionet 1966).

4. Decomposition problems connected with the investigation of absorption spectra were described e.g. by Doetsch (1928); Sticker (1930a), (1930b); Schellenberg (1932); Szigeti (1947); Kiss, Sándorfy (1948); Berencz (1955a), (1955b); Noble, Hayes, Jr., Eden (1959).

5. From the literature concerning the electrophoretical separation of proteins mention can be made of Svedberg, Pedersen (1940) p. 265; Wiedemann (1947); Wallner, Ulke (1952); Medgyessy (1953), (1954c); Wuhrmann, Wunderly (1957) III. B. To all this a more detailed description is given e.g. in Medgyessy (1953), (1954b), (1955a), (1955b), (1955c), (1957), (1961a) pp. 18—21.

6. The decomposition of superpositions of density functions plays a role not only in spectroscopy but in mathematical statistics, mathematical economics, biochemistry, etc.; see e.g. Harding (1949); Cassie (1954); Daeves, Beckel (1958); Carnahan (1964); Inouye (1964); Varga (1966); Bhattacharya (1967); Larson, Kenneth (1967); Meszéna (1968); Varga (1968); Židkov, Ščedrin, Rambidi, Egorova (1968); Gregor (1969).

7. Literature of the method of counter current distribution in chemical technology is given e.g. by Jantzen (1932); Craig (1944); Stene (1945); Craig (1951). A survey or supplement can be found, in addition, in Medgyessy (1954a), (1954c); Medgyessy, Rényi, Tettamanti, Vincze (1954); Medgyessy (1961a) pp. 22—23.

8. Besides chemical technology the decomposition of superpositions of discrete distributions also plays a role in mathematical statistics as well as in other sciences; see e.g. McPhee (1953); Thionet (1966).

II. SOME MATHEMATICAL TOOLS

§ 1. UNIMODAL DENSITY FUNCTIONS

This paragraph summarizes results concerning unimodal density functions to be used later. Their unimodality can be most easily defined by means of the corresponding distribution function.

Definition 1. A distribution function $F(x)$ is called **unimodal** at the point $x=a$ (the *mode* of the distribution function) — in short: (a) **unimodal** — if $F(x)$ is convex in the interval $(-\infty, a)$ and is concave in (a, ∞).

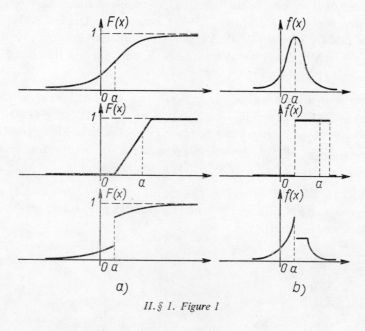

II.§ 1. Figure 1

If instead of being convex (concave) it is strictly convex (concave), then in the definition it is said to be **strictly** (a) **unimodal** (Hinčin 1938; Gnedenko, Kolmogorov 1954 § 32).

$F(x)$ may be discontinuous at the point $x=a$ only.

Figure 1a shows possible types of graphs of unimodal distribution functions.

Definition 2. A density function $f(x)$ is called **(strictly)** (a) **unimodal** at the point $x=a$ (the *mode*) if the corresponding distribution function $F(x)$ is (strictly) (a) unimodal.

Then $f(x)$ is monotone non-decreasing (increasing) in $(-\infty, a)$ and monotone non-increasing (decreasing) in (a, ∞). At $x=a$ (the mode) $f(x)$ is not defined in every case (Feller 1966 p. 115).

Figure 1b shows the possible types of graphs of unimodal density functions.

Distribution or density functions symmetrical around some point $x=b$ will also occur in the following; they will be called shortly (b) **symmetrical**.

The terms "unimodal", "symmetrical", etc., will be used also for the *graphs* of the corresponding functions.

The point $(a, f(a))$ of the graph of a strictly (a) unimodal density function $f(x)$ is called the *peak* of the graph of $f(x)$.

Testing for the unimodality is, evidently, simple if a differential equation of type $f'(x)=\Phi(x)f(x)$ for $f(x)$ holds — for instance, if the function $\Phi(x)$ has a single change of sign; cf. the case of the Pearson family of density functions (Pearson 1894).

Supplements and problems to Ch. II, § 1

1. The concepts in the main part of the present paragraph can also be extended to the function equal to distribution (density) function of the above type multiplied by a constant.

2. Since in applying the ideas and methods of decomposition as well as in other fields it is frequently necessary to decide whether a distribution (density) function is unimodal, some important **theorems on the unimodality of distribution (density) functions** will be listed below.

This survey may also be of interest in its own right.

Theorem 1. *A function* $\varphi(t)$ *is the characteristic function of an* (a) *unimodal distribution function* $F(x)$ *if and only if*

$$\varphi(t) = \frac{e^{iat}}{t} \int_0^t \psi(u)\, du = e^{iat} \int_0^1 \psi(tu)\, du$$

where $\psi(u)$ *is some characteristic function* (Hinčin 1938; Gnedenko, Kolmogorov 1954 § 32; Feller 1966 p. 501).

The theorem can also be deduced (Medgyessy 1963) from the representation theorem of the convex characteristic functions of G. Pólya (Pólya 1949; Dugué, Girault 1955 p. 292); related theorems: Isii (1958); Olshen, Savage (1970).

Corollary 1. A differentiable function $\varphi(t)$ is the characteristic function of a (0) unimodal distribution function if and only if $\dfrac{d}{dt}[t\varphi(t)] = \psi(t)$ where $\varphi(t)$ is some characteristic function (Girault 1955).

Theorem 2. A density function $f(x)$ is (0) unimodal if and only if

$$f(x) = \begin{cases} \displaystyle\int_x^\infty \frac{dP(y)}{y} & (x > 0) \\[3mm] \displaystyle -\int_{-\infty}^x \frac{dP(y)}{y} & (x < 0) \end{cases}$$

where $P(y)$ is a distribution function. At $x = 0$ $f(x)$ is not always defined (Girault 1955 pp. 252—258).

Evidently, if $p(y)$ is some density function then the function

$$g(x) = \begin{cases} \displaystyle\int_x^\infty \frac{p(y)}{y}\,dy & (x > 0) \\[3mm] \displaystyle -\int_{-\infty}^x \frac{p(y)}{y}\,dy & (x < 0) \end{cases}$$

(at $x = 0$ $g(x)$ is not defined) is a (0) unimodal density function.

Theorem 3. If some sequence of unimodal distribution functions converges to a distribution function, then the latter is also unimodal (Wintner 1936; Lapin 1947; Gnedenko, Kolmogorov 1954 § 32; Ibragimov, Linnik 1965 p. 79).

Theorem 4. If $F_1(x)$ is an (a_1) symmetrical, (a_1) unimodal distribution function and $F_2(x)$ is an (a_2) symmetrical, (a_2) unimodal distribution function, then $F_1(x) * F_2(x)$ is an $(a_1 + a_2)$ symmetrical, $(a_1 + a_2)$ unimodal distribution function (Feldheim 1939, pp. 21—22; Wintner 1936 pp. 47, 77—78; Ibragimov, Linnik 1965 pp. 80—81).

An analogous theorem can be established for density functions.

*Definition 3. A distribution function $F(x)$ is called **strongly unimodal** if, for an arbitrary unimodal distribution function $H(x)$, $F(x) * H(x)$ is unimodal* (Ibragimov 1956).

Then $F(x)$ itself is also unimodal.

Theorem 5. The convolution of two strongly unimodal distribution functions is strongly unimodal (Ibragimov 1956).

Extensions to several strongly unimodal distribution functions and convolution powers are obvious.

Theorem 6. If a sequence of strongly unimodal distribution functions converges weakly to some distribution function, then the limit distribution function will also be strongly unimodal (Ibragimov 1956).

Theorem 7. A non-degenerate unimodal distribution function $F(x)$ is strongly unimodal if and only if $F(x)$ is continuous and the function $\log F'(x)$, where $F'(x)$ denotes the (everywhere existing) left (or right) hand derivative of $F(x)$, is concave over that set of points where neither the left nor the right derivative of $F(x)$ is zero (Ibragimov 1956).

Thus we can speak, analogously, also of *strongly unimodal density functions* with properties such as those listed in Theorems 5 and 6.

Evidently the mirror image of a strongly unimodal distribution or density function is also strongly unimodal. Further, translation does not affect the strong unimodality.

By Theorems 5, 6 and 7 if M ($M=1, 2, ...$) continuous distribution functions have derivatives that are logarithmically concave in the sense described in Theorem 7, then the derivative of their convolution is also logarithmically concave in the above-mentioned sense. Furthermore, if this convolution converges weakly to a distribution function when $M \to \infty$, then the derivative of this limit distribution function is also logarithmically concave in the above-mentioned sense.

Generally in statements the terms "strongly unimodal distribution function" and "distribution function whose derivative is logarithmically concave" (in the above sense) can be used equivalently.

Example 1. The *normal distribution function* with density function

$$f(x) = \frac{e^{-\frac{x^2}{4\gamma}}}{\sqrt{4\pi\gamma}}$$

is, by virtue of Theorem 7, strongly unimodal.

This has been proved also separately by Ibragimov (1956).

Example 2. The *exponential distribution function* with density function

$$f(x) = \begin{cases} \dfrac{e^{-\frac{x}{\delta}}}{\delta} & (x > 0) \\ 0 & (x \leq 0) \end{cases} \quad (\delta > 0)$$

is, by virtue of Theorem 7, strongly unimodal.

Example 3. The *Gamma distribution function* with density function

$$f(x) = \begin{cases} \dfrac{x^{p-1} e^{-\frac{x}{\delta}}}{\delta^p \, \Gamma(p)} & (x > 0) \\[3mm] 0 & (x \leqq 0) \end{cases} \qquad (\delta > 0, \quad p > 0)$$

is, for $p=1, 2, \ldots$, being a convolution of exponential distribution functions, strongly unimodal in view of Example 2 and Theorem 5. Moreover it is, by virtue of Theorem 7, strongly unimodal for any p, $p \geqq 1$. For $0 < p < 1$ the criterion fails (Kubik 1966).

Example 4. Let $f_1(x)$ be a translated normal density function and $f_j(x)$ $(j=2, 3, \ldots)$ translated exponential density functions or mirror images of such ones. Then by Examples 1 and 2 and Theorems 5 and 6

$$f(x) = f_1(x) * \lim_{M \to \infty} [f_2(x) * f_3(x) * \ldots * f_M(x)]$$

is, if the limit exists, a strongly unimodal density function. Passing over to characteristic functions it follows from the preceding that

if the characteristic function $\lambda(t)$ of a density function $f(x)$ has the form

$$\lambda(t) = e^{-\gamma t^2} e^{i\delta t} \prod_{\nu=1}^{\infty} \frac{e^{-i\delta_\nu t}}{(1 - i\delta_\nu t)}$$

where $\gamma \geqq 0$ and δ, δ_ν are real, $0 < \gamma + \sum_{\nu=1}^{\infty} \delta_\nu^2 < \infty$, then $f(x)$ is strongly unimodal.

The character of the corresponding density function is obvious.

$\lambda(t)$ is, at the same time, the unique representation of the characteristic function of a so-called *Pólya frequency function* (Schoenberg 1951), i.e. a function $\dfrac{\Lambda^*(x)}{\int_{-\infty}^{\infty} \Lambda^*(x)\,dx}$, where $\Lambda^*(x)$ is a non-monotone, *totally positive* function, integrable on $(-\infty, \infty)$. We recall that a measurable real function $\Lambda(x)$ $(-\infty < x < \infty)$ is called *totally positive* if $\Lambda(x) \neq 0$ for at least two different x-values, and if, for any $n=1, 2, \ldots$ and arbitrary systems $x_1 < \ldots < x_n$ and $y_1 < \ldots < y_n$,

$$\begin{vmatrix} \Lambda(x_1-y_1) & \Lambda(x_1-y_2) & \ldots & \Lambda(x_1-y_n) \\ \Lambda(x_2-y_1) & \Lambda(x_2-y_2) & \ldots & \Lambda(x_2-y_n) \\ \vdots & \vdots & & \vdots \\ \Lambda(x_n-y_1) & \Lambda(x_n-y_2) & \ldots & \Lambda(x_n-y_n) \end{vmatrix} \geqq 0$$

(Schoenberg 1947, 1951).

$\lambda(t)$ — more exactly, a certain integral function $\Psi(s)$ for which $\Psi(-it)=\lambda(t)$ and, as it can be shown (Schoenberg 1947, 1951) $\dfrac{1}{\Psi(s)} \displaystyle\int_{-\infty}^{\infty} e^{-xs}\Lambda(x)\,dx$ where $\Lambda(x)$ is a density function multiplied by a constant — appeared for the first time in an article of G. Pólya (1915). This is the explanation of the terminology.

Examples of Pólya frequency functions with characteristic function $\lambda(t)$ are e.g.

$$\Lambda^*(x) = C_1 e^{-|x|}$$

or

$$\Lambda^*(x) = \begin{cases} C_2 \displaystyle\sum_{v=0}^{\infty}(-1)^{v+1}v^2 e^{-xv^2} & (x > 0) \\ 0 & (x \leq 0). \end{cases}$$

These are important for mathematical statistics; C_1, C_2 are norming factors applied to the totally positive functions (Schoenberg 1947, 1951 pp. 343, 372).

The essence of the preceding is worth formulating as

Theorem 8. The Pólya frequency functions are strongly unimodal (Medgyessy 1968, 1971b, 1972a).

The unimodality of the Pólya frequency functions was established earlier (Schoenberg 1947; Hirschman, Widder 1949; Hirschman 1950).

The totally positive functions and Pólya frequency functions have a vast literature; besides the quoted articles there is also a book on them (Karlin 1968). Their properties are, however, *not* needed here.

If for the above function $\Lambda(x)$ and for arbitrary $x_1 < x_2$, $y_1 < y_2$

$$\begin{vmatrix} \Lambda(x_1-y_1) & \Lambda(x_1-y_2) \\ \Lambda(x_2-y_1) & \Lambda(x_2-y_2) \end{vmatrix} \geq 0,$$

then $\Lambda(x)$ will be called *twice positive*; a generalization to k times positive functions is obvious (Schoenberg 1947, 1951). Evidently, every totally positive function is k times positive for each $k \equiv 1, 2, \ldots$ and so, in particular, it is also twice positive.

A measurable continuous real function $\Lambda(x)$ is twice positive if and only if

$$\Lambda(x) = e^{-\psi(x)}$$

where $-\psi(x)$ is concave and continuous for $\alpha < x < \beta$ ($-\infty \leq \alpha < \beta \leq +\infty$), $\psi(x) = = +\infty$ for $x < \alpha$ and $x > \beta$, and, for finite α and β, $\psi(\alpha+0) \leq \psi(\alpha) \leq +\infty$ and $\psi(\beta-0) \leq \psi(\beta) \leq +\infty$ (Schoenberg 1947, 1951). We can therefore establish

Theorem 9. A twice positive function $\Lambda(x)$ is, except for a constant factor, a strongly unimodal density function (Medgyessy 1968, 1971b, 1972a).

Proof. By our assumptions $\Lambda(x)=e^{-\psi(x)}\neq0$ if $\alpha<x<\beta$ and so $\Lambda(x)$ is, except for a constant factor, a density function continuous in (α,β). Let the corresponding distribution function be $F(x)$. $F(x)$ is continuous, $F'(x)=\Lambda(x)\neq0$ and $\log F'(x)=-\psi(x)$ is concave if $\alpha<x<\beta$. Thus, by Theorem 7, $F(x)$ is a strongly unimodal distribution function. This is equivalent to the assertion of the theorem.

Hence it follows that the characteristic function of a strongly unimodal density function represented by a twice positive function has the form $\varphi(t)=C\int_\alpha^\beta e^{-\psi(x)}e^{itx}\,dx$, — the further investigation of which is an unsolved **problem**.

The form of $\lambda(t)$ shows that a convolution of Pólya frequency functions, being strongly unimodal, belongs to the same type; this follows also from the general theory of Pólya frequency functions (Schoenberg 1947).

Except for the Pólya frequency functions, the analytical representation, e.g. by a characteristic function, of the twice positive functions is an unsolved **problem** (Schoenberg 1955). If it were not so we should get e.g. the general form of the characteristic function of a great class of strongly unimodal density functions.

Now we shall consider *sufficient* conditions for the unimodality of distribution functions. In general the knowledge of the mode of the distribution function is not needed in them.

Theorem 10. *The so-called stable distribution functions* $S_A(x,B,c)$ (Lévy 1924, 1925 pp. 252—277; Hinčin, Lévy 1937; Hinčin 1938; Gnedenko, Kolmogorov 1954 § 34) *possessing characteristic functions*

$$\varphi(t)=e^{-c\,|t|^A\{1+iB\,\text{sgn}\,t\cdot\omega(t,A)\}}$$

$$\left(c>0,\ 0<A\leqq2,\ |B|\leqq1,\ \omega(t,A)=\tan\frac{\pi A}{2}\ (A\neq1),\ \omega(t,1)=\frac{2}{\pi}\log|t|\right)$$

are unimodal (Ibragimov, Černin 1959).

Since these distribution functions have derivatives of all orders (Gnedenko, Kolmogorov 1954 § 36), the preceding means that the so-called *stable density functions* $s_A(x,B,c)=\dfrac{d}{dx}S_A(x,B,c)$ belonging to the characteristic function $\varphi(t)$ are well-defined and their graphs are — intuitively — unimodal.

Evidently, $s_A(x,B,c)=s_A(-x,-B,c)$.

Examples. By the theorem the elementary stable density functions are unimodal (this can be seen directly also). These functions are:

1. the *normal density function, which will be written in the present book in the form*

$$f(x) = \frac{e^{-\frac{x^2}{4c}}}{\sqrt{4\pi c}}$$

for, in this case, its characteristic function $\varphi(t) = e^{-ct^2}$ corresponds to the above form (with $A=2$, B arbitrary);

2. the *Cauchy density function* which has the form

$$f(x) = \frac{1}{\pi c} \frac{1}{1 + x^2/c^2},$$

(here $A=1$, $B=0$);

3. the *density function of the type* V *of Pearson* which has the form

$$f(x) = \begin{cases} \dfrac{c}{\sqrt{2\pi}} \dfrac{e^{-\frac{c^2}{2x}}}{x^{3/2}} & (x > 0) \\[2ex] 0 & (x \leqq 0) \end{cases}$$

(here $A=1/2$, $B=-1$). It has interesting applications (Ovseevič, Jaglom 1954; Feller 1966 pp. 170—172).

As to the analytical form of the other — unimodal — stable density functions see e.g. Wintner (1941); Pollard (1946); Bergström (1953); Linnik (1954); Skorohod (1954); Zolotarev (1954), (1956); Ibragimov, Linnik (1965) III, §3; for a survey on them, with tables and graphs see Holt, Crow (1973).

Definition 4. A distribution function $F(x)$ is said to be *infinitely divisible* if its characteristic function $\psi(t)$ can be represented in the form

$$\psi(t) = \exp\left[i\gamma t - \frac{\sigma^2 t^2}{2} + \int_{-\infty}^{-0} \left(e^{itu} - 1 - \frac{itu}{1+u^2} \right) dM(u) + \int_{+0}^{\infty} \left(e^{itu} - 1 - \frac{itu}{1+u^2} \right) dN(u) \right]$$

(the canonical representation of P. Lévy; Gnedenko, Kolmogorov 1954 § 18) where γ, δ are real, $M(u)$ and $N(u)$ are non-vanishing and non-decreasing in the intervals $(-\infty, 0)$ and $(0, \infty)$, respectively; $M(-\infty) = N(+\infty) = 0$ and $\int_{-\varepsilon}^{0} u^2\, dM(u) + \int_{0}^{\varepsilon} u^2\, dN(u) < \infty$ for any $\varepsilon > 0$.

Definition 5. **A** distribution function $F(x)$ is said to belong to the *class* $\mathscr{L}$ if its characteristic function $\lambda(t)$ satisfies the functional equation

$$\lambda(t) = \lambda(\alpha t)\lambda_\alpha(t)$$

for every α $(0<\alpha<1)$, where $\lambda_\alpha(t)$ is some characteristic function. The distribution functions belonging to the class $\mathscr{L}$ are infinitely divisible (Lévy 1937; Gnedenko, Kolmogorov 1954 § 29).

Theorem 11. *The symmetrical distribution functions belonging to the class $\mathscr{L}$ are all unimodal* (Wintner 1956 p. 840).

Analogous statements are true for the corresponding density functions, too.

Theorem 12. *The distribution functions belonging to the class $\mathscr{L}$, in the P. Lévy canonical representation of which either $M(u)\equiv 0$ or $N(u)\equiv 0$, are unimodal* (Wolfe 1971).

Theorem 13. *Let $\{F(x, y)\}$ $(c<y<d)$ be a one-parametric family of* (a) *unimodal distribution functions; let the corresponding family of characteristic functions be $\{\varphi(t, y)\}$. If $A(y)$ is some distribution function then the mixture distribution function $G(x) = \int_c^d F(x, y)dA(y)$ is also* (a) *unimodal and its characteristic function is $\omega(t) = \int_c^d \varphi(t, y)dA(y)$. If, in addition, $F(x, y)$ is also* (a) *symmetrical then so is $G(x)$* (Medgyessy 1967d).

Examples. Let ξ be a stochastic variable with a (0) unimodal distribution function, and let η be another variable independent of ξ with the distribution function $F(x)$. If $\varphi_\xi(t)$, $\varphi_\eta(t)$ are the characteristic functions of ξ and η, respectively, then the characteristic function of $\xi\eta$ is $\varphi_{\xi\eta}(t) = \int_{-\infty}^{\infty} \varphi_\xi(tu)dF(u)$ and that of $\dfrac{\xi}{\eta}$ is

$$\varphi_{\frac{\xi}{\eta}}(t) = \int_{-\infty}^{\infty} \varphi_\xi\left(\frac{t}{u}\right) dF(u) \quad \text{(Girault 1955 pp. 290—292).}$$

By the theorem, the distribution functions of $\xi\eta$ and of $\dfrac{\xi}{\eta}$, respectively are also (0) unimodal.

Theorem 14. *Let $\Phi(t)$ be the characteristic function of a* (0) *unimodal,* (0) *symmetrical distribution function and $\{p_r\}$ $(r=0, 1, ...)$ a discrete distribution with generating function $g(z)$. Then*

$$\sum_{r=0}^{\infty} p_r[\Phi(t)]^r = g[\Phi(t)]$$

will be the characteristic function of a (0) *unimodal,* (0) *symmetrical distribution function* (Medgyessy 1967*d*).

The analogous theorems for density functions are obvious.

Example 1. If $\Phi(t)$ is the characteristic function of a (0) unimodal, (0) symmetrical density function then the characteristic function of the type of B. de Finetti (de Finetti 1930) $\varphi(t) = e^{\lambda[\Phi(t)-1]}$ $(\lambda > 0)$ will belong to a (0) unimodal, (0) symmetrical density function.

Example 2. Let $\Phi(t)$ be as above. Lukács (1957, 1964 pp. 215—216) proved that

$$\varphi(t) = \frac{p-1}{p-\Phi(t)} \quad (p > 1)$$

is the characteristic function of some infinitely divisible distribution function $F(x)$. By Theorem 13, $F(x)$ is then (0) unimodal and (0) symmetrical.

Definition 6. Let $\mathcal{M}$ denote the class of (γ) symmetrical, infinitely divisible distribution functions for which it holds that in P. Lévy's canonical representation of their characteristic functions $\psi(t)$, in which, by the symmetry, $N(u) = M(-u)$, $M(u)$ is convex in $(-\infty, 0)$ (Medgyessy 1967*d*).

Evidently, these distribution functions are (γ) symmetrical.

A *new class of unimodal distribution functions* has been introduced by the next theorem.

Theorem 15. *Any distribution function belonging to the class $\mathcal{M}$ is unimodal* (Medgyessy 1967*d*).

Example 1. Evidently the symmetrical stable density functions belong to this class; their unimodality has been proved, however, in another way.

Example 2. $\gamma = \sigma^2 = 0$,

$$M(u) = \int_{-\infty}^{u} M'(y)\,dy,$$

where

$$M'(u) = \begin{cases} \dfrac{\beta}{|u|^{1+\beta}} & (-1 < u \leq 0) \\[2ex] \beta e^{-\frac{1}{2}(|u|-1)} & (u \leq -1) \end{cases} \quad (u < 0, \quad 0 < \alpha < 1),$$

satisfy all conditions of the theorem. Thus the distribution function $W(x, \beta)$ with the characteristic function

$$\psi(t) = \exp\left[i\gamma t + 2\beta \int_{-\infty}^{-0} (\cos tu - 1) M'(u) \, du\right]$$

is (γ) unimodal, which is a valuable piece of information, as $W(x, \beta)$ cannot be written in a closed form. We notice that $W(x, \beta)$ does *not* belong to the class $\mathscr{L}$.

Theorem 16. *Let* $\psi(t)$ *be the characteristic function of a* (γ) *symmetrical,* (γ) *unimodal, infinitely divisible distribution function belonging to the class* $\mathscr{M}$. *Then* $[\psi(t)]^c$ $(c > 0)$ *will be the characteristic function of a* $(c\gamma)$ *unimodal,* $(c\gamma)$ *symmetrical infinitely divisible distribution function which also belongs to* $\mathscr{M}$ (Medgyessy 1967d).

Theorem 17. *Let* $\psi(t)$ *be the characteristic function of a* (0) *symmetrical,* (0) *unimodal infinitely divisible distribution function belonging to the class* $\mathscr{M}$ *and let* $B(x)$ *be a distribution function. Then the function*

$$\omega(t) = \int_0^\infty [\psi(t)]^x \, dB(x) = L[-\log \psi(t)]$$

where $L(s)$ *denotes the Laplace–Stieltjes transform of* $B(x)$ *will be the characteristic function of a* (0) *unimodal,* (0) *symmetrical infinitely divisible distribution function* (Medgyessy 1967d).

Analogous theorems and definitions can be established for density functions in the same way.

Families E of unimodal distribution (density) functions having the property that the convolution of any two of their members belongs to the same family are worth investigating. Such families are — by virtue of their definitions — those of 1. the Pólya frequency functions (Schoenberg 1951) — this is seen from their characteristic functions; 2. the stable distribution functions with characteristic functions having the same parameters A and B; 3. the symmetrical distribution functions of the class $\mathscr{L}$; 4. the distribution functions of the class $\mathscr{M}$.

We notice that strong unimodality need not be supposed (this implies no contradiction because, in case of the strongly unimodal distribution functions, convolutions with an arbitrary unimodal distribution function occur while here the other distribution function is *not* arbitrary. However, the determination of all such families E is an unsolved **problem**.

Furthermore, one should consider first the unsolved **problem**: to determine all such families of unimodal distribution functions in which the convolution (not contained eventually in the family) of any two members of the family is itself unimodal.

Theorem 18. *Let the density function $f(x)$ vanishing for $x \leqq 0$ be derivable and its derivative be representable in form of a Laplace transform, i.e. let* $f'(x) = \int\limits_0^\infty e^{-xt} \varphi(t) dt$ $(x>0)$. *If $\varphi(t)$ possesses a single change of sign in $(0, \infty)$ then $f(x)$ is unimodal* (Medgyessy 1968).

Proof. It is known (Doetsch 1950, p. 149) that if a real function $F(t)$ has, for $t>0$, n changes of sign, then the Laplace transform

$$f(s) = \int\limits_0^\infty e^{-st} F(t) dt$$

possesses at most n different zeros for the real s-values of the half plane of convergence, i.e. n changes of sign at most. Consequently, the function $f'(x)$ will have at most one change of sign in $(0, \infty)$, i.e. $f(x)$ will possess at most one extremum. Since $f(x)$ is a density function, there will be exactly one maximum, thus $f(x)$ is unimodal.

Analogous theorems can be established if instead of e^{-xt} there stands a kernel representing a Pólya frequency function (Schoenberg 1951; Hirschman, Widder 1955 p. 108).

Sometimes the unimodality of a density function can be decided by means of the fact that its characteristic function satisfies a certain differential equation. A relevant criterion is provided by

Theorem 19. *Let the characteristic function $\varphi(t)$ of a density function $f(x)$ satisfy the differential equation*

$$\sum_{\mu=0}^M B_\mu \frac{d^\mu}{dt^\mu} \varphi(t) + \sum_{v=0}^N A_v \frac{d^v}{dt^v} [t \varphi(t)] = 0,$$

where $M > 0, N > 0$, A_v, B_μ are constants. Further let

$$\lim_{|t| \to \infty} \frac{d^\varrho}{dt^\varrho} \varphi(t) = 0 \qquad (\varrho = 0, 1, \ldots M-1),$$

$$\lim_{|t| \to \infty} \frac{d^\tau}{dt^\tau} [t \varphi(t)] = 0 \qquad (\tau = 0, 1, \ldots N-1),$$

$$\int\limits_{-\infty}^\infty |\varphi(t)| dt < \infty, \qquad \int\limits_{-\infty}^\infty |t \varphi(t)| dt < \infty$$

and the function

$$\Phi(x) = \frac{\displaystyle\sum_{\mu=0}^{M} B_\mu i^\mu x^\mu}{\displaystyle\sum_{\nu=0}^{N} A_\nu i^{\nu-1} x^\nu} \qquad (-\infty < x < \infty)$$

have a single change of sign. Then $f(x)$ is unimodal (Medgyessy 1972*a*).

Proof. From the conditions of the theorem we find, by constructing the inverse Fourier transform of our differential equation, that $f'(x)=\Phi(x)f(x)$; hence the assertion follows.

The mode of $f(x)$ can be calculated immediately from the above. For $N=0$, $M=1$, the condition that $\Phi(x)$ has a single change of sign is fulfilled from the beginning. (It is interesting to relate this theorem with the family of Pearson density functions (Pearson 1894).)

In investigations on unimodality one can apply also the following theorems.

Theorem 20. Let $\varphi(t)$ be a real-valued even function which satisfies the following conditions: 1. $\varphi(0)=1$; 2. $\lim\limits_{t\to 0} \varphi(t)=0$; 3. *$t\,\varphi'(t)$ exists and is continuous for every real t and, in addition,* $\lim\limits_{t\to\infty} t\,\varphi'(t) = \lim\limits_{|t|\to\infty} t\,\varphi'(t)=0$; 4. *the function $g(t)=\varphi(t)+t\,\varphi'(t)$ is convex for $t>0$. Then $\varphi(t)$ is the characteristic function of a* (0) *symmetrical,* (0) *unimodal distribution function* (Laha 1961; Lukács 1961 p. 311).

This theorem is a combination of the theorem of Pólya (1949) concerning convex characteristic functions with Corollary 1 to Theorem 1 above. In this context see also Medgyessy (1974*a*) for generalizations.

Theorem 21. Let $\varphi(t)$ be a continuous real-valued and even function of the real variable t such that $\varphi(0)=1$ and $\varphi(t)=A(t)$ for $t>0$ where the function $A(t)$ satisfies the following conditions: 1. *$A(z)$ defined as a function of the complex variable z ($z=re^{i\theta}=u+iv$, u and v both real) is regular in the region $D(r>0, -\varepsilon_1<\theta<\pi/2+\varepsilon_2$ where ε_1 and ε_2 are arbitrarily small positive numbers) of the complex z-plane;* 2. *$|A(z)|=O(1)$ as $|z|\to 0$, $|A(z)|=O(|z|^{-\delta})$ as $|z|\to\infty$ ($\delta>1$);* 3. *Im $A(iv)\leqq 0$ for $v>0$. Then $\varphi(t)$ is the characteristic function of a* (0) *symmetrical,* (0) *unimodal and absolutely continuous density function* (Laha 1961).

Example. By the aid of these theorems it can be proved that $\varphi(t)=\dfrac{1}{1+|t|^\alpha}$
($0<\alpha<2$) is the characteristic function of a (0) *symmetrical,* (0) *unimodal, absolutely continuous density function*; it is known that this statement holds also for $\alpha=2$. (The fact that $\varphi(t)$ is a characteristic function was proved earlier by Linnik

1953 p. 271.) From all this a new proof can be obtained for the old theorem, a particular case of Theorem 10 (Wintner 1936 p. 78; Feldheim 1939 p. 21): *All symmetrical stable density functions are unimodal* (Laha 1961).

Theorem 22. Let $\varphi(t)$ be a continuous, real-valued, even function which satisfies the following conditions: 1. $\varphi(0) = 1$; 2. $\lim\limits_{t \to \infty} \varphi(t) = 0$ *and* 3. $-\varphi'(t)$ *is convex for* $t > 0$. *Then* $\varphi(t)$ *is the characteristic function of a* (0) *symmetrical,* (0) *unimodal density function* (Askey 1975).

Historical remarks. The investigation of unimodal density functions started at the end of the 19th century, when it had become necessary to approximate the unimodal graph of empirical density functions by similarly *unimodal* functions (Jordan 1927 p. 242). It was in the course of these investigations that the family of unimodal density functions known nowadays as the family of Pearson density functions (Pearson 1894) was introduced. It contained many important types. Later it became clear that one can also get the most characteristic ones from a completely different background: the investigation of the limit distribution of sums of independent stochastic variables. This led to the exploration of the infinitely divisible distribution functions in the thirties; their general characteristics did not, however, reveal anything of unimodality relations at all. As a consequence of this, the investigation of the questions of unimodality on the basis of characteristic functions was started (Hinčin 1938). It gained a great impetus as a result of the observation of Chung (1953) that the convolution of unimodal distribution functions — in contrast to the earlier erroneous statement (Lapin 1947) — is *not* necessarily unimodal. Together with this the statement, earlier considered to be true, that the stable distribution functions are all unimodal also remained to be proved. However, it was not proved until later (Ibragimov, Černin 1959); meanwhile the question of the unimodality of the convolution of unimodal distribution functions was cleared up by Ibragimov (1956). The investigation of the unimodality of distribution functions also continued since then (Laha 1961; Lukács 1961; Fisz 1962; Medgyessy 1963, 1967d, 1968; Wolfe 1971; Askey 1975).

On the other hand it is interesting that researches concerning the unimodality of discrete distributions were very scarce. It was due to this that several authors hoped to correct the balance (Medgyessy 1968, 1971b, 1972a, 1972c; Keilson, Gerber 1971); the results are summarized in Ch. II, §2.

Remark. In the footnote on p. 856 in Wintner (1956) one reads that the observation of Chung (1953) that the convolution of unimodal distribution functions is not necessarily unimodal is, originally, due to P. Lévy. However, this statement is *false*. P. Lévy presented merely examples of (0) unimodal, non-symmetrical distribution functions, whose convolution was unimodal, but not (0) unimodal (cf. Feldheim 1939 p. 22).

The following **problems** are worth mentioning.

1. Can a functional equation be given for the characteristic functions of the distribution functions of the class $\mathcal{M}$ — analogous to that valid for the distribution functions belonging to the class $\mathcal{L}$?

2. Let the characteristic function $\varphi(t)$ of a distribution function $F(x)$ be identically zero outside of some finite interval. What are the conditions to be imposed on $\varphi(t)$ that $F(x)$ be unimodal? (Examples show that there exist such $\varphi(t)$ which yield unimodal $F(x)$.)

3. Under what conditions can the convolution of two (0) unimodal, *non-symmetrical* distribution functions be again (eventually, (0)) unimodal? (The specialization for convolution powers is obvious.)

It is known that two (0) unimodal non-symmetrical stable density functions with the same parameters A, B in their characteristic functions have these properties.

4. Let $F(x)$ be a strongly unimodal distribution function. Then $[F(x)]^{*n}$ ($n = 2, 3, \ldots$) is, evidently, also strongly unimodal. If $F(x)$ is only (0) unimodal, when will $[F(x)]^{*n}$ be also (0) unimodal?

5. Let $F(x)$ be a strongly unimodal infinitely divisible distribution function. The Lévy–Hinčin canonical representation of its characteristic function has to be determined.

6. In what circumstances can the convolution of a unimodal distribution function and a non-unimodal distribution function be unimodal?

7. When can a mixture of *non* (0) unimodal distribution functions be unimodal?

8. The following conjecture should be examined: Let $F(x)$ be a distribution function. Then there is a number n_0 such that if $n > n_0$ then $[F(x)]^{*n}$ will be a unimodal distribution function. This conjecture is suggested by a theorem of B. V. Gnedenko (Gnedenko 1954 p. 255) stating that the limit distribution function of an appropriately normed sum of equally distributed, independent random variables is normal — i.e. unimodal. This conjecture is the analogue of that formulated for discrete distributions by A. Rényi (see *Supplements and problems to* Ch. II, §2, Problem 7 at the end of Section 1).

§ 2. UNIMODAL DISCRETE DISTRIBUTIONS

This paragraph summarizes results concerning unimodal discrete distributions, needed in the book.

With discrete distributions, unimodality cannot be defined in the same way as for continuous distribution functions because their distribution functions are step functions. However, the following definition suggests itself.

Definition 1. A discrete distribution $\{p_n\}$ $(n=...-1, 0, 1, ...)$ is called *unimodal* at the point $n=a$ — in short: (a) **unimodal** — if in the sequence

$$..., p_{-1}-p_{-2}, \quad p_0-p_{-1}, \quad p_1-p_0, ..., \quad p_n-p_{n-1}, ...$$

exactly one change of sign occurs — more precisely, if p_n-p_{n-1} becomes negative *for the first time* at $n=a+1$ (Medgyessy 1968, 1971b, 1972a, 1972c).

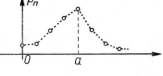

Figure 1 shows the possible types of graphs of (a) unimodal discrete distributions. (Auxiliary dotted conjunction lines were applied.)

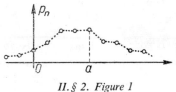

The term "unimodal" will be used with its above meaning also when speaking of the *graphs* of discrete distributions. Related, equivalent definitions have been given by Holgate (1970) and Keilson, Gerber (1971).

II.§ 2. Figure 1

The point (a, p_a) of the graph of an (a) unimodal discrete distribution $\{p_n\}$ $(n=-1, 0, 1, ...)$ is called the *peak* of the graph of $\{p_n\}$.

Testing for the unimodality is, evidently, simple if there holds a recursive relation type $p_{n+1}=\Phi(n)p_n$ between the members of $\{p_n\}$, for instance if there is a single change of sign in the sequence $\{\Phi(n)-1\}$.

Supplements and problems to Ch. II, § 2

1. Since in applying the ideas and methods described in the present work it is frequently necessary to decide whether a density function is unimodal, some important **theorems on the unimodality of discrete distributions** will be listed below. These theorems are due to a great extent to J. Keilson, H. Gerber and P. Medgyessy and have much in common with those on unimodal distribution functions summarized in Ch. II, § 1.

This survey may be of interest on its own, too, since this field of research has been somewhat neglected in probability theory.

Theorem 1. *The necessary and sufficient condition that a discrete distribution* $\{p_n\}$ $(n=\ldots, -1, 0, 1, \ldots)$ *be* (a) *unimodal is that if*

$$(a-n+\theta)(p_n-p_{n-1}) = q_n \quad (n = \ldots, -1, 0, 1, \ldots)$$

where θ $(0<\theta<1)$ *is arbitrary, then* $\{q_n\}$ *is some discrete distribution with* $q_{a+1}>0$ (Medgyessy 1968, 1971*b*, 1972*a*, 1972*c*).

Theorem 2. *A discrete distribution* $\{p_n\}$ $(n=\ldots, -1, 0, 1, \ldots)$ *is* (a) *unimodal if and only if its characteristic function* $\gamma(t)$ *has the form*

$$\gamma(t) = \frac{e^{i(a+\theta)t}}{i(1-e^{it})} \int_0^t \varkappa(u) e^{-i(a+\theta)u} du$$

where θ $(0<\theta<1)$ *is arbitrary,* $\varkappa(t)=\sum_{n=-\infty}^{\infty} q_n e^{int}$ *is the characteristic function of some discrete distribution* $\{q_n\}$ $(n=\ldots, -1, 0, 1, \ldots)$ *and* $q_{a+1}>0$ (Medgyessy 1968, 1971*b*, 1972*a*, 1972*c*).

The analogue of this for generating functions — if $\{p_n\}$ $(n=0, 1, \ldots)$ is written for $\{p_n\}$ $(n=\ldots, -1, 0, 1, \ldots; \ p_n=0$ if $n<0)$ — is

Theorem 3. *A discrete distribution* $\{p_n\}$ $(n=0, 1, \ldots)$ *is* (a) *unimodal* $(a\geq 0)$ *if and only if its generating function* $\pi(z)$ *has, for* $0<\varepsilon\leq x<1$, *the form*

$$\pi(x) = \frac{x^{a+\theta}}{(1-x)} \int_x^1 \frac{q(u)}{u^{a+\theta+1}} du \quad (0 < \varepsilon \leq x < 1)$$

where θ $(0<\theta<1)$ *is arbitrary,* $q(z)=\sum_{n=0}^{\infty} q_n z^n$ $(|z|\leq 1)$ *is the generating function of a discrete distribution* $\{q_n\}$ $(n=0, 1, \ldots)$ *for which* $q_{a+1}>0$, *and* ε *is arbitrarily small* (Medgyessy 1968, 1971*b*, 1972*a*, 1972*c*).

By means of these theorems we can thus generate unimodal distributions.

In Theorem 3

$$\pi'(x)x(x-1)+\pi(x)[a+\theta-(a+\theta-1)x] = q(x) \quad (0 <\varepsilon \leq x < 1)$$

can be taken also as a necessary and sufficient condition.

The case $a=0$ can also be treated more easily.

Theorem 4. *A discrete distribution* $\{p_n\}$ $(n=0, 1, \ldots)$ $(p_0>0)$ *is* (0) *unimodal if and only if*

$$\left\{\frac{p_{n-1}-p_n}{p_0}\right\} = \{q_n\} \quad (n = 1, 2, \ldots)$$

is some discrete distribution for which $q_1 > 0$ (Medgyessy 1968, 1971*b*, 1972*a*, 1972*c*).

Theorem 5. A discrete distribution $\{p_n\}$ ($n = 0, 1, ...; p_0 > 0$) *with generating function* $\pi(z)$ *is* (0) *unimodal if and only if at the notation*

$$\frac{\pi(z)(1-z)}{\pi(0)} + 1 = q(z) \qquad (|z| \leq 1)$$

$$q(z) = \sum_{n=1}^{\infty} q_n z^n \qquad (|z| \leq 1)$$

is the generating function of some distribution $\{q_n\}$ ($n = 1, 2, ...$) for which $q_1 > 0$ (Medgyessy 1968, 1971*b*, 1972*a*, 1972*c*).

The statements in these two theorems can also be extended to a distribution $\{p_n\}$ ($n = 0, 1, ..., N$) automatically.

Theorem 6. If a sequence of unimodal discrete distributions converges to some discrete distribution, then the limit distribution will also be unimodal (Medgyessy 1968, 1971*b*, 1972*a*, 1972*c*; Keilson, Gerber 1971).

Definition 2. A discrete distribution $\{p_n\}$ ($n = ..., -1, 0, 1, ...$) is called **strongly unimodal** if, for an arbitrary unimodal distribution $\{h_n\}$, $\{p_n\} * \{h_n\}$ is unimodal (Keilson, Gerber 1971).

Then $\{p_n\}$ itself is also unimodal.

Theorem 7. The convolution of two strongly unimodal discrete distributions is strongly unimodal (Keilson, Gerber 1971).

Extensions to several strongly unimodal discrete distributions and convolution powers are obvious.

Theorem 8. If a sequence of strongly unimodal discrete distributions converges to some discrete distribution, then the limit distribution will also be strongly unimodal (Keilson, Gerber 1971).

Evidently, the mirror image of a strongly unimodal discrete distribution is also strongly unimodal. Further, translation does not affect the strong unimodality.

Theorem 9. A unimodal discrete distribution $\{p_n\}$ ($n = ..., -1, 0, 1, ...$) *is strongly unimodal if and only if* $\{p_n\}$ *is logarithmically concave, i.e.* $p_n^2 - p_{n-1}p_{n+1} \geq 0$ *for all n* (Keilson, Gerber 1971).

Then the quantities $\dfrac{p_{n-1}}{p_n}$ form a monotone non-decreasing sequence, i.e.

$\dfrac{p_{n-1}}{p_n} \leqq \dfrac{p_n}{p_{n+1}}$, and the quantities

$$\frac{p_n^2 - p_{n-1}p_{n+1}}{\sum\limits_{n=-\infty}^{\infty}(p_n^2 - p_{n-1}p_{n+1})} = q_n \qquad (n = \dots, -1, 0, 1, \dots)$$

form a discrete distribution if $0 < \sum\limits_{n=-\infty}^{\infty}(p_n^2 - p_{n-1}p_{n+1}) < \infty$.

One realizes that the concept "strongly unimodal discrete distribution" is the counterpart of the concept "strongly unimodal distribution function" introduced in Ch. II, § 1.

By Theorems 7, 8 and 9, if the members of M ($M = 1, 2, \dots$) discrete distributions form logarithmically concave sequences, then the members of their convolution also form a logarithmically concave sequence. (This is also a consequence of a famous theorem of M. Fekete (Fekete, Pólya 1912).) Furthermore if this convolution converges to a distribution when $M \to \infty$, then this limit distribution is also logarithmically concave.

Generally in statements the terms "strongly unimodal discrete distribution" and "discrete distribution such that the sequence of its members is logarithmically concave" can be used equivalently.

Theorem 10. *If* $\{p_n\}$ ($n = 0, 1, \dots$) *is a discrete distribution with generating function* $\pi(z)$, *if* $0 < \sum\limits_{n=1}^{\infty}(p_n^2 - p_{n-1}p_{n+1}) < \infty$, *if* $q(x)$ *is defined by*

$$\frac{\dfrac{1}{2\pi}\int\limits_{-\pi}^{\pi}|\Pi(\sqrt{x}\,e^{i\theta})|^2(1 - e^{2i\theta})\,d\theta - \Pi(0)^2}{\dfrac{1}{2\pi}\int\limits_{-\pi}^{\pi}|\Pi(e^{i\theta})|^2(1 - e^{2i\theta})\,d\theta - \Pi(0)^2} = q(x) \qquad (0 < x < 1)$$

and if $q(z) = \sum\limits_{n=1}^{\infty} q_n z^n$ ($|z| \leqq 1$) *is the generating function of some discrete distribution* $\{q_n\}$ ($n = 1, 2, \dots$), *then* $\{p_n\}$ *is logarithmically concave, i.e. strongly unimodal* (Medgyessy 1968, 1971b, 1972a, 1972c).

Example 1. The members of the *binomial distribution*

$$\left\{\binom{N}{n}p^n q^{N-n}\right\} \qquad (n = 0, 1, \dots N; \ p > 0, \ q > 0, \ p + q = 1, \ N = 1, 2, \dots)$$

form a logarithmically concave sequence. Indeed

$$\left[\binom{N}{n}p^n q^{N-n}\right]^2 - \binom{N}{n-1}p^{n-1}q^{N-n+1}\binom{N}{n+1}p^{n+1}q^{N-n-1} =$$

$$\binom{N}{n}^2 p^{2n}q^{2N-2n}\left(1 - \frac{n(N-n)}{(n+1)(N-n+1)}\right) \geqq 0 \quad (n = 0, 1, ..., N)$$

Example 2. The members of the *Poisson distribution*

$$\left\{e^{-\lambda}\frac{\lambda^n}{n!}\right\} \qquad (n = 0, 1, ...; \ \lambda > 0)$$

form a logarithmically concave sequence. Indeed

$$\left(e^{-\lambda}\frac{\lambda^n}{n!}\right)^2 - e^{-\lambda}\frac{\lambda^{n-1}}{(n-1)!}e^{-\lambda}\frac{\lambda^{n+1}}{(n+1)!} = e^{-2\lambda}\frac{\lambda^{2n}}{(n!)^2}\left(1 - \frac{n}{n+1}\right) \geqq 0 \quad (n = 0, 1, ...).$$

Example 3. The members of the *geometric distribution*

$$\{(1-p)p^n\} \qquad (n = 0, 1, ...; \ 0 < p < 1)$$

form a logarithmically concave sequence. Indeed

$$[(1-p)p^n]^2 - (1-p)p^{n-1}(1-p)p^{n+1} = 0 \qquad (n = 0, 1, 2, ...).$$

Consequently, the strong unimodality of these discrete distributions follows from Theorem 9. (For further examples and results see Keilson, Gerber 1971).

Example 4. By Theorem 7 the Rth convolution power of a geometric distribution $\{(1-p)p^n\}$ $(n=0, 1, ...)$, $\{(1-p)p^n\}^{*R}$ $(R=1, 2, ...)$ i.e. the *negative binomial distribution* $\left\{\binom{R-1+n}{R-1}(1-p)^R p^n\right\}$ $(n=0,1,...)$ is also strongly unimodal.

Example 5. Since a binomial distribution and a geometric distribution are strongly unimodal, Theorem 7 shows that *the convolution of a binomial distribution and a geometric distribution* is strongly unimodal.

Example 6. Let $\{p_n^{(1)}\}$ be a Poisson distribution and $\{p_n^{(j)}\}$ $(j=2, 3, ...)$ a convolution of a binomial distribution and a geometric distribution. Then by Examples 2 and 5 and by Theorems 7 and 8

$$\{p_n\} = \{p_n^{(1)}\} * \lim_{M \to \infty} [\{p_n^{(2)}\} * \{p_n^{(3)}\} * ... * \{p_n^{(M)}\}]$$

is, if the limit exists, a strongly unimodal discrete distribution. Passing over to generating functions, it follows from the preceding that *if a generating function $\pi(z)$ of a discrete distribution $\{p_n\}$ $(n=0, 1, ...)$ has the form*

$$\pi(z) = Ce^{c_1 z} \frac{\prod_{v=1}^{\infty} (1+\alpha_v z)}{\prod_{v=1}^{\infty} (1-\beta_v z)} \qquad (|z| \leq 1)$$

where C is a convenient norming constant, $c_1 > 0$, $\alpha_v \geq 0$, $\beta_v \geq 0$, $\sum_{v=1}^{\infty}(\alpha_v + \beta_v) < \infty$, then $\{p_n\}$ is strongly unimodal.

The character of the corresponding discrete distribution is obvious.

$\pi(z)$ is, at the same time, the unique representation of the power series $\sum_{n=0}^{\infty} a_n z^n$ whose coefficients a_n form a so-called *totally positive sequence,* i.e. every minor of arbitrary order of the matrix

$$\begin{pmatrix} a_0 & a_1 & a_2 & a_3 \ldots \\ 0 & a_0 & a_1 & a_2 \ldots \\ 0 & 0 & a_0 & a_1 \ldots \\ 0 & 0 & 0 & a_0 \ldots \\ \vdots & \vdots & \vdots & \vdots \end{pmatrix}$$

is non-negative; such sequences are analogues of the totally positive functions (see Ch. II, § 1) (Schoenberg 1948).

The essence of the preceding is worth formulating as

Theorem 11. *The discrete distributions represented (up to a constant factor) by the members of totally positive sequences are strongly unimodal* (Medgyessy 1968, 1971b, 1972a).

The totally positive sequences have a vast literature nowadays (Aissen, Schoenberg, Whitney 1952; Edrei 1952, 1953a, 1953b; Schoenberg 1953, 1955; see also the book of Karlin 1968). The results of these studies are, however, *not* needed here.

A particular case of $\pi(z)$ was already known earlier (Lipka 1938).

If merely the minors of 1st and 2nd order of the above matrix are non-negative, then the *sequence* of the coefficients will be called *twice positive;* a generalization to *k times positive sequences* is obvious (Fekete, Pólya 1912; Schoenberg 1955). Evidently, every totally positive sequence is k times positive at the same time and, consequently, is also twice positive.

Since from

$$p_n^2 - p_{n-1} p_{n+1} \geq 0 \ (n = 1, 2, ...), \text{ i.e. } \frac{p_{n-1}}{p_n} \leq \frac{p_n}{p_{n+1}},$$

it follows that

$$\begin{vmatrix} p_n & p_{n+r} \\ p_{n-s} & p_{n-s+r} \end{vmatrix} \geqq 0 \quad (r = 1, 2, \ldots; \quad s = 1, 2, \ldots; \quad p_n = 0 \quad \text{if} \quad n < 0)$$

the members of a logarithmically concave discrete distribution $\{p_n\}$ $(n=0, 1, \ldots)$ form also a twice positive sequence. Up to a constant factor, the converse is obvious. Thus the concepts "*logarithmically concave sequence*" and "*twice positive sequence*" can be regarded as equivalent, and, in view of Theorem 9, we can establish

Theorem 12. *A discrete distribution generated by a twice positive sequence is strongly unimodal* (Medgyessy 1968, 1971*b*, 1972*a*).

The form of $\pi(z)$ shows that the convolution of distributions given by totally positive sequences and being, consequently, strongly unimodal, belongs to the same type; this follows also from the general theory of totally positive sequences (Schoenberg 1948).

Unfortunately it is an unsolved **problem** how to give such a unique representation of power series whose coefficients form a twice positive sequence as the unique representation of power series whose coefficients present a totally positive sequence (Schoenberg 1955). If it were not so we should get the general form of the generating function of a great class of strongly unimodal distributions.

In the following, *sufficient* conditions will be given for the unimodality of discrete distributions. In general the knowledge of the place of the peak of the distribution is not needed in them.

We notice that the statements in Theorems 7, 8, 9 and 11 can be extended automatically to a distribution $\{p_n\}$ $(n=0, 1, \ldots, N)$.

Theorem 13. *If* $\{p_n\}$ $(n=0, 1, \ldots)$ *is a discrete distribution and for some* α *and* β $(\alpha>0, \beta>0, \alpha+\beta=1)$

$$p_n \geqq \alpha p_{n-1} + \beta p_{n+1} \quad (n = 1, 2, \ldots)$$

(*if* $\beta p_1 - \alpha \beta_0 \neq 0$, *this is equivalent to saying that*

$$\left\{ \frac{p_n - \alpha p_{n-1} - \beta p_{n+1}}{\sum\limits_{n=1}^{\infty} (p_n - \alpha p_{n-1} - \beta p_{n+1})} \right\} = \left\{ \frac{p_n - \alpha p_{n-1} - \beta p_{n+1}}{\beta p_1 - \alpha p_0} \right\} = \{q_n\} \quad (n = 1, 2, \ldots)$$

is some discrete distribution), then $\{p_n\}$ *is unimodal* (Medgyessy 1968, 1971*b*, 1972*a*, 1972*c*).

In the case of $\alpha = \beta = 1/2$, the sequence $\{p_n\}$ is called *concave*.

Theorem 14. If the generating function of a discrete distribution $\{p_n\}$ $(n = 0, 1, ...)$ is $\pi(z)$ and for some α, β $(\alpha > 0, \beta > 0, \alpha + \beta = 1)$, $\beta p_1 - \alpha p_0 \neq 0$, and

$$\frac{1}{\beta p_1 - \alpha p_0}\left[\pi(z)\left(1 - \alpha z - \frac{\beta}{z}\right) + p_0\left(\frac{\beta}{z} - 1\right) + \beta p_1\right] = q(z) = \sum_{n=1}^{\infty} q_n z^n \qquad (|z| \leq 1)$$

is the generating function of some discrete distribution for which $q_0 = 0$, then $\{p_n\}$ is unimodal (Medgyessy 1968, 1971*b*, 1972*a*, 1972*c*).

Theorem 15. If $\{p_n\}$ $(n = 0, 1, ...; p_n \neq 0)$ is a discrete distribution and for some α and β $(\alpha > 0, \beta > 0)$

$$\left(\frac{p_{n-1}}{p_n}\right)^{\alpha} \leq \left(\frac{p_n}{p_{n+1}}\right)^{\beta} \qquad (n = 1, 2, ...)$$

(if $0 < \sum_{n=1}^{\infty} (p_n^{\alpha+\beta} - p_{n-1}^{\alpha} p_{n+1}^{\beta}) < \infty$, this is equivalent to saying that

$$\left\{\frac{p_n^{\alpha+\beta} - p_{n-1}^{\alpha} p_{n+1}^{\beta}}{\sum_{n=1}^{\infty} (p_n^{\alpha+\beta} - p_{n-1}^{\alpha} p_{n+1}^{\beta})}\right\} = \{q_n\} \qquad (n = 1, 2, ...)$$

is some discrete distribution), then $\{p_n\}$ is unimodal (Medgyessy 1968, 1971*b*, 1972*a*, 1972*c*).

If the preceding inequality holds for $\alpha = \beta = 1$, $\{p_n\}$ is logarithmically concave; this case has been investigated above.

Theorem 16. If $\{p_n\}$ $(n = 0, 1, ...)$ is a discrete distribution and

$$p_0 \leq p_1,$$

$$p_n > \min(p_{n-1}, p_{n+1}) \qquad (n = 1, 2, ...)$$

i.e.

$$p_0 \leq p_1,$$

$$p_n > \frac{1}{2}[p_{n-1} + p_{n+1} - |p_{n-1} - p_{n+1}|]$$

(if $0 < \sum_{n=1}^{\infty} [p_n - \min(p_{n-1}, p_{n+1})] < \infty$, this is equivalent to saying that

$$\left\{\frac{p_n - \min(p_{n-1}, p_{n+1})}{\sum_{n=1}^{\infty} [p_n - \min(p_{n-1}, p_{n+1})]}\right\} = \{q_n\} \qquad (n = 1, 2, ...)$$

is some discrete distribution), then $\{p_n\}$ is unimodal (Medgyessy 1968, 1971*b*, 1972*a*, 1972*c*).

The condition of the theorem expresses that the members of $\{p_n\}$ form a so-called *strictly quasi-concave* sequence. The last theorem is the analogue of the one on the unimodality of the so-called *quasi-concave functions* (Kovács 1963).

The composition of this theorem for the corresponding generating functions is an unsolved **problem**.

Further *sufficient* conditions for the unimodality of discrete distributions can be obtained on the basis of relations, studied long ago, which exist between the number of changes of sign in the sequence of the coefficients of a polynomial or a power series (representing, in general, an integral function) and the properties of the zeros of the polynomial or power series. *While in the earlier investigations the properties of the zeros were deduced from the number of changes of sign in the sequence of coefficients, here these relations will be used in the "opposite direction"; mainly those relations will be considered which imply the equality of the number of changes of sign and the number of zeros.* Such theorems are, in connection with Theorems 14—16, special cases of Descartes' rule of sign (Lipka 1938, 1942) or the theorem that if the zeros of the polynomial $P(z)=a_0 z^N + a_1 z^{N-1} + \dots + a_N$ with real coefficients are contained in the sector

$$\pi - \frac{\pi}{k+1} \leqq \arg z \leqq \pi + \frac{\pi}{k+1}$$

then the sequence $a_0, \dots, a_N$ is k times positive (Schoenberg 1954; see also Karlin 1968 p. 415). It is known that the distribution is defined as unimodal if, in the sequence of the first differences of its members, only a single change of sign occurs.

Some theorems built on those ideas will be presented here.

Theorem 17. If the generating function $P(z)$ of a discrete distribution

$$\{p_0, p_1, \dots p_n, \dots p_N\} \quad (p_n > 0; \quad n = 0, \dots, N)$$

has only real roots, then this distribution is unimodal (Medgyessy 1968, 1971*b*, 1972*a*, 1972*c*).

Example. As the generating function

$$P(z) = (pz + q)^N \quad (|z| \leqq 1, \quad p > 0, \quad q > 0, \quad p + q = 1, \quad N = 1, 2, \dots)$$

of a binomial distribution has only real roots, by virtue of the theorem the distribution is unimodal. (This is a curious proof of a well-known property.)

Theorem 18. If the generating function $P(z)$ of a discrete distribution $\{p_0, p_1, \dots, p_n, \dots, p_N\}$ $(p_0 > 0, p_N > 0)$ possesses only one pair of conjugate complex roots $\xi \pm i\eta$ with $\xi < -2$ and the other roots are real, then this distribution is unimodal (Medgyessy 1968, 1971*b*, 1972*a*, 1972*c*).

The evident unimodality of the distribution {45/52, 6/52, 1/52} cannot be deduced from any of the former theorems (e.g. its members do not form a twice positive sequence). The roots of its generating function $P(z)=\dfrac{1}{52}(45+6z+z^2)$ are $-3\pm6i$ and there is no real root; hence, the unimodality of the distribution follows from Theorem 18.

Theorem 19. *If the generating function* $P(z)$ *of a discrete distribution* $\{p_n\}$ $(n=0, 1, ...; p_0>0)$ *has the form*

$$P(z) = \frac{e^{\alpha z}}{P(1)} \prod_{v=1}^{\infty} \left(1-\frac{z}{\alpha_v}\right) \qquad (|z| \leq 1)$$

it can be defined for any z and among its roots α_v *there are M complex ones, and there is no positive; further, if the complex roots lie in a circle of radius r whose centre is in the point* $(-d, 0)$ *where* $d > r\sqrt{M+1}$, *then this distribution is unimodal* (Medgyessy 1968, 1971b, 1972a, 1972c).

The next theorem is connected with a theorem concerning moments (Fejér 1914).

Theorem 20. *Let* $\{p_n\}$ $(n=0, 1, ...)$ *be a discrete distribution with the generating function* $\Pi(z)$. *If* $\Pi(-s)(1+s)$ *can be defined for any* $s>0$ *and equals the Laplace transform of some continuous function* $f(x)$ *vanishing for* $x<0$ *with* $f(0)\geqq0$ *and all moments of* $f(x)$ *existing, and if, moreover,* $f(x)$ *has a single zero in* $(0, \infty)$, *then this distribution* $\{p_n\}$ *is unimodal* (Medgyessy 1968, 1971b, 1972a, 1972c).

Example. Let us consider the negative binomial distribution $\{(n+1)(1-p)^2p^n\}$ $(n=0, 1, ...; 0<p<1)$ with the generating function $\Pi(z)=\left(\dfrac{1-p}{1-pz}\right)^2$. It is easily seen that

$$\Pi(-s)(1+s) = \frac{1-p}{(1+ps)^2}(1+s) = \int_0^\infty e^{-sx}f(x)\,dx,$$

where

$$f(x) = \left(\frac{1-p}{p}\right)e^{-\frac{x}{p}}\left[\left(\frac{1-p}{p}\right)-x\right].$$

This function has a single zero; as the other conditions of the theorem are also fulfilled, we find in this way that the considered distribution is unimodal.

Sometimes the unimodality of a discrete distribution can be inferred from its generating function satisfying a certain differential equation. A relevant criterion is provided by

Theorem 21. *Let the generating function* $g(z)$ *of a discrete distribution* $\{p_n\}$ $(n=0, 1, ...; p_0 > 0)$ *satisfy the differential equation*

$$\sum_{v=0}^{M} (A_v z^v + B_v z^{v+1}) \frac{d^v}{dz^v} g(z) = 0 \qquad (0 < |z| \leqq 1)$$

where $M > 0$, A_v, B_v *are constants. Further let*

$$\sum_{v=0}^{M} A_v \binom{n+1}{v} v! \neq 0 \quad \text{if} \quad n = 0, 1, ...$$

and

$$\sum_{v=0}^{M} B_v \binom{n}{v} v! \neq 0 \qquad \text{if} \quad n = 0, 1, ..., N \qquad (N \leqq +\infty)$$

and let there be at most one change of sign in the sequence

$$\left\{ -\frac{\sum\limits_{v=0}^{M} B_v \binom{n}{v} v!}{\sum\limits_{v=0}^{M} A_v \binom{n+1}{v} v!} - 1 \right\} \qquad (n=0, 1, ...).$$

Then $\{p_n\}$ *is unimodal* (Medgyessy 1972a).

Proof. Inserting the power series of $g(z)$ in our differential equation and comparing coefficients we get $p_{n+1} = \Phi(n)p_n$ $(n=0, 1, ...)$ where

$$\Phi(n) = -\frac{\sum\limits_{v=0}^{M} B_v \binom{n}{v} v!}{\sum\limits_{v=0}^{M} A_v \binom{n+1}{v} v!}.$$

From the conditions of the theorem we have $p_n \neq 0$ if $n=0, 1, ..., N$ $(N \leqq +\infty)$. Thus in the sequence of the quantities $p_{n+1} - p_n = [\Phi(n) - 1] p_n$ the number of changes of sign is equal to that in the sequence of the quantities $\Phi(n) - 1$, that is at most one; consequently the number of changes of sign of the sequence $p_0, p_1 - p_0, ..., p_{n+1} - p_n, ...$ is exactly one; hence, by Definition 1, the unimodality of $\{p_n\}$ follows.

The mode of $\{p_n\}$ can be calculated immediately from the above. When $M=1$ the condition that at most one change of sign occurs in the sequence $\{\Phi(n) - 1\}$ is fulfilled from the beginning.

We mention also

Theorem 22. Let $f(\lambda)$ be a unimodal density function being identically 0 for $\lambda \leqq 0$. Then the discrete (mixture) distribution

$$\left\{ \frac{1}{n!} \int_0^\infty e^{-\lambda} \lambda^n f(\lambda) d\lambda \right\} \qquad (n = 0, 1, \ldots)$$

is unimodal (Holgate 1970).

The following unsolved **problems**, possessing some analogy with certain problems concerning distribution functions are worth mentioning.

1. When will the convolution of (0) unimodal discrete distributions be also (0) unimodal? (The specialization for convolution powers is obvious.)

2. We know that the generating function of an infinitely divisible discrete distribution always has the form

$$g(z) = e^{\lambda[f(z)-1]} \qquad (\lambda > 0)$$

where $f(z)$ is some generating function (Feller 1957 p. 271). What conditions on $f(z)$ will ensure that $g(z)$ is the generating function of a unimodal discrete distribution? (Conjecture: if the coefficients of $f(z)$ form a logarithmically concave sequence then the solution is $f(z) = z$.)

3. Under what general conditions will the convolution of a unimodal distribution and a non-unimodal distribution be unimodal?

4. When will the mixture of unimodal discrete distributions be unimodal?

5. Those families of unimodal discrete distributions should be determined in which the convolution of any two members is unimodal. (This raises the question of the *discrete analogues of the stable distribution functions*.)

6. A conjecture by A. Rényi: Let $\{p_m\}$ $(m = 0, 1, \ldots)$ be a discrete distribution. Then there is a number n_0 such that if $n \geqq n_0$, $\{p_m\}^{*n}$ will be a unimodal discrete distribution. This conjecture was suggested by the theorem that the limit density function of an appropriately normed sum of equally distributed independent discrete random variables is normal, i.e. unimodal (Gnedenko 1954 p. 252).

7. Is it possible to *discretize*, appropriately, distribution functions so that their unimodality can be investigated with the aid of the theorems valid for discrete distributions?

§ 3. IDENTIFICATION PROBLEMS CONNECTED WITH MIXTURES OF DENSITY FUNCTIONS AND DISCRETE DISTRIBUTIONS

The decomposition of a superposition $k(x)=\sum_{k=1}^{N} p_k f(x, \alpha_k, \beta_k)$ has meaning only if the superposition is *"identifiable"* (Teicher 1960), that is, if it is identical with another superposition $k^*(x)=\sum_{k=1}^{N'} p'_k f(x, \alpha'_k, \beta'_k)$ if and only if $N'=N$, $p'_k=p_k$, $\alpha'_k=\alpha_k$, $\beta'_k=\beta_k$ $(k=1, 2, ..., N)$. Thus in case of a given superposition the first step is to decide whether it is identifiable or not.

Since a superposition is, except for constant factor, a so-called *finite mixture* (Teicher 1963), only a mixture of a finite number of components occurs. The results concerning the identifiability of such mixtures (Teicher 1960, 1961, 1963; Barndorff-Nielsen 1965; Yakowitz, Spragins 1968; Tallis 1969) can be used in deciding the identifiability of a concrete superposition. Their criteria will not be given here; only those superpositions will be enumerated which can be stated to be *identifiable* on account of the results of the above-mentioned authors. These are the *superposition of*

1. *stable (e.g. normal, Cauchy) density functions;*

2. *Gamma density functions;*

3. ch *density functions of type* $f(x)=\dfrac{1}{2\,\mathrm{ch}(\pi x/2)}$;

4. *negative binomial distributions;*

5. *binomial distributions belonging to the type* $\left\{ \sum_{k=1}^{N} p_k \binom{\alpha_k}{n} \beta_k^n (1-\beta_k)^{\alpha_k-n} \right\}$ $(n=0, 1, ...)$ *where the* α_k *are different integers,* β_k $(0<\beta_k<1)$ *is given.*

At the same time, the *superpositions of binomial distributions belonging to the type* $\left\{ \sum_{k=1}^{N} p_k \binom{\alpha}{n} \beta_k^n (1-\beta_k)^{\alpha-n} \right\}$ $(n=0, 1, ..., \alpha)$ *where* α *is a given integer,* $0<\beta_k<1$ *and the* β_k *are different, are identifiable only if* $\alpha \geqq 2N-1$.

In the following, any investigated superposition is always assumed to be identifiable. For the examples in the present book this is true without exception.

Supplements and problems to Ch. II, § 3

1. A problem of identification — in case of finite mixtures of normal density functions — was investigated much earlier than the previously mentioned ones (Pearson 1894; Lonn 1932).

2. In the application of certain decomposition methods, it will automatically

happen that the superposition to be decomposed is identifiable, e.g. in the decomposition of a superposition of stable density functions (cf. Teicher 1963). However, a general method of proof for identification problems cannot be built up from these facts.

3. The proof technique of theorems on identifiability, based in contrast to the usual ones on the utilization of continuity properties, is well shown with the notations of Ch. I, § 1 by

Theorem 1. Let $s_A(x, B, c_k)$ $(c_k > 0)$ be a stable density function (cf. Ch. II, § 1) with characteristic function

$$\varphi_k(t) = e^{-c_k |t|^A \{1 + iB \operatorname{sgn} t \cdot \omega(t, A)\}}$$

$$\left(c_k > 0, \ 0 < A \leqq 2, \ |B| \leqq 1, \ \omega(t, A) = \tan\frac{\pi A}{2} \ (A \neq 1), \ \omega(t, 1) = \frac{2}{\pi} \log|t|\right).$$

Then the superposition

$$k(x) = \sum_{k=1}^{N} p_k s_A(x - \gamma_k, B, c_k)$$

$$(p_k > 0; \ (\gamma_i, c_i) \not\equiv (\gamma_j, c_j) \ (i \neq j); \ 0 < c_1 \leqq c_2 \leqq \ldots \leqq c_N)$$

is identifiable (Medgyessy 1962, Teicher 1963).

Proof. We have to prove that $k(x)$ is identical with a superposition

$$k_1(x) = \sum_{k=1}^{N} p_k' s_A(x - \gamma_k', B, c_k')$$

$$(p_k' > 0; \ (\gamma_i', c_i') \not\equiv (\gamma_j', c_j') \ (i \neq j); \ 0 < c_1' \leqq \ldots \leqq c_N')$$

only if

$$N = N', \ p_k = p_k', \ (\gamma_i, c_i) \equiv (\gamma_i', c_i').$$

Let us pass on to characteristic functions. Since the stable distribution functions can be differentiated arbitrarily many times (Hinčin 1938a p. 101; Gnedenko, Kolmogorov 1954 p. 183), this is a one-to-one correspondence. Now let us suppose identity; let

$$\sum_{k=1}^{N} p_k e^{i\gamma_k t} e^{-c_k |t|^A W(t)} \equiv \sum_{k=1}^{N'} p_k' e^{i\gamma_k' t} e^{-c_k' |t|^A W(t)}$$

$$\left(W(t) = 1 + iB \operatorname{sgn} t \cdot \omega(t, A), \ \omega(t, A) = \tan\frac{\pi A}{2} \ (A \neq 1), \ \omega(t, 1) = \frac{2}{\pi} \log|t|\right).$$

Let an ordering be introduced for the characteristic functions. We make the convention that $e^{i\Gamma_1 t} e^{-C_1 |t|^A W(t)}$ "precedes" the characteristic function $e^{i\Gamma_2 t} e^{-C_2 |t|^A W(t)}$ (Γ_1, Γ_2 are from among the quantities γ_i, γ_i', and C_1, C_2 are from among the c_i, c_i'), if $C_1 < C_2$ or, if $C_1 = C_2$, then $\Gamma_1 < \Gamma_2$. Without any restriction of generality it can also be assumed that if $v<\mu$ then $e^{i\gamma_v t}e^{-c_v|t|^A W(t)}$ "precedes" the function $e^{i\gamma_\mu t}e^{-c_\mu|t|^A W(t)}$ and similarly in case of the " ′ " parameters, further, that $e^{i\gamma_1 t}e^{-c_1|t|^A W(t)}$ "precedes" the function $e^{i\gamma_1' t}e^{-c_1'|t|^A W(t)}$ if they are not identical. Consequently $e^{i\gamma_1 t}e^{-c_1|t|^A W(t)}$ "precedes" $e^{i\gamma_v' t}e^{-c_v'|t|^A W(t)}$ ($1\leq v'\leq N'$). Dividing the identity between the characteristic functions by $e^{i\gamma_1 t}e^{-c_1|t|^A W(t)}$ (when doing so, the first term in the right-hand side reduces to p_1) and passing to the distribution functions one finds that the sum consisting of discontinuous function(s) and of continuous function(s) is identical with a continuous function. Since $p_1>0$ this contradiction can be raised only if $e^{i\gamma_1 t}e^{-c_1|t|^A W(t)}\equiv e^{i\gamma_1' t}e^{-c_1'|t|^A W(t)}$ that is $\gamma_1=\gamma_1'$, $c_1=c_1'$ and $p_1=p_1'$. Carrying out the simplification allowed by this fact and then repeating analogously the whole procedure, one gets $p_k=p_k'$, $\gamma_k=\gamma_k'$, $c_k=c_k'$ if $1\leq k\leq \min(N, N')$. If $N\neq N'$ — where we can assume $N>N'$ — it follows by cancelling identical terms that $\sum_{k=N'+1}^{N} p_k s_A(x-\gamma_k, B, c_k)\equiv 0$, i.e. $p_k=0$, if $N'+1\leq k\leq N$, which yields a contradiction unless $N=N'$; hence the statements of the theorem follow.

A plausible generalization of the method is given by Teicher (1963), who does not use the characteristic function and continuity considerations (Medgyessy 1962), but applies another transformation and makes use of the behaviour of the transforms at infinity. Evidently, the different tools can be applied in further combinations also; generalizations to types other than the stable density functions are also possible.

§ 4. THE CHARACTERIZATION OF THE SHAPE OF THE GRAPH OF A DENSITY FUNCTION

In comparing the graphs of density functions we often say that the graph of the one is "narrower" than that of the other. The "narrowness" has a fundamental role in our investigations; therefore we have to give an exact definition of these intuitive concepts. We shall make use of an earlier paper (Medgyessy 1954).

We notice that unimodality is not necessarily supposed for the moment in the following.

The "narrowness" of the graph of a density function can be defined most easily with the aid of the corresponding distribution function.

In the present book it will be sufficient to restrict the considerations to distribution functions and density functions obtained by means of a translation (which may be an 0-translation also) from one of the following types of distribution function and density function, respectively (see Fig. 1).

(A_1) **distribution function:** a strictly monotone increasing, differentiable (0) symmetrical distribution function (i.e. $F(-x)=1-F(x)$). Then the value of the function at $x=0$ is $1/2$, and in case of unimodality the mode appears at $x=0$.

(A_1) **density function:** a density function generating an (A_1) distribution function. Then it is (0) symmetrical, and in case of unimodality it is (0) unimodal.

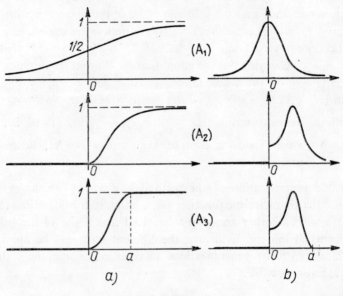

II.§ 4. Figure 1

(A_2) **distribution function:** a distribution function identically zero in $(-\infty, 0]$, which is strictly monotone increasing in $(0, \infty)$, differentiable and possessing the value 0 at $x=+0$.

(A_2) **density function:** a density function generating an (A_2) distribution function. Then it is identically 0 in $(-\infty, 0]$.

(A_3) **distribution function:** the same as the (A_2) distribution function with the difference that its value is 1 not at infinity but from some point A ($A>0$) onwards.

(A_3) **density function:** the same as the (A_2) density function, with the difference that it is identically zero also in (A, ∞).

Definition 1. Let $F_1(x)$ and $F_2(x)$ be (A_1) (or (A_2)) distribution functions. Let the graphs of $F_1(x)$ and $F_2(x)$ be $\mathcal{G}F_1(x)$ and $\mathcal{G}F_2(x)$ respectively. We say that the graph of $F_2(x)$ is "**narrower**" than the graph of $F_1(x)$ — in notation: $\mathcal{G}F_2(x) \prec \mathcal{G}F_1(x)$ — if the graph of $F_2(x)$ can be generated from the graph of $F_1(x)$ by such a compression along the x-axis and directed toward the origin, at which the distance of the abscissae of any two points of ordinates y_1 and y_2 of the graph of $F_2(x)$ is less than the distance between the abscissae of the points of the same ordinates y_1 and y_2 of the graph of $F_1(x)$ (see Fig. 2a) (Medgyessy 1971b, 1972a).

The relation denoted by "$\prec$" is, evidently, transitive.

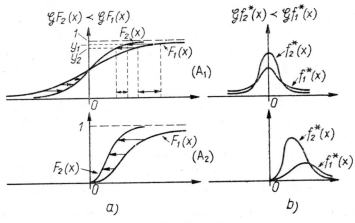

II. § 4. Figure 2

Of the curves of distribution functions $F_2^*(x)$ and $F_1^*(x)$, generated from $F_2(x)$ and $F_1(x)$ respectively by a translation we say that the same narrowness relation exists between them as $\mathscr{G}F_2(x)\prec\mathscr{G}F_1(x)$; we write $\mathscr{G}F_2^*(x)\prec\mathscr{G}F_1^*(x)$.

In the following the distribution functions will always be reduced to (A_1) (or (A_2) or (A_3)) distribution functions by a suitable translation.

The graphs of distribution functions generated merely by a translation from each other will be considered as equally "narrow".

Theorem 1. $\mathscr{G}F_2(x)\prec\mathscr{G}F_1(x)$ *is equivalent to any one of the following statements:*

(A) $F_1^{-1}(y)-F_2^{-1}(y)$ *is a monotone increasing function whose value is 0 at* $y=0$ (Medgyessy 1964).

(B) $F_2(x)=F_1(\varphi(x))$ *where* $\varphi(x)$ *is defined on* $(-\infty, \infty)$, $\varphi(0)=0$, $\varphi'(x)$ *exists, and* $\varphi'(x)>1$, *i.e.* $\varphi(x)$ *is a monotone increasing function* (Medgyessy 1971b, 1972a).

(C) *The function* $F(x)=F_1(x+a)-F_2(x)$, *where in the case of* (A_1) *distribution functions* $-\infty<a<\infty$, *and in the case of* (A_2) *distribution functions* $a>0$, *has a single change of sign, at* $x=x_a$ *say, for any permitted* a *and* $F(x)<0$ *if* $x>x_a$ (Medgyessy 1971b, 1972a).

Proof. (A) This is dealt with in Medgyessy (1964). (B) The sufficiency: Since $\varphi'(x)>1$ and $\varphi(x)$ is monotone increasing, it follows that $\varphi(x_2)-\varphi(x_1)> x_2-x_1$ if $x_2-x_1>0$. Let $x_2=F_2^{-1}(y_2)$, $x_1=F_2^{-1}(y_1)$; then $y_2>y_1$. Since $F_2(x)= F_1(\varphi(x))$ we have $F_1^{-1}(x)=\varphi(F_2^{-1}(x))$. All this yields $[F_1^{-1}(y_2)-F_2^{-1}(y_2)]- [F_1^{-1}(y_1)-F_2^{-1}(y_1)]>0$, which proves, by (A), the assertion. The proof of the necessity proceeds similarly. (C) A way of proving the sufficiency is, in the case of (A_1) distribution functions (cf. van Zwet 1964a, pp. 60—61), as follows: By the monotonicity properties $(F_1^{-1}(x)$ is *strictly monotone* etc.) the function $F_1^{-1}(F_1(x+a))-F_1^{-1}(F_2(x))$ has the same number of changes of sign as $F_1(x+a)- F_2(x)$. Thus the function $G(x)=x+a-F_1^{-1}(F_2(x))$ has a unique change of sign. As $F_1^{-1}[F_2(x)]=\varphi(x)$ is odd, either $\varphi'(x)>1$ or $\varphi'(x)<1$. At the same time, for a sufficiently large x, when $F_1(x+a)-F_2(x)<0$ then $x+a-\varphi(x)<0$ i.e. $\varphi'(x)>1$. Hence the assertion follows by (B). The necessity can be proved similarly. In the case of an (A_2) distribution function its image obtained by a mirroring around the origin and 1 are added to it; finally the result is multiplied by $1/2$. All this yields an (A_1) distribution function, which with the preceding considerations gives the proof. The necessity can be proved easily.

In addition it is easily seen that the graph of $F_2(x)$ passes over the graph of $F_1(x)$ if $x>x_a$ and the converse; further $\max F_2'(x) > \max F_1'(x)$.

The translation of the distribution functions does not essentially influence the foregoing statements.

They can be expressed also for (A_1) or (A_2) density functions, by an appropriate rephrasing of the statements valid for the corresponding distribution functions.

Theorem 2. If $F_1(x)$, $F_2(x)$ are (A_1) distribution functions which possess expectations and $\mathscr{G}F_2(x)<\mathscr{G}F_1(x)$, further if $p(x)$ is a (0) symmetrical Pólya frequency function such that for any given $\omega>0$, $\int_{-\infty}^{-\xi} p(x)dx<\omega\,p(-\xi)$ if $\xi>\xi(\omega)>0$, where $\xi(\omega)$ is a threshold number depending on ω, then

$$\mathscr{G}\left[\int_{-\infty}^{\infty} p(y-x)F_2(x)\,dx\right] < \mathscr{G}\left[\int_{-\infty}^{\infty} p(y-x)F_1(x)\,dx\right]$$

i.e. the graph of the distribution function obtained from $F_2(x)$ by this convolution transformation is narrower than the graph of the distribution function obtained from $F_1(x)$ (Medgyessy 1971b, 1972a).

Proof. Because of (C) in Theorem 1, it suffices to prove that the function

$$F(y) = \int_{-\infty}^{\infty} p(y-x+a)F_1(x)\,dx - \int_{-\infty}^{\infty} p(y-x)F_2(x)\,dx$$

has a single change of sign, say, at $y=y_a$ and $F(y)<0$ if $y>y_a$. Since

$$F(y) = \int_{-\infty}^{\infty} p(y-x)[F_1(x+a)-F_2(x)]\,dx,$$

by the theorem of Schoenberg (1951) to be cited in Theorem 10 of Ch. II, § 8, the number of changes of sign of $F(y)$ is at most equal to that of $F_1(x+a)-F_2(x)$ i.e. it is at most one. If it can be shown that for sufficiently large values of y $F(y)<0$ and for sufficiently small values of y $F(y)>0$, then the proof is complete because in this case $F(y)$ possesses at least one change of sign, that is, by what has been said above, exactly one change of sign, at $y=y_0$ say, and $F(y)<0$ if $y>y_a$. Only $F(y)<0$ will be proved here; the rest proceeds analogously. Let the change of sign of $F_1(x+a)-F_2(x)$ take place at $x=x_a$ and $y>x_a$. Then, since $p(y-x)=p(x-y)$ and $p(x_a-y)\leq p(x-y)$ if $x_a\leq x\leq 2y-x_a$ $(y>x_a)$,

$$F(y) = \int_{-\infty}^{x_a} [F_1(x+a)-F_2(x)]\,p(x-y)\,dx + \int_{x_a}^{\infty} [F_1(x+a)-F_2(x)]\,p(x-y)\,dx \leq$$

$$\int_{-\infty}^{x_a} p(x-y)\,dx - p(x_a-y)\left|\int_{x_a}^{|2y-x_a|} [F_1(x+a)-F_2(x)]\,dx\right| <$$

$$\int_{-\infty}^{x_a} p(x-y)\,dx - p(x_a-y)\,\delta\left|\int_{x_a}^{\infty} [F_1(x+a)-F_2(x)]\,dx\right|$$

where δ $(0<\delta<1)$ is fixed, if $y>x_a$ and $y>y_1(\delta)$ where $y_1(\delta)$ depends on δ and is sufficiently great. We have here, that $\left|\int_{x_a}^{|2y-x_a|} [F_1(x+a)-F_2(x)\,dx\right|$ is a monotone increasing function of y, and that because of the existence of the expectations, $\left|\int_{x_a}^{\infty} [F_1(x+a)-F_2(x)]\,dx\right|=\omega$ also exists. By the condition imposed on $p(x)$, nevertheless,

$$F(y) \leq \int_{-\infty}^{x_a-y} p(x)\,dx - \omega\delta p(x_a-y) < 0$$

if $y-x_a>y_2(\omega\delta)$ where $y_2(\omega\delta)$ is a threshold number depending on $\omega\delta$. Thus if $y>\max\,(y_1(\delta),\,x_a+y_2(\omega\delta))$, then $F(y)<0$ and the proof is complete.

The extension of this theorem to (A_1) density functions is trivial.

Definition 2. Let $f_1^*(x)$, and $f_2^*(x)$ be density functions generated by some translations from certain (A_1) (or (A_2)) density functions. Let the graphs of $f_1^*(x)$ and $f_2^*(x)$ be $\mathscr{G}f_1^*(x)$ and $\mathscr{G}f_2^*(x)$ respectively. We say that $\mathscr{G}f_2^*(x)$ is **"narrower"** than $\mathscr{G}f_1(x)$ — in notation: $\mathscr{G}f_2^*(x) \prec \mathscr{G}f_1^*(x)$ — if for the corresponding distribution functions $F_1^*(x)$, $F_2^*(x)$ the relation $\mathscr{G}F_2^*(x) \prec \mathscr{G}F_1^*(x)$ holds (Medgyessy 1971b, 1972a).

Intuitively $\mathscr{G}f_2^*(x) \prec \mathscr{G}f_1^*(x)$ means that $\mathscr{G}f_2^*(x)$ originates from $\mathscr{G}f_1^*(x)$ by a compression and a subsequent translation accompanied, however, by an area-preserving vertical dilatation (see Fig. 2b).

The above statement on distribution functions can be automatically extended to density functions *via* the distribution functions belonging to these density functions.

By the preceding definitions and considerations it will often be possible and also simpler to work with distribution functions.

In the following narrowness comparisons are given for the graphs of the members of a family $\{G(x, \lambda)\}$ $(\Lambda_1 \leqq \lambda \leqq \Lambda_2)$ of (A_1) (or (A_2)) distribution functions, — or, for the graphs of the members of a family $\{g(x, \lambda)\}$ of (A_1) (or (A_2)) density functions. According to the above statements, it suffices to investigate these functions for $x > 0$.

Definition 3. If λ_i, λ_j $(\Lambda_1 \leqq \lambda_i < \lambda_j \leqq \Lambda_2)$ are two different values of the parameter λ, and $\mathscr{G}G(x, \lambda_i) \prec \mathscr{G}G(x, \lambda_j)$, then we say that λ is the (Λ_2, Λ_1) **monotone formant** of the graph of $G(x, \lambda)$; if, conversely, $\mathscr{G}G(x, \lambda_j) \prec \mathscr{G}G(x, \lambda_i)$, then we say that λ is the (Λ_1, Λ_2) **monotone formant** of the graph of $G(x, \lambda)$ (Medgyessy 1971b, 1972a).

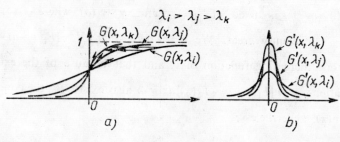

II. § 4. *Figure 3*

The notations (Λ_2, Λ_1) and (Λ_1, Λ_2) indicate the *direction* of the monotone change of λ. The intuitive meaning of the monotone formant is clear; as λ varies monotonely the corresponding graph becomes more and more narrow; the narrowness is characterized by a single number (see Fig. 3a).

Retaining the preceding definitions, we often say, to be brief: "λ is the **monotone formant** of $G(x, \lambda)$" and so on.

Definition 4. We say that λ is the (Λ_2, Λ_1) (or (Λ_1, Λ_2)) **monotone formant** of the graph of an (A_1) (or (A_2)) density function $g(x, \lambda)$ $(\Lambda_1 \leqq \lambda \leqq \Lambda_2)$ if it is the

(A_2, A_1) (or (A_1, A_2)) monotone formant of the distribution function $\int_{-\infty}^{x} g(y, \lambda)\, dy$ belonging to $g(x, \lambda)$ (Medgyessy 1971b, 1972a).

The intuitive meaning of this is evident (see Fig. 3b).

The statements in connection with the relation $\mathcal{G}F_2(x) \prec \mathcal{G}F_1(x)$ can be expressed also in relation with the monotone formant.

For instance we can say that λ *is the* (A_2, A_1) *monotone formant of an* (A_1) *distribution function* $G(x, \lambda)$ if the function $H(x) = G(x+a, \lambda_i) - G(x, \lambda_j)$ $(A_1 < \lambda_i < \lambda_j < A_2)$ has, for any a, a unique change of sign at some $x = x_a$ and if $x > x_a$, then $H(x) < 0$ — and so on, automatically.

Example 1. *Distribution functions (density functions) obtained by changing a scale parameter.* Let $G(x, \lambda) = F\left(\dfrac{x}{\lambda^\omega}\right)$ $(0 \leq \lambda < A_2,\ \omega > 0)$ where $F(x)$ is an (A_1) (or (A_2)) distribution function. As the diminishing of the scale parameter λ represents the compression described in Definition 1, $\mathcal{G}G(x, \lambda_j) \prec \mathcal{G}G(x, \lambda_i)$ if $0 < \lambda_j < \lambda_i < A_2$ i.e. λ is a $(A_2, 0)$ monotone formant of the graph of $F\left(\dfrac{x}{\lambda^\omega}\right)$. Evidently λ is the $(A_2, 0)$ monotone formant of $\mathcal{G}F\left(\dfrac{x-\alpha}{\lambda^\omega}\right)$, too, where α is arbitrary.

If we start from an (A_1) (or (A_2)) density function $\dfrac{1}{\lambda^\omega} f\left(\dfrac{x}{\lambda^\omega}\right)$ then from the definition of the monotone formant of density functions it immediately follows that λ is the $(A_2, 0)$ monotone formant of $\mathcal{G}\,\dfrac{1}{\lambda^\omega} f\left(\dfrac{x-\alpha}{\lambda^\omega}\right)$ $(0 < \lambda < A_2,\ \omega > 0)$.

It is also clear that in case of a density function $\dfrac{1}{(\beta-\lambda)^\omega} f\left(\dfrac{x-\alpha}{(\beta-\lambda)^\omega}\right)$ $(0 < \lambda < \beta$, $\beta,\ \alpha$ are given) λ will be the $(0, \beta)$ monotone formant of the graph of $\dfrac{1}{(\beta-\lambda)^\omega} f\left(\dfrac{x-\alpha}{(\beta-\lambda)^\omega}\right)$. The properties of the graphs of the corresponding distribution functions are obvious (see Fig. 4).

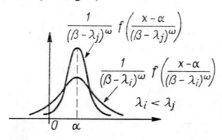

II. § 4. Figure 4

Example 2. Gamma distribution functions (density functions) of different order.
Let

$$G(x, \lambda) = \begin{cases} \int\limits_0^x \dfrac{B^\lambda y^{\lambda-1} e^{-By}}{\Gamma(\lambda)}\, dy & (x > 0) \\ 0 & (x \leq 0) \end{cases} \qquad (1 \leq \lambda \leq \Lambda_2;\ B > 0).$$

Then λ is a $(\Lambda_2, 1)$ monotone formant of $\mathscr{G}G(x, \lambda)$. Since $G(x, \lambda)$ is an (A_2) distribution function, it is sufficient to prove that if $\Lambda_2 > \lambda_i > \lambda_j > 1$, then the function $H(x) = G(x+a, \lambda_i) - G(x, \lambda_j)$ $(a>0)$ possesses a single change of sign. However, this is a particular case of a proof published in the book of van Zwet (1964a, pp. 60—61). It is easily seen also that for some x_a, $H(x) < 0$, if $x > x_0$; thus $\mathscr{G}G(x, \lambda_j) \prec \mathscr{G}G(x, \lambda_i)$. By our definition λ is also a $(\Lambda_2, 1)$ monotone formant of the Gamma density function

$$g(x, \lambda) = \begin{cases} \dfrac{B^\lambda x^{\lambda-1} e^{-Bx}}{\Gamma(\lambda)} & (x > 0) \\ 0 & (x \leq 0). \end{cases}$$

Example 3. The distribution function with characteristic function $\Phi(t) = \dfrac{\operatorname{ch} \lambda t}{\operatorname{ch} \beta t}$

$(\beta > 0, 0 \leq \lambda < \beta)$. $\Phi(t)$ is, in fact, a characteristic function; see Ch. II, § 8 below. The corresponding inverse Fourier transform,

$$\frac{1}{2\pi} \int\limits_{-\infty}^{\infty} \frac{\operatorname{ch} \lambda t}{\operatorname{ch} \beta t} e^{-ixt}\, dt = \frac{1}{\beta} \frac{\operatorname{ch} \dfrac{\pi x}{2\beta} \cdot \cos \dfrac{\pi \lambda}{2\beta}}{\operatorname{ch} \dfrac{\pi x}{\beta} + \cos \dfrac{\pi \lambda}{\beta}}$$

is, as it can be shown, an (A_1) density function; thus the corresponding distribution function is

$$G(x, \lambda) = \int\limits_{-\infty}^{x} \frac{1}{\beta} \frac{\operatorname{ch} \dfrac{\pi y}{2\beta} \cdot \cos \dfrac{\pi \lambda}{2\beta}}{\operatorname{ch} \dfrac{\pi y}{\beta} + \cos \dfrac{\pi \lambda}{\beta}}\, dy = \frac{1}{\pi} \left[\arctan \left(\frac{\operatorname{sh} \dfrac{\pi x}{2\beta}}{\cos \dfrac{\pi x}{2\beta}} \right) + \frac{\pi}{2} \right].$$

To show that λ is a $(0, \beta)$ monotone formant of $\mathscr{G}G(x, \lambda)$ it is sufficient to prove that if $0 < \lambda_i < \lambda_j < \beta$, then the function

$$H(x) = G(x+a, \lambda_i) - G(x, \lambda_j) =$$

$$\frac{1}{\pi} \left[\arctan \left(\frac{\operatorname{sh} \dfrac{\pi(x+a)}{2\beta}}{\cos \dfrac{\pi \lambda_i}{2\beta}} \right) - \arctan \left(\frac{\operatorname{sh} \dfrac{\pi x}{2\beta}}{\cos \dfrac{\pi \lambda_j}{2\beta}} \right) \right] \qquad (x > 0)$$

has a single change of sign. The number of changes of sign of $H(x)$ is, however, the same as that of the function

$$A(x) = \frac{\text{sh}\dfrac{\pi(x+a)}{2\beta}}{\cos\dfrac{\pi\lambda_i}{2\beta}} - \frac{\text{sh}\dfrac{\pi x}{2\beta}}{\cos\dfrac{\pi\lambda_j}{2\beta}}$$

(cf. the considerations in Theorem 1 (C)). From the properties of the function sh x it can be deduced easily that this number equals one. It is easily seen also that for some x_a $H(x)<0$ if $x>x_0$. Thus λ is the $(0,\beta)$ monotone formant of $\mathcal{G}G(x,\lambda)$.

Thus we can also say that λ is the $(0,\beta)$ monotone formant of the graph of the (A_1) density function

$$g(x,\lambda) = \frac{1}{\beta} \frac{\text{ch}\dfrac{\pi x}{2\beta}\cdot\cos\dfrac{\pi\lambda}{2\beta}}{\text{ch}\dfrac{\pi x}{\beta} + \cos\dfrac{\pi\lambda}{\beta}}.$$

From the viewpoint of analytical treatment the definition of the concept "narrower" introduced in the present paragraph is not convenient because it needs implicitly the investigation of the inverse functions. However, it seemed to be the most natural and the most suitable of the possible definitions for the change of shape properties. Its advantage is that the abscissa of the peak of a graph does not need to be determined; on the other hand it can be applied only in the case of (A_1) or (A_2) distribution (density) functions and it is generally useless in the case of (A_3) distribution functions or appropriate distribution function types defined by their characteristic functions (e.g. of infinitely divisible distribution functions) because it cannot be brought into relation with the characteristic function.

Thus another definition of "*narrower*" has become necessary.

From among the consequences of the relation $\mathcal{G}F_2(x)\prec\mathcal{G}F_1(x)$ there are two on which a new definition can be built *making use of the fact of unimodality*. This new definition can be handled more easily and, in addition, it can be applied also to (A_3) distribution (density) functions.

Definition 5. Let $F_1(x)$ and $F_2(x)$ be strictly unimodal (A_1) (or (A_2) or (A_3)) distribution functions. We say that the graph of $F_2(x)$ is "**narrower in the wider sense**" than the graph of $F_1(x)$ if (1) in the case of (A_1) or (A_2) distribution functions the graph of $F_2(x)$ passes, for $x>0$, over the graph of $F_1(x)$ — while in the case of (A_3) distribution functions the graph of $F_2(x)$ passes either everywhere over or everywhere under the graph of $F_1(x)$ at points x where $F_1(x) \neq 0$ or $F_1(x) \neq 1$; (2) the maximum of the derivative of $F_2(x)$ is greater than the maximum of the derivative of $F_1(x)$ (see Fig. 5a). This relation of "narrowness" will be denoted by $\mathscr{G}F_2(x) \overset{w}{\prec} \mathscr{G}F_1(x)$ (Medgyessy 1971b, 1972a).

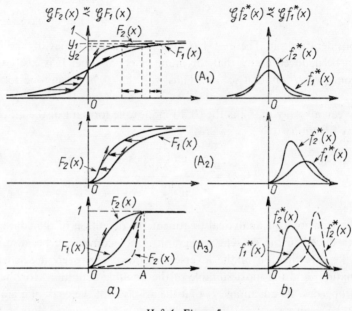

II. § 4. Figure 5

In the case of (A₁) or (A₂) distribution functions the relation $\overset{w}{\prec}$ is transitive.

Of the graphs of distribution functions $F_2^*(x)$ and $F_1^*(x)$ generated from $F_2(x)$ and $F_1(x)$ by means of a translation we say that there exists the same narrowness relation between them as that described above; this will be denoted also by $\mathscr{G}F_2^*(x) \overset{w}{\prec} \mathscr{G}F_1^*(x)$.

In the following such distribution functions will be considered from the viewpoint of the present narrowness relation by reducing them to the above (A₁) (or (A₂) or (A₃)) distribution functions by some translation.

Graphs of the above type, each generated from the other by some translation, will be called equally "narrow" (in the wider sense).

If the present "narrowness" relation holds, then e.g. in case of (A₁) or (A₂) distribution functions $\mathscr{G}F_2(x)$ originates from $\mathscr{G}F_1(x)$ by such a compression in which the abscissae of the points of the first graph tend to 0 — or, in case of (A₃) distribution functions, to another point — but the mutual distance of the abscissae of any two points of the ordinates y_1 and y_2 does not decrease necessarily;

the increase of the maximal value of the derivative is accompanied by an increase of the peak height of the corresponding density functions; this is a rather natural characteristic of the "narrowing" of the graph of the distribution function.

Definition 6. Let $f_1(x)$, $f_2(x)$ be strictly unimodal (A_1) $\big($or (A_2) or $(A_3)\big)$ density functions and $f_1^*(x)$, $f_2^*(x)$ be density functions having originated from the former ones by a translation (0-translation means the lack of translation). Let $\mathscr{G}f_1^*(x)$, $\mathscr{G}f_2^*(x)$ be the graphs of $f_1^*(x)$, $f_2^*(x)$ respectively. We say that $\mathscr{G}f_2^*(x)$ is "**narrower in the wider sense**" than $\mathscr{G}f_1^*(x)$ — in notation: $\mathscr{G}f_2^*(x) \overset{w}{\prec} \mathscr{G}f_1^*(x)$ — if $\mathscr{G}F_2^*(x) \overset{w}{\prec} \mathscr{G}F_1^*(x)$ holds for the distribution functions $F_1^*(x)$ and $F_2^*(x)$ belonging to $f_1^*(x)$ and $f_2^*(x)$ respectively (Medgyessy 1971*b*, 1972*a*).

The intuitive meaning of this is clear if the above remark is taken into account (see Fig. 5b).

Clearly, if the relation $\prec$ holds, then $\overset{w}{\prec}$ also holds.

Statements concerning density functions will in future be made *via* the corresponding distribution functions automatically.

"Narrowness" comparisons concerning the graphs of the members of one-parameter families $\{G(x, \lambda)\}$ and $\{g(x, \lambda)\}$ $(A_1 \leqq \lambda \leqq A_2)$ of strictly unimodal (A_1) $\big($or (A_2) or $(A_3)\big)$ distribution functions and density functions will be introduced here in the same way as in dealing with the former concept of "narrowness". It suffices to write the sign $\overset{w}{\prec}$ instead of $\prec$ everywhere at the appropriate place. Thus the terms: (A_2, A_2) $\big($or $(A, A_2)\big)$ **monotone formant in the wider sense** — or, in short: *monotone formant in the wider sense* — and the extension of the earlier statements concerning the monotone formant, etc. can be interpreted using this revised concept of narrowness.

For instance the statement "λ is the monotone formant in the wider sense of the (A_1) distribution function $G(x, \lambda)$" will mean that if $\lambda_i > \lambda_j > A_1$, then $\mathscr{G}G(x, \lambda_j) \overset{w}{\prec} \mathscr{G}G(x, \lambda_i)$ i.e. the graph of $G(x, \lambda_j)$ will pass over the graph of $G(x, \lambda_i)$ if $x > 0$ and the peak height of $G'(x, \lambda_j)$ will be greater than the peak height of

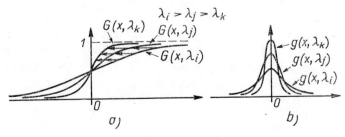

II. § 4. *Figure 6*

$G'(x, \lambda_i)$. The situation is similar for the monotone formant in the wider sense of an (A_1) density function, and so on (see Fig. 6).

The criteria for finding the monotone formant in the wider sense of a strictly unimodal (A_1) (or (A_2) or (A_3)) distribution function are significantly simpler than the criteria for the former monotone formant. One has for instance

Lemma 1. *If in the case of a strictly unimodal* (A_1) *distribution function* $G(x, \lambda)$ $(\Lambda_1 < \lambda < \Lambda_2)$ $\dfrac{\partial}{\partial \lambda} G(x, \lambda) < 0$ *for* $x > 0$, *and the height of peak of* $G'_x(x, \lambda)$ *increases as* λ *diminishes, then* λ *is a* (Λ_2, Λ_1) *monotone formant in the wider sense of the graph of* $G(x, \lambda)$ (Medgyessy 1971b, 1972a).

Analogous theorems can be made mechanically.

Example 1. The monotone formant λ in the former examples is, evidently, also a monotone formant in the wider sense in the case of strictly unimodal distribution functions.

Example 2. The *Beta distribution functions of type*

$$F(x, \beta) = \begin{cases} \displaystyle\int_0^x \dfrac{x^{\beta-1}(1-x)^{c-\beta-1}}{B(\beta, c-\beta)}\, dx & (0 < x \le 1) \\ 0 & \text{otherwise} \end{cases} \qquad (0 \le \beta \le 1/2)$$

$(c - \beta > 0, c$ is given). This is a strictly unimodal (A_3) distribution function. By simple differentiation it can be shown that β is a $(1/2, 0)$ monotone formant in the wider sense of the graph of $F(x, \beta)$. *A monotone formant in the sense of the former definition cannot be defined here from the first stage as it is connected with* (A_1) (*or* (A_2)) *distribution functions.*

Example 3. *The distribution function*

$$F(x, \beta, \lambda) = \int_{-\infty}^x \frac{1}{2}\left[\frac{1}{\pi(\beta-\lambda)}\frac{1}{1+y^2/(\beta-\lambda)^2} + \frac{1}{\pi(\beta+\lambda)}\frac{1}{1+y^2/(\beta+\lambda)^2}\right] dy$$

$$(0 < \lambda < \beta).$$

This is the superposition of a compressed Cauchy distribution function and a dilated one and is a strictly unimodal (A_1) distribution function. An elementary calculation shows that λ is the $(0, \beta)$ monotone formant in the wider sense of the graph of $F(x, \beta, \lambda)$, but *it is not a* $(0, \beta)$ *monotone formant of this graph in the former sense.*

Supplements and problems to Ch. II, § 4

1. The characterization of the shape of the graphs of distribution and density functions was also investigated by Medgyessy (1967a) mainly in that case when the variance or the expectation does not exist.

2. The concept "narrower" can be defined easily also in connection with the graphs of non-(A_1) (or (A_2) or (A_3)) distribution functions and density functions, in the same way as in Definitions 1 or 2 (cf. Medgyessy 1964).

3. In connection with (B) in Theorem 1 we notice that if $\varphi(x)$ is *convex* in $(0, \infty)$, too, then the transformation represented by it is identical with that introduced by van Zwet (1964a, 1964b). However, this restriction is not needed.

We also notice that the definition of the concept "narrowness" in Theorem 1 (C) is related to the definition of *peakedness* of Birnbaum (1948).

According to this definition a distribution function $F_2(x)$ is, at some point ξ, "more peaked" than the distribution function $F_1(x)$ at the point η if for any arbitrary $T > 0$ $F_2(\xi + T) - F_2(\xi - T) > F_1(\eta + T) - F_1(\eta - T)$. Our criterion in Theorem 1 (C) implies this if x_a is taken for both ξ and η; for at the distance T to the right and to the left from x_a $\mathscr{G}F_2(x)$ passes over $\mathscr{G}F_1(x)$ and under $\mathscr{G}F_1(x)$, respectively, and this means that $F_2(x)$ is "more peaked" at $x = x_a$ than $F_1(x)$.

4. In connection with Theorem 2 we notice that the more precise determination of the Pólya frequency function satisfying the conditions of the theorem is an unsolved **problem.**

5. If λ is a monotone formant of the graph $\mathscr{G}$ of an (A_1) (or (A_2)) distribution function $G(x, \lambda)$, then it is also a monotone formant of $G(\varphi(x), \lambda)$ where $\varphi(x)$ is the monotone function appearing in Theorem 1 (B) given that $G(\varphi(x), \lambda)$ is again an (A_1) (or (A_2)) distribution function, since a compression or extension does not influence the way in which $\mathscr{G}$ changes when λ increases (or decreases).

It is also clear that when the parameter λ^* is introduced instead of λ, where λ^* is a strictly monotone function of λ, this λ^* will also be a monotone formant of $G(x, \lambda)$ (with an appropriate domain of definition).

6. In principle the notion of monotone formant can be extended to a *discrete-valued* parameter λ. However, such an extension is not needed here (cf. *Supplements and problems to* Ch. IV, § 1, 25).

7. The present concept of monotone formant is *not* identical with the "monotone formant" introduced by Medgyessy (1961a p. 122). In the latter work a parameter c of a characteristic function $\varphi(t, c)$ $(c_1 \leqq c \leqq c_2)$ of a distribution function $F(x, c)$ was called "monotone formant" if $\varphi(t, c) \rightarrow e^{iAt}$ as $c \downarrow c_1$ (or $c \uparrow c_2$) where A is

some constant. Evidently, the graph of $F(x, c)$ did not become "narrower" simultaneously in such a simple manner as in the case of the above definition. We could say only that the P. Lévy distance of $F(x, c)$ and some degenerate distribution (Gnedenko, Kolmogorov 1954 p. 33) tended to zero if $c \downarrow c_1$ or $c \uparrow c_2$.

In certain cases the two definitions coincide, of course. This happens, for example, if $\varphi(t, c) = e^{izt} \Phi(ct)$ or, further, if $\varphi(t, c) = e^{izt}[\Phi(t)]^c$ $(c_1 \leq c \leq c_2)$ (cf. Theorem 4 in 12 below) where c is a $(c_2, 0)$ monotone formant, in the present sense, of the graph of $F(x, c)$. Then the monotone formant can be read off from the analytical form. *In such cases the monotone formants of the factors of a convolution may often be added to yield the monotone formant of that convolution, too* (cf. Medgyessy 1964).

8. The present definition of narrowness is independent of the existence of moments. However, in some cases — e.g. in case of a density function belonging to the type $\dfrac{1}{\lambda^\omega} f\left(\dfrac{x-\alpha}{\lambda^\omega}\right)$ $(\omega > 0)$ the corresponding *variance* (if it exists) stands in a one-to-one correspondence with the monotone formant λ; consequently it may play the role of that.

9. When $F_1(x) \neq 0$ in $(-\infty, 0)$ further $F_1(0) = F_2(0)$ and $F_1(x)$ is (0) unimodal, then the fact that $F_2(x)$ is narrower in the wider sense than $F_1(x)$ is equivalent to the statement "$F_2(x)$ is 'more peaked' at the origin than $F_1(x)$", in the sense given by Birnbaum (1948).

10. The definition given above of "narrower in the wider sense" is independent of the existence of the moments of the density function. However, in some cases e.g. in case of a strictly unimodal density function belonging to the type $\dfrac{1}{\lambda^\omega} f\left(\dfrac{x-\alpha}{\lambda^\omega}\right)$ $(\omega > 0)$ the *variance* — if it exists — is in a one-to-one correspondence with the monotone formant λ in the wider sense and, consequently, it may play the role of that.

11. Example 3 for the monotone formant in the wider sense is a consequence of the following

Theorem 3. If $f(x)$ is a strictly (0) unimodal (A_1) density function, then λ will be the $(0, \beta)$ monotone formant in the wider sense of the graph of the strictly (0) unimodal (A_1) density function

$$g(x, \lambda) = \frac{1}{2}\left[\frac{1}{(\beta-\lambda)^\omega} f\left(\frac{x}{(\beta-\lambda)^\omega}\right) + \frac{1}{(\beta+\lambda)^\omega} f\left(\frac{1}{(\beta+\lambda)^\omega}\right)\right] \quad (0 \leq \lambda < \beta, \ \omega > 0)$$

if for $x > 0$

$$\frac{f'(x)}{f(x)} > -\frac{\omega+1}{\omega x}$$

(Medgyessy 1971*b*, 1972*a*).

Proof. It suffices to prove only that the ordinates of $\int\limits_{-\infty}^{x} g(y, \lambda)\, dy$ increase monotonically if $\lambda \uparrow \beta$, i.e. that

$$\frac{\partial}{\partial\lambda} \int\limits_{-\infty}^{x} g(y, \lambda)\, dy = \frac{\partial}{\partial\lambda}\left[F\left(\frac{x}{(\beta-\lambda)^\omega}\right) + F\left(\frac{x}{(\beta+\lambda)^\omega}\right)\right] > 0$$

where $F(x) = \int\limits_{-\infty}^{x} f(y)\, dy$. A sufficient condition of this is that

$$\frac{1}{(\beta+\lambda)^{\omega+1}} f\left(\frac{x}{(\beta+\lambda)^\omega}\right) < \frac{1}{(\beta-\lambda)^{\omega+1}} f\left(\frac{x}{(\beta-\lambda)^\omega}\right) \qquad (x > 0).$$

This holds if $\dfrac{1}{\beta^{\omega+1}} f\left(\dfrac{x}{\beta^\omega}\right)$ is a monotone decreasing function of β, i.e.

$\dfrac{\partial}{\partial\beta}\left[\dfrac{1}{\beta^{\omega+1}} f\left(\dfrac{x}{\beta^\omega}\right)\right] < 0$; this follows, however, from the inequality $\dfrac{f'(x)}{f(x)} > $

$-\dfrac{\omega+1}{\omega x}$ $(x > 0)$.

If $f(x)$ is a (0) symmetrical density function of the Pearson family, i.e. if $\dfrac{f'(x)}{f(x)} = \dfrac{x}{A+Cx^2}$ (A, C are constants) then $\dfrac{f'(x)}{f(x)} > -\dfrac{\omega+1}{\omega x}$ holds if $\dfrac{x}{A+Cx^2} > $

$-\dfrac{\omega+1}{\omega x}$ $(x > 0)$. This cannot be fulfilled in the case of a normal density function ($A < 0, C = 0$); in the case of $A \neq 0, C \neq 0$, when $A = C < 0$ also holds, this will already be fulfilled if $|A| \geqq \dfrac{\omega}{\omega+1}$. The Pearson density functions corresponding to these conditions have the form $f(x) = \dfrac{R}{(1+x^2)^{1/2|A|}}$ (R is a constant). As $f(x)$ is a density function, $\dfrac{1}{2|A|} > 1/2$ i.e. $|A| < 1$ i.e. $\dfrac{\omega}{\omega+1} \leqq |A| < 1$. $|A| = 1/2$, $\omega \leqq 1$ yields the case of the distribution function derived from the Cauchy density function, figuring in Example 3.

12. On the basis of the characteristic function it is, generally, difficult to decide even whether some parameter is a monotone formant of the corresponding distribution function in the wider sense or not. A relevant criterion is yielded by

Theorem 4. *Let the strictly unimodal* (A_2) *distribution function* $F(x, \lambda)$ *have the characteristic function belonging to the type of B. de Finetti* (de Finetti 1930)

$$\varphi(t, \lambda) = e^{\lambda[\psi(t)-1]} \qquad (0 < \lambda < \Lambda_2)$$

where $\int\limits_{-\infty}^{\infty} |\varphi(t, \lambda)|\,dt < \infty$ *and* $\psi(t)$ *is the characteristic function of some* (A_2) *distribution function* $G(x)$. *Let the height of the peak of* $F_x'(x, \lambda)$ *increase when* $\lambda\downarrow 0$. *Then* λ *is a* $(\Lambda_2, 0)$ *monotone formant in the wider sense of the graph of* $F(x, \lambda)$ (Medgyessy 1971b, 1972a).

Proof. We have $F(0, \lambda) = 0$, $G(0) = 0$. Further, by the inversion formula

$$F(x, \lambda) = \frac{1}{2\pi} \int\limits_{-\infty}^{\infty} \frac{1 - e^{-ixt}}{it}\, \varphi(t, \lambda)\,dt$$

under our conditions, and we have

$$\frac{\partial F}{\partial \lambda} = \frac{1}{2\pi} \int\limits_{-\infty}^{\infty} \frac{1 - e^{-ixt}}{it}\, \varphi_\lambda'(t, \lambda)\,dt.$$

As $\varphi_\lambda'(t, \lambda) = \varphi(t, \lambda)[\psi(t) - 1]$,

$$\frac{\partial F}{\partial \lambda} = \frac{1}{2\pi} \int\limits_{-\infty}^{\infty} \frac{1 - e^{-ixt}}{it}\, \varphi(t, \lambda)\psi(t)\,dt - \frac{1}{2\pi} \int\limits_{-\infty}^{\infty} \frac{1 - e^{-ixt}}{it}\, \varphi(t, \lambda)\,dt =$$

$$\int\limits_0^x F(x - y, \lambda)\,dG(y) - F(x, \lambda) \leqq \int\limits_0^x F(x - y, \lambda)\,dG(y) - F(x, y) \int\limits_0^x dG(y) =$$

$$\int\limits_0^x [F(x - y, \lambda) - F(x, y)]\,dG(y) < 0,$$

as $F(x - y, \lambda) - F(x, \lambda) < 0$.

Thus $F(x, \lambda)$ increases at a fixed x, if $\lambda\downarrow 0$. Consequently, if the peak height of $F_x'(x, \lambda)$ increases as $\lambda\downarrow 0$, λ will be a $(\Lambda_2, 0)$ monotone formant (in the wider sense) of $\mathscr{G}F(x, \lambda)$.

13. Naturally, it is possible to give further definitions of monotone formant. We saw that in certain cases the variance could be taken as such, if it depended on a single parameter only. This can be generalized for (A_1) (or (A_2)) distribution functions of a family $\{G(x, \lambda)\}$ $(\Lambda_1 < \lambda < \Lambda_2)$ in the following definition of "monotone formant" (and of "narrowness" respectively) (Medgyessy 1971b, 1972a).

If the graph over $(0, \infty)$ of such a distribution function $G(x, \lambda)$ always lies over the graph of some function $\gamma(x, \lambda)$ of x and λ, which approaches "monotonely" both the y axis and the straight line $y = 1$ as e.g. $\lambda\downarrow\Lambda_1$ then we say that λ is the (Λ_2, Λ_1) "monotone formant", *in this newer sense*, of the graph of $G(x, \lambda)$. Clearly $G(x, \lambda)$ tends to a degenerate distribution function if $\lambda\downarrow\Lambda_1$; this justifies the definition.

The classical inequalities of probability theory as well as their generalizations (cf. Godwin 1964) provide the basis for the introduction of this new type of monotone formant because they give bounds for a distribution function i.e. types of $\gamma(x, \lambda)$. Let for example $G(x, \lambda)$ be an (A_2) distribution function for which the variance exists. Then from the basic idea of the Čebyšev inequality we have $1 - G(x, \lambda) \leqq \dfrac{1}{x^2} \int\limits_0^\infty y^2 dG(y, \lambda)$ i.e. $G(x, \lambda) \geqq 1 - \dfrac{m_2(\lambda)}{x^2}$ $\left(m_2(\lambda) = \int\limits_0^\infty y^2 dG(x, \lambda)\right)$ where $m_2(\lambda)$ depends on λ. The right-hand side corresponds to the function $\gamma(x, \lambda)$. If $m_2(\lambda) \downarrow 0$ for $\lambda \downarrow \Lambda_1$ then the graph of $1 - \dfrac{m_2(\lambda)}{x^2}$ approaches "monotonely" the y-axis and the straight line $y = 1$; consequently $G(x, \lambda)$ tends to the degenerate distribution function with jump at $x = 0$. Thus, in this case λ can be considered as the "monotone formant" in the newer sense of the graph of $G(x, \lambda)$.

Also the variance can be introduced, in the same way, into the function $\gamma(x, \lambda)$.

If λ is a monotone function of the variance then the latter can, clearly, be taken also for the "monotone formant". This supports the general view of regarding the variance as a measure of the "narrowness" of the graph of a distribution (or, rather, density) function in such cases when the variance is not proportional to some scale parameter. Evidently, the Markov inequality gives another possibility of introducing a function $\gamma(x, \lambda)$ and a "monotone formant" in case of (A_2) distribution functions if the variance does not exist.

It is possible to construct a function $\gamma(x, \lambda)$, independently of the existence of moments of any order. However, it is an unsolved **problem** whether a function $\gamma(x, \lambda)$ can be constructed based only on the characteristic function $\varphi(t)$ of $G(x, \lambda)$ e.g. by the measure of dispersion of Hinčin, $-\int\limits_0^a \log |\varphi(t)| dt$ ($|\varphi(t)| > 0$ if $0 \leqq t \leqq a$) (Hinčin 1937; Linnik 1960 pp. 52—53; Lukács 1964 p. 116), or by using the measure of dispersion of Ito ("the measure of concentratedness of the distribution function"), $\dfrac{1}{\pi} \int\limits_{-\infty}^\infty \dfrac{|\varphi(t)|^2}{1 + t^2} dt$ (Ito 1960 pp. 42—49).

§ 5. THE CHARACTERIZATION OF THE SHAPE OF THE GRAPH OF A DISCRETE DISTRIBUTION

In our investigations the "narrowness" of graphs of discrete distributions also has a fundamental role. Thus we have to establish this intuitive concept, too.

We wish to establish the characterization of the shape of the graph of a discrete distribution in full analogy of what has been said in § 4 in connection with the concept of "narrower in the wider sense".

5*

The "narrowness" of the graph of a discrete distribution $\{q_n\}$ $(n=\ldots, -1, 0, 1, \ldots)$ can be defined most easily with the aid of the corresponding *cumulative distribution* $\{h_n\} = \left\{ \sum\limits_{v=-\infty}^{n} q_v \right\}$ $(n=\ldots, -1, 0, 1, \ldots)$.

In the following only such discrete distributions and cumulative distributions will be considered which are generated by a 0 or positive integer-valued translation (the lack of translation being considered as a translation of 0 value) from one of the following types of distribution and cumulative distribution, respectively (see Fig. 1).

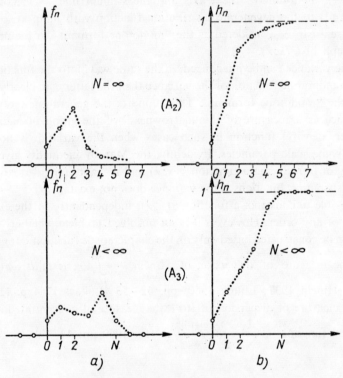

II. § 5. *Figure 1*

(A_2) **distribution:** A unimodal discrete distribution $\{f_n\}$ which is not 0 at $n=0, 1, \ldots$ and is 0 at points of negative integer abscissae.

(A_2) **cumulative distribution:** a discrete cumulative distribution $\{h_n\} = \left\{ \sum\limits_{v=0}^{n} f_v \right\}$ $(n=0, 1, \ldots)$ generated by an (A_2) distribution and defined at integer points. Thus it is $\neq 0$ at $n=0, 1, \ldots$, and it is 0 for $n=\ldots, -2, -1$.

(A_3) **distribution:** the same as the (A_2) distribution with the difference that it is 0 also at $n=N+1, N+2, \ldots$ where N $(N\geqq 0)$ is given.

(A_3) **cumulative distribution:** the same as the (A_2) cumulative distribution with the difference that it is 1 not at infinity but already at $n = N, N+1, \ldots$ where $N(N \geqq 0)$ is given.

We notice that here and in what follows the points of a distribution or a cumulative distribution are joined to aid the imagination by a dotted line.

Definition 1. Let $\{h_n^{(1)}\}$ and $\{h_n^{(2)}\}$ be (A_2) (or (A_3)) cumulative discrete distributions. Let the graphs of $\{h_n^{(1)}\}$ and $\{h_n^{(2)}\}$ be $\mathscr{G}\{h_n^{(1)}\}$ and $\mathscr{G}\{h_n^{(2)}\}$ respectively. We say that $\mathscr{G}\{h_n^{(2)}\}$ is "*narrower*" than $\mathscr{G}\{h_n^{(1)}\}$ — in notation: $\mathscr{G}\{h_n^{(2)}\} \overset{w}{\prec} \mathscr{G}\{h_n^{(1)}\}$ — if 1. the graph of $\{h_n^{(2)}\}$ in the case of an (A_2) cumulative distribution lies above the graph of $\{h_n^{(1)}\}$, in the case of an (A_3) cumulative distribution either everywhere above the graph of $\{h_n^{(1)}\}$, or everywhere under the graph of $\{h_n^{(2)}\}$ where $h_n^{(1)} \neq 0$ or $h_n^{(1)} \neq 1$; 2. the maximum of the first differences of the values of $\{h_n^{(2)}\}$ is greater than the maximum of the first differences of $\{h_n^{(1)}\}$ i.e. the peak height of the distribution belonging to the former is greater than that of the latter (see Figs 2a, b) (Medgyessy 1971*b*, 1972*a*).

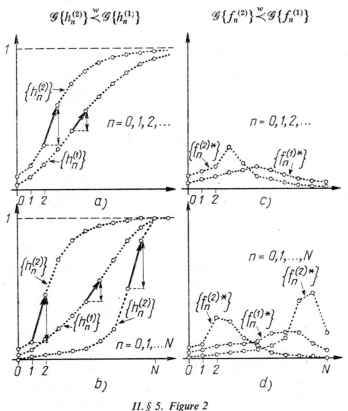

II. § 5. *Figure 2*

In case of (A_2) cumulative distributions the relation $\overset{w}{\prec}$ is transitive. The narrowness relation described above is also said to exist between the graphs of the cumulative distributions $\{h_n^{(2)*}\}$ and $\{h_n^{(1)*}\}$ generated from $\{h_n^{(2)}\}$ and $\{h_n^{(1)}\}$ respectively by means of a translation of 0 or positive integer value. This will be denoted also by $\mathcal{G}\{h_n^{(2)*}\} \overset{w}{\prec} \mathcal{G}\{h_n^{(1)*}\}$.

In the following, cumulative distributions will be reduced if possible to the above (A_2) $\big($or $(A_3)\big)$ cumulative distributions by means of an appropriate translation, when under examination from the viewpoint of this narrowness relation.

Graphs of cumulative distributions that can be derived from each other by some translation will be considered equally narrow.

If the present "narrowness" relation holds, the relation between the two graphs is, intuitively, the same *in broad lines* as the one between the two related graphs in case of the $\overset{w}{\prec}$ relation introduced in § 4 for graphs of distribution functions. Joining the points of the graphs of the cumulative distributions by a dotted line the resulting picture will readily show all this. Thus our definition is reasonable.

Definition 2. Let $\{f_n^{(1)}\}$, $\{f_n^{(2)}\}$ be unimodal (A_2) $\big($or $(A_3)\big)$ distributions and $\{f_n^{(1)*}\}$, $\{f_n^{(2)*}\}$ be discrete distributions originated by a 0 or integer valued translation from the former ones (0-translation means the lack of translation). Let the graphs of $\{f_n^{(1)*}\}$ and $\{f_n^{(2)*}\}$ be $\mathcal{G}\{f_n^{(1)*}\}$ and $\mathcal{G}\{f_n^{(2)*}\}$ respectively. We say that $\mathcal{G}\{f_n^{(2)*}\}$ is "*narrower*" than $\mathcal{G}\{f_n^{(1)*}\}$ — in notation: $\mathcal{G}\{f_n^{(2)*}\} \overset{w}{\prec} \mathcal{G}\{f_n^{(1)*}\}$ — if $\mathcal{G}\{h_n^{(2)*}\} \overset{w}{\prec} \mathcal{G}\{h_n^{(1)*}\}$ holds for the cumulative discrete distributions $\{h_n^{(1)*}\} = \left\{ \sum_{v=0}^{n} f_v^{(1)*} \right\}$ and $\{h_n^{(2)*}\} = \left\{ \sum_{v=0}^{n} f_n^{(2)*} \right\}$ (Medgyessy 1971b, 1972a).

The intuitive meaning of this is obvious on account of the above (see Figs 2c, d).

Statements concerning distributions will be composed, also here, *via* the corresponding cumulative distributions automatically.

In the following, narrowness comparisons will be made among the graphs of the members of a one-parameter family $\{h_n(\lambda)\}$ and $\{f_n(\lambda)\}$ $(A_1 \leqq \lambda \leqq A_2)$ of (A_2) $\big($or $(A_3)\big)$ cumulative distributions and (A_2) $\big($or $(A_3)\big)$ distributions respectively.

Definition 3. If λ_i, λ_j are two values of the parameter λ $(A_1 < \lambda_i < \lambda_j < A_2)$ and $\mathcal{G}\{h_n(\lambda_i)\} \overset{w}{\prec} \mathcal{G}\{h_n(\lambda_j)\}$ $(n = 0, 1, \ldots)$ then we say that λ is the (A_2, A_1) **monotone formant** of the graph of the (A_2) $\big($or $(A_3)\big)$ cumulative distribution $\{h_n(\lambda)\}$; if, on the contrary, $\mathcal{G}\{h_n(\lambda_j)\} \overset{w}{\prec} \mathcal{G}\{h_n(\lambda_i)\}$, then we say that λ is the

(Λ_1, Λ_2) **monotone formant** of the graph of the (A_2) (or (A_3)) *cumulative distribution* $\{h_n(\lambda)\}$ (Medgyessy 1971b, 1972a).

The notations (Λ_1, Λ_2) and (Λ_2, Λ_1) respectively, indicate the direction of the monotone change of λ. The two kinds of formant are called, here also, by the common name *monotone formant*; naturally keeping the above definitions.

The intuitive meaning of the monotone formant is clear: as λ varies monotonely the corresponding graph becomes more and more narrow (this is shown well by the graph of the cumulative distribution). The narrowness can thus be characterized by a single number (see Fig. 3a).

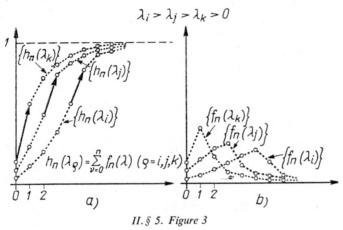

II.§ 5. *Figure 3*

Definition 4. We say that λ is the (Λ_2, Λ_1) (or (Λ_1, Λ_2)) **monotone formant** of the graph of an (A_2) (or (A_3)) distribution $\{f_n(\lambda)\}$ if it is a (Λ_2, Λ_1) (or (Λ_1, Λ_2)) monotone formant of the (A_2) (or (A_3)) cumulative distribution belonging to $\{f_n(\lambda)\}$ (Medgyessy 1971b, 1972a).

Its intuitive meaning is clear after the preceding (see Fig. 3b).

The criteria for the monotone formant of an (A_2) (or (A_3)) cumulative distribution are much simpler than those given in § 4 for the monotone formant introduced there. Such a criterion is given e.g. by

Lemma 1. If in case of an (A_2) cumulative distribution $\{h_n(\lambda)\}$, $\dfrac{d}{d\lambda} h_n(\lambda) < 0$ and the peak height of $h_{n+1}(\lambda) - h_n(\lambda)$, i.e. of the (A_2) distribution corresponding to $\{h_n(\lambda)\}$ increases if λ diminishes, then λ is a (Λ_2, Λ_1) monotone formant of the graph of $\{h_n(\lambda)\}$.

A similar lemma for the other type of monotone formant can be given automatically.

In what follows we shall be dealing with how to decide from the generating function $g(z, \lambda)$ of a distribution $\{f_n(\lambda)\}$ ($n=0, 1, ...$) whether λ is a monotone formant — supposing that we already know that the peak height of $\{f_n(\lambda)\}$ does not decrease with a corresponding monotone change of λ. The following two theorems are useful in making such a decision.

Theorem 1. *Let* $g(z, \lambda)$ *($|z| < 1$) be the generating function of an* (A_2) *(or* (A_3)) *distribution* $\{f_n(\lambda)\}$ *($n=0, 1, ..., N; N \leqq +\infty; \Lambda_1 \leqq \lambda \leqq \Lambda_2$). If in the case of* $\lambda \uparrow \Lambda_2$ *(or* $\lambda \downarrow \Lambda_1$*) the peak height of* $\{f_n(\lambda)\}$ *increases and, in addition,* $g'_\lambda(z, \lambda)$ *exists and all the coefficients of the power series of the function* $\dfrac{g'_\lambda(z, \lambda)}{1-z}$ *are positive (or negative) in the case of an* (A_2) *distribution and of the same sign in the case of an* (A_3) *distribution, then* λ *is the* (Λ_1, Λ_2) *(or* (Λ_2, Λ_1)) *monotone formant of the graph of* $\{f_n(\lambda)\}$ (Medgyessy 1971*b*, 1972*a*).

Proof. The generating function of the quantities $\sum\limits_{\nu=0}^{n} f_\nu(\lambda)$ i.e. of the cumulative distribution $\{h_n(\lambda)\}$ belonging to $\{f_n(\lambda)\}$ is $\dfrac{g(z, \lambda)}{1-z}$ ($|z| < 1$). Thus $\dfrac{g'_\lambda(z, \lambda)}{1-z} = \sum\limits_{n=0}^{\infty} \left(\dfrac{d}{d\lambda} h_n(\lambda)\right) z^n$. Hence, and from Lemma 1, the assertion already follows from the definition of the monotone formant.

Theorem 2. *Let* $w(z)$ *be the generating function of an infinitely divisible* (A_2) *(or* (A_3)) *distribution* $\{f_n\}$ *($n=0, 1, ...$) with the canonical representation* $w(z)=e^{B[H(z)-1]}$ *($B>0$) where* $H(z)=\sum\limits_{r=1}^{\infty} H_r z^r$ *is a generating function in which* $H_r \geqq 0$ *($r=1, 2, ...$). Let* $\{g_n(\lambda)\}$ *($0 < \lambda < \Lambda_2$) be an infinitely divisible* (A_2) *(or* (A_3)) *distribution whose generating function is* $g(z, \lambda)=w(z)^\lambda$. *If the peak height of the graph of* $\{g_n(\lambda)\}$ *increases if* $\lambda \downarrow 0$, *then* λ *is the* ($\Lambda_2, 0$) *monotone formant of the graph of* $\{g_n(\lambda)\}$ (Medgyessy 1971*b*, 1972*a*).

Proof. The proof of the canonical representation $w(z)=e^{B[H(z)-1]}$ ($B>0$) can be found in Feller's book (1957 p. 271); the proof that $w(z)^\lambda=g(z, \lambda)$ ($\lambda>0$) is a generating function is to be found in Gnedenko, Kolmogorov (1954 § 17). By Theorem 1 it suffices to prove that the coefficients of the power series of $\dfrac{g'_\lambda(z, \lambda)}{1-z}$ are negative. Now

$$\frac{g'_\lambda(z, \lambda)}{1-z} = \frac{e^{B\lambda[H(z)-1]}B[1-H(z)]}{1-z} \qquad (|z| < 1).$$

Here the first factor is a generating function. In

$$\frac{1-H(z)}{1-z} = 1 + \sum_{v=1}^{\infty} \left(1 - \sum_{r=1}^{v} H_r\right) z^v$$

the coefficients are non-negative. This proves that the coefficients of $\dfrac{g_\lambda'(z,\lambda)}{1-z}$ are negative.

The investigation of the change of the peak height of some (A_2) (or (A_3)) distribution $\{f_n(\lambda)\}$ $(n=0, 1, ...)$ is supported e.g. by the useful but almost trivial

Lemma 2. If in the case of an (A_2) (or (A_3)) distribution $\{f_n(\lambda)\}$, when λ decreases, the members of the (A_2) (or (A_3)) distribution belonging to the new value of λ increase up to the abscissa of the peak of the original distribution or to another point lying to the right of this, then the peak height of $\{f_n(\lambda)\}$ increases when λ decreases (Medgyessy 1971b, 1972a).

By this and similar simple statements the generally difficult determination of the peak height can be avoided.

In cases when the changing of the peak height cannot be stated exactly because of analytical difficulties, *instead of the increasing of the peak height we shall make the convention that we accept the fact that some lower bound of the peak height increases, when e.g. λ decreases.*

Example 1. The Poisson distribution. Let us consider the Poisson distribution $\left\{e^{-\lambda}\dfrac{\lambda^n}{n!}\right\}$ $(n=0, 1, ...; 0<\lambda<\Lambda_2)$ with generating function $g(z,\lambda)=e^{\lambda(z-1)}$. It is unimodal. Taking in Theorem 2 $B=1$, $h(z)=z$, we see that λ is a $(\Lambda_2, 0)$ monotone formant of the graph of $\left\{e^{-\lambda}\dfrac{\lambda^n}{n!}\right\}$, supposing that the peak height of this graph increases as $\lambda \downarrow 0$. However, this can be proved by the help of Lemma 1. Firstly for a fixed λ-value λ_0, the abscissa of the peak of $\mathscr{G}\left\{e^{-\lambda_0}\dfrac{\lambda_0^n}{n!}\right\}$ is at $[\lambda_0]$ ([ ]: integer part). Secondly,

$$\frac{d}{d\lambda} e^{-\lambda}\frac{\lambda^n}{n!} \begin{cases} > 0 & \text{if } \lambda < n, \\ = 0 & \text{if } \lambda = n, \\ < 0 & \text{if } \lambda > n; \end{cases}$$

thus the members of our Poisson distribution increase when λ decreases, if $n \le \lambda_0$, i.e. if $n \le [\lambda_0]+1$. These n-values are at most equal to the abscissa of the peak; hence the assertion follows.

A direct proof of the latter assertion is easy also.

Example 2. *The negative binomial distribution.* Let us consider the negative binomial distribution $\left\{\begin{pmatrix} R-1+n \\ R-1 \end{pmatrix}(1-\lambda)^R \lambda^n\right\}$ $(n=0, 1, ..., R=1, 2, ...; 0<\lambda<1)$ with generating function $g(z, \lambda) = \left[\dfrac{(1-\lambda)}{(1-\lambda z)}\right]^R$. It is unimodal. Theorem 1 can be applied: $\dfrac{g'_\lambda(z, \lambda)}{1-z} = -R \dfrac{(1-\lambda)^{R-1}}{(1-\lambda z)^{R+1}}$ and all the coefficients of this power series are negative. Thus λ is a $(1, 0)$ monotone formant of the graph of $\left\{\begin{pmatrix} R-1+n \\ R-1 \end{pmatrix}(1-\lambda)^R \lambda^n\right\}$, supposing that the peak height of this graph increases as $\lambda \downarrow 0$, which holds by Lemma 2, as for a fixed λ-value, λ_0, the peak of this graph lies at $\left[\dfrac{R\lambda_0-1}{1-\lambda_0}\right]+1$. On the other hand

$$\frac{d}{d\lambda}\begin{pmatrix} R-1+n \\ R-1 \end{pmatrix}(1-\lambda)^R \lambda^n \begin{cases} >0 & \text{if } \lambda < \dfrac{n}{R+n}, \\[2mm] =0 & \text{if } \lambda = \dfrac{n}{R+n}, \\[2mm] <0 & \text{if } \lambda > \dfrac{n}{R+n}; \end{cases}$$

thus the members of our distribution increase with the decrease of λ if $n \leqq \dfrac{R\lambda_0}{1-\lambda_0}$ i.e. if $n \leqq \left[\dfrac{R\lambda_0}{1-\lambda_0}\right]+1$. These n-values are at most equal to the abscissa of the peak; hence the assertion follows.

Example 3. *The binomial distribution.* Let us consider the binomial distribution $\left\{\begin{pmatrix} R \\ n \end{pmatrix}e^{-\lambda n}(1-e^{-\lambda})^{R-n}\right\}$ $(n=0, 1, ..., R; R=1, 2, ...; 0<\lambda<\log 2)$ with the generating function $g(z, \lambda)=[e^{-\lambda}(z-1)+1]^R$. It is unimodal. Again Theorem 2 can be applied: $\dfrac{g'_\lambda(z, \lambda)}{1-z} = e^{-\lambda} R[e^{-\lambda}(z-1)+1]^{R-1}$ and all the coefficients of this power series are positive. Thus λ is a $(\log 2, 0)$ monotone formant of the graph of $\left\{\begin{pmatrix} R \\ n \end{pmatrix}e^{-\lambda n}(1-e^{-\lambda})^{R-n}\right\}$, given that the peak height of this graph does not decrease at $\lambda \downarrow 0$. By Lemma 2, this also holds; for a fixed λ-value, λ_0, the peak of this graph lies at $[(R+1)e^{-\lambda_0}]$, further

$$\frac{d}{d\lambda}\binom{R}{n}e^{-\lambda n}(1-e^{-\lambda})^{R-n} \begin{cases} >0 & \text{if } \lambda < \log\dfrac{R}{n}, \\[2mm] =0 & \text{if } \lambda = \log\dfrac{R}{n}, \\[2mm] <0 & \text{if } \lambda > \log\dfrac{R}{n}; \end{cases}$$

thus the members of the distribution decrease with the decrease of λ if $n < Re^{-\lambda_0}$ i.e. if

$$n = \begin{cases} [Re^{-\lambda_0}]-1 & \text{if } Re^{-\lambda_0} \text{ is an integer,} \\ [Re^{-\lambda_0}] & \text{otherwise,} \end{cases}$$

and increase otherwise. These n-values are equal to or greater than the abscissa of the peak; hence the assertion follows.

Example 4. *The discrete distribution of generating function*

$$g(z, \lambda) = \left(\frac{1-p}{1-\lambda}\right)^R\left(\frac{1-\lambda z}{1-pz}\right)^R \qquad (0 < \lambda < p < 1; \quad R = 1, 2, \ldots).$$

This generating function is the quotient of two generating functions each belonging to a negative binomial distribution with parameters (p, R) and (λ, R) respectively. In § 8 we shall see that $g(z, \lambda)$ is actually a generating function of an infinitely divisible distribution $\{q_n(p, \lambda)\}$; the latter is unimodal, too. This can be seen by making use of the transformation

$$\left(\frac{1-p}{1-\lambda}\right)^R\left(\frac{1-\lambda z}{1-pz}\right)^R = \left(\frac{1-p}{1-\lambda}\right)^R\left(\frac{\lambda}{p}+\frac{1-\lambda/p}{1+pz}\right)^R =$$

$$\left(\frac{1-p}{1-\lambda}\right)^R\left(\frac{\lambda}{p}\right)^R\left[1+\sum_{v=1}^{R}\binom{R}{v}\left(\frac{p/\lambda-1}{1-p}\right)^v\left(\frac{1-p}{1-pz}\right)^v\right]$$

(presenting the generating function $g(z, \lambda)$ as a mixture of generating functions of negative binomial distributions of different order). Hence the corresponding distribution is

$$\{q_n(p, \lambda)\} = \left\{\left(\frac{1-p}{1-\lambda}\right)^R\left(\frac{\lambda}{p}\right)^R\left[1+p^n\sum_{v=1}^{R}\binom{R}{v}\left(\frac{p}{\lambda}-1\right)^v\binom{v-1+n}{v-1}\right]\right\}.$$

In this $\displaystyle\sum_{v=1}^{R}\binom{R}{v}\left(\frac{p}{\lambda}-1\right)^v\binom{v-1+n}{v-1}$ is a polynomial $P_{R-1}(n)$ in n of the $(R-1)$th degree, with *positive* coefficients. For a moment, let $p^n P_{R-1}(n) = V(n)$ be considered as a continuous function of n. Its derivative is, as $P_{R-1}(n)$ is a convex function, either negative or has a single change of sign; thus the function has only one extremum. Returning to discrete n-values we have from this that $\{q_n(p, \lambda)\}$ is unimodal. — We can make use of Theorem 1. We have

$$\frac{g'_\lambda(z, \lambda)}{1-z} = \frac{R}{(1-\lambda)^2} \left(\frac{1-p}{1-\lambda}\right)^{R-1} \left(\frac{1-\lambda z}{1-pz}\right)^{R-1} \left(\frac{1-p}{1-pz}\right)$$

which has, being a product of generating function and positive constants, a power series expansion with positive coefficients. Thus by Theorem 1 λ is a $(0, p)$ monotone formant of the graph of $\{q_n(p, \lambda)\}$, supposing that the peak height of this graph does not decrease as $\lambda \uparrow p$. Here Lemma 2 cannot be applied because of the technical difficulties: we can only make use of the convention given there and showing that *some lower bound of the peak height increases when* $\lambda \uparrow p$. We shall use the inequality

$$\sum_{n=0}^{\infty} [q_n(p, \lambda)]^2 \leq \max_{0 \leq n < \infty} q_n(p, \lambda).$$

The right-hand side is equal to the peak height. The left-hand side can be expressed by means of the generating function:

$$\sum_{n=0}^{\infty} [q_n(p, \lambda)]^2 = \frac{1}{2\pi} \int_0^{2\pi} |g(e^{it}, \lambda)|^2 dt =$$

$$\frac{1}{2\pi} \int_0^{2\pi} \left(\frac{1-p}{1-\lambda}\right)^{2R} \left(\frac{1-2\lambda \cos t + \lambda^2}{1-2p \cos t + p^2}\right)^R dt.$$

For the derivative with respect to λ of the right-hand side we have

$$\frac{\partial}{\partial \lambda} \frac{1}{2\pi} \int_0^{2\pi} \left(\frac{1-p}{1-\lambda}\right)^{2R} \left(\frac{1-2\lambda \cos t + \lambda^2}{1-2p \cos t + p^2}\right)^R dt =$$

$$\frac{R}{\pi} \frac{(1-p)^{2R}(1+\lambda)}{(1-\lambda)^{2R+1}} \int_0^{2\pi} \frac{(1-2\lambda \cos t + \lambda^2)^{R-1}}{(1-2p \cos t + p^2)^R} (1-\cos t) dt > 0,$$

since the integrand is non-negative. Thus this lower bound of the peak height increases as $\lambda \uparrow p$; which was to be proved.

Supplements and problems to Ch. II, § 5

1. The concept "narrower" is not defined in case of a *non*-(A_2) (or (A_3)) distribution or (A_2) (or (A_3)) cumulative distribution; our considerations cannot be extended to this case.

2. It is clear that if λ is the monotone formant of an (A_2) (or (A_3)) cumulative distribution $\{h_n(\lambda)\}$ and we introduce a new parameter λ^* instead of λ, where λ^* is a strictly monotone function of λ and $\lambda^* > \lambda$, then λ^* will be a monotone formant of the resulting (A_2) (or (A_3)) cumulative distribution $\{h_n(\lambda^*)\}$, with a convenient domain of definition.

3. The notion of monotone formant can be extended to a *discrete-valued* parameter λ, too. However, this extension can be neglected here. (Cf. *Supplements and problems to* Ch. IV, § 1, 21.)

4. The concept of the monotone formant, used here, is *not* identical with the "monotone formant" introduced earlier by the author (Medgyessy 1961a p. 122). In that book a parameter c of the characteristic function $\varphi(t, c)$ belonging to a discrete distribution $\{p_n(c)\}$ ($c_1 \leqq c \leqq c_2$) was called a "monotone formant" if $\varphi(t, c) \to e^{iMt}$ as $c \downarrow c_1$ (or $c \uparrow c_2$) where M is some integer. At any rate, the graph of the distribution function belonging to $\{p_n(c)\}$ did not become "narrower", simultaneously, in such a manner, as in case of our above definition. We can only say that the P. Lévy distance of the distribution function belonging to the distribution $\{p_n(c)\}$ and of some degenerate distribution function (Gnedenko, Kolmogorov 1954 p. 33) tends to zero if $c \downarrow c_1$ (or $c \uparrow c_2$). In certain cases the two definitions coincide, of course, e.g. if $\varphi(t, c) = e^{iMt}[\Phi(t)]^c$ where c is a $(c_2, 0)$ monotone formant, in the present sense, of the graph of $\{p_n(c)\}$. The Poisson distribution $(\varphi(t, c) = [e^{(e^{it}-1)}])$ and other infinitely divisible discrete distributions also belong to this class. In addition it is easily seen that the two definitions also coincide for the binomial distribution with $\varphi(t, c) = [e^{-c}(e^{it}-1)+1]^R$ if $c \downarrow 0$, and for the negative binomial distribution with $\varphi(t, c) = \dfrac{1-c}{1-ce^{it}}$ if $c \downarrow 0$, too.

5. The present definition of narrowness is independent of the existence of moments. However, the *variance,* if it exists, is sometimes in a one-to-one correspondence with the monotone formant λ and the former can thus play the role of the latter.

6. Naturally further definitions of "monotone formant" are also possible in case of discrete distributions. In practice the variance is often taken as a "monotone formant" if it contains a single parameter. This may support, in case of

a family $\{h_n(\lambda)\}$ $(\Lambda_1 \leqq \lambda \leqq \Lambda_2)$ of (A_2) (or (A_3)) cumulative distributions, the introduction of a newer concept of monotone formant and narrowness in the same way as described in *Supplements and problems to* Ch. II, § 4, 13, because the statements given there can be applied also to the graph of the distribution function of a discrete distribution. Consequently, the examples there concerning the use of the moments or the variance as a monotone formant (in the present sense) will be valid also in our case *via* generalizations of Čebyšev's inequality or other relations. All this makes it possible to find several new types of "monotone formant"; however, they will not be investigated here.

§ 6. UNIMODAL TRANSFORMS OF INCREASED NARROWNESS OF DENSITY FUNCTIONS

Definition 1. By a **unimodal transform of increased narrowness** of a strictly unimodal (A_1) (or (A_2) or (A_3)) density function $f(x)$ we mean a strictly unimodal (A_1) (or (A_2) or (A_3)) density function $g(y)$ which is in a well-defined analytical relation with $f(x)$ and for which the narrowness relation (cf. § 4) $\mathscr{G}g(y) \prec \mathscr{G}f(x)$ (or $\mathscr{G}g(y) \overset{w}{\prec} \mathscr{G}f(x)$) holds (Medgyessy 1971b, 1972a).

In many instances it is easier to find an appropriate $g(y)$ to a given $f(x)$ starting from the characteristic function $\varphi(t)$ of $f(x)$, on the basis of some very simple transformation of $\varphi(t)$.

Examples for all this will be seen in Chapter III.

§ 7. UNIMODAL TRANSFORMS OF INCREASED NARROWNESS OF DISCRETE DISTRIBUTIONS

Definition 1. By a **unimodal transform of increased narrowness** of a (unimodal) (A_2) (or (A_3)) distribution $\{f_n\}$ $(n = ..., -1, 0, 1, ...; f_v < 0,$ if $v < 0)$ we mean a unimodal (A_2) (or (A_3)) distribution $\{g_m\}$ $(m = ..., -1, 0, 1, ...; g_\mu = 0$ if $\mu < 0)$ which is in a well-defined analytical relation with $\{f_n\}$ and for which the narrowness relation (cf. § 5) $\mathscr{G}\{g_m\} \overset{w}{\prec} \mathscr{G}\{f_n\}$ holds (Medgyessy 1971b, 1972a).

In many instances it is easier to find an appropriate $\{g_m\}$ to a given $\{f_n\}$ starting from the generating function $g(z)$ of $\{f_n\}$, on the basis of some very simple transformation of $g(z)$.

Examples are given in Chapter IV.

§ 8. SOME SPECIAL THEOREMS CONCERNING DENSITY FUNCTIONS AND DISCRETE DISTRIBUTIONS

Here we enumerate various theorems which are referred to in different parts of the book, but cannot be included in any of the preceding points.

1. *The quotient of two characteristic functions.*

According to the (unique) Lévy–Hinčin canonical representation of the characteristic function $\varphi(t)$ of a non-degenerate infinitely divisible distribution function,

(1) $$\varphi(t) = \exp\left[i\gamma t + \int\limits_{\infty}^{\infty} \left(e^{itu} - 1 - \frac{itu}{1+u^2}\right) \frac{1+u^2}{u^2}\, dG(u)\right]$$

where γ is real, $G(u)$ that may be called the *spectral function*, is a non-decreasing function of bounded variation and for $u=0$ $\left(e^{itu} - 1 - \dfrac{itu}{1+u^2}\right) \dfrac{1+u^2}{u^2}$ is defined as equal to 0 (Gnedenko, Kolmogorov 1954 § 18).

Theorem 1. Let $\varphi_1(t)$ and $\varphi_2(t)$ be characteristic functions of non-degenerate infinitely divisible distribution functions whose canonical representations of type (1) contain the parameter and spectral function $\gamma_1, G_1(u)$ and $\gamma_2, G_2(u)$ respectively. If $G_1(u) - G_2(u)$ is a non-decreasing function of bounded variation, then $\dfrac{\varphi_1(t)}{\varphi_2(t)}$ is the characteristic function of a non-degenerate infinitely divisible distribution function. The condition is necessary and sufficient (Medgyessy 1961a pp. 168— 171).

Proof. The somewhat trivial proof is given in the book by P. Medgyessy (1961a pp. 168—171) and follows from the explicit expression of $\dfrac{\varphi_1(t)}{\varphi_2(t)}$ which is the same as that in (1) merely with $\gamma_1 - \gamma_2$ and $G_1(u) - G_2(u)$ instead of γ and $G(u)$, respectively; and from the uniqueness of the canonical representation of the characteristic function of an infinitely divisible distribution function.

Example. Let $\varphi_1(t) = \dfrac{1}{\operatorname{ch}\beta_1 t}$ and $\varphi_2(t) = \dfrac{1}{\operatorname{ch}\beta_2 t}$ $(0 < \beta_2 < \beta_1 < \infty)$ be the characteristic functions of the ch density functions $f_1(x) = \dfrac{1}{2\beta_1 \operatorname{ch}(\pi x/2\beta_1)}$ and $f_2(x) = \dfrac{1}{2\beta_2 \operatorname{ch}(\pi x/2\beta_2)}$ respectively. $\varphi_1(t)$ and $\varphi_2(t)$ are characteristic functions of infinitely divisible density functions: it can be proved that $\varphi_i(t)$ has a canonical representation belonging to type (1) with the parameter and the spectral function

$$\gamma_i = -\int\limits_{-\infty}^{\infty} \frac{y^2}{2(1+y^2)\,\mathrm{sh}\,(\pi y/2\beta_i)}\,dy \quad \text{and} \quad G_i(u) = \int\limits_{-\infty}^{u} \frac{y}{2(1+y^2)\,\mathrm{sh}\,(\pi y/2\beta_i)}\,dy$$

$$(i = 1, 2).$$

$G_1(u)-G_2(u)$ is non-decreasing (see Medgyessy 1961a pp. 64—66). The other condition of the theorem being fulfilled, we finally have that $\dfrac{\mathrm{ch}\,\beta_2 t}{\mathrm{ch}\,\beta_1 t}$ is the characteristic function of an infinitely divisible density function $f(x)$ for which routine calculation gives (cf. § 4, Example 3)

$$f(x) = \frac{1}{\beta_1}\,\frac{\mathrm{ch}\dfrac{\pi x}{2\beta_1}\cdot\cos\dfrac{\pi\beta_2}{2\beta_1}}{\mathrm{ch}\dfrac{\pi x}{\beta_1}+\cos\dfrac{\pi\beta_2}{\beta_1}}.$$

2. *The quotient of two generating functions.*

Although the relevant results can be obtained by the preceding theorem, it is worthwhile giving a more adequate theorem. First we recall that the necessary and sufficient condition that a generating function $g(z)$ be the generating function of some infinitely divisible discrete distribution sounds as follows: $g(z)$ has the form

(2) $$g(z) = e^{\lambda[h(z)-1]}$$

where $\lambda > 0$ and $h(z)$ is some generating function (Feller 1957 p. 271). We have, then, the somewhat trivial

Theorem 2. Let $g_1(z)$ and $g_2(z)$ be generating functions of non-degenerate infinitely divisible discrete distributions whose canonical forms belonging to type (2) contain the parameter and generating function $\lambda_1, h_1(z)$ and $\lambda_2, h_2(z)$ respectively, where $\lambda_1 > 0, \lambda_2 > 0$. If $\dfrac{\lambda_1 h_1(z)-\lambda_2 h_2(z)}{\lambda_1-\lambda_2}$ is a generating function then $\dfrac{g_1(z)}{g_2(z)}$ is the generating function of an infinitely divisible discrete distribution. The condition is necessary and sufficient (Medgyessy 1971b, 1972a).

Proof. The proof follows immediately from the explicit form of $\dfrac{g_1(z)}{g_2(z)}$ and from the uniqueness of the representation (2). Translations of integer value of the distributions belonging to $g_1(z)$ and $g_2(z)$ do not affect the validity of the theorem.

Example. Let $g_1(z) = \left(\dfrac{1-p_1}{1-p_1 z}\right)^R$, $g_2(z) = \left(\dfrac{1-p_2}{1-p_2 z}\right)^R$ be the generating functions of order R $(R = 1, 2, ...)$ of the negative binomial distributions of parameters (R, p_1) and (R, p_2) respectively, where $0 < p_2 < p_1 < 1$. Evidently

$$g_i(z) = \exp\left[\log\left(\frac{1}{1-p_i}\right) \cdot \left\{\frac{\log\left(\dfrac{1}{1-p_i z}\right)}{\log\left(\dfrac{1}{1-p_i}\right)} - 1\right\}\right] \qquad (i = 1, 2).$$

We have

$$\lambda_i = \log\left(\frac{1}{1-p_i}\right), \quad h_i(z) = \frac{\log\left(\dfrac{1}{1-p_i z}\right)}{\log\left(\dfrac{1}{1-p_i}\right)}$$

and

$$\frac{\lambda_1 h_1(z) - \lambda_2 h_2(z)}{\lambda_1 - \lambda_2} = \frac{\displaystyle\sum_{n=1}^{\infty} \frac{(p_1^n - p_2^n) z^n}{n}}{\log\left(\dfrac{1-p_2}{1-p_1}\right)}$$

which is the generating function of the discrete distribution

$$\{q_n\} = \left\{\frac{p_1^n - p_2^n}{n \log\left(\dfrac{1-p_2}{1-p_1}\right)}\right\} \qquad (n = 1, 2, ...).$$

Thus $\dfrac{g_1(z)}{g_2(z)} = \left(\dfrac{1-p_1}{1-p_2}\right)^R \left(\dfrac{1-p_2 z}{1-p_1 z}\right)^R$ is the generating function of an infinitely divisible discrete distribution (the distribution $\{q_n\}$).

3. *Partial differential equations for stable density functions.*

Let $s_A(x, B, c)$ be a stable density function (cf. Ch. II, § 1) with parameters A, B whose characteristic function has the form

$$\varphi(t) = e^{-c|t|^A\{1+iB\,\text{sgn}\,t \cdot \Omega(A)\}}$$

where $c > 0$, $0 < A \leq 2$, $|B| \leq 1$, $\Omega(A) = \tan\dfrac{\pi A}{2}$ $(A \neq 1)$, $\Omega(1) = 0$ and $A = \dfrac{m}{n}$ where m, n are relative primes. (Evidently $s_A(x-\gamma, B, c) = s_A(-x+\gamma, -B, c)$.) These stable density functions form, evidently, only a subset of the family of stable density functions. Then we have

Theorem 3. *If for some* $\varkappa$ *and* M, $\varkappa=0, \pm 1, \ldots$; $M=1, 2, \ldots$ *and for given values of the parameters* $A=\dfrac{m}{n}$ *and* B *the conditions*

$$\frac{\pi}{2} \frac{2\varkappa - Mm}{Mn} = \begin{cases} \arctan\left(B\tan\dfrac{\pi m}{2n}\right) & (A \neq 1) \\ 0 & (A = 1) \end{cases}, \qquad \left|\frac{2x - Mm}{Mm}\right| \leq 1$$

are fulfilled, then $s_A(x, B, c)=f(x, c)$ *satisfies the partial differential equation*

$$(3) \qquad (-1)^{Mn}(1+P^2)^{\frac{Mn}{2}} \frac{\partial^{Mm} f}{\partial x^{Mm}} - (-1)^{\varkappa} \frac{\partial^{Mn} f}{\partial c^{Mn}} = 0$$

where

$$P = \begin{cases} B\tan\dfrac{\pi A}{2} = B\tan\dfrac{\pi m}{2n} & (A \neq 1) \\ 0 & (A = 1) \end{cases}$$

(Medgyessy 1956).

The proof as well as other theorems concerning partial differential equations for stable density functions can be found in the works of Medgyessy (1956, 1961a pp. 188—199).

The theorem and the subsequent statements are, evidently, true even if the occurring stable density functions are translated by a constant value.

Let us emphasize that A was rational and, as it can be proved, for $A=1$ only the Cauchy density function can occur.

Example 1. From Theorem 3 it follows that if $Mn=1$ is fixed then only *the normal density function* $f(x, c)=\dfrac{e^{-\frac{x^2}{4c}}}{\sqrt{4\pi c}}$ can be the solution of the partial differential equation (3). Here $\varphi(t)=e^{-ct^2}$, $A=2$, B is arbitrary, $\Omega(A)=0$. Thus $m=2$, $n=1$, $P=0$, $M=1$, $\varkappa=1$ and the partial differential equation (3) becomes the partial differential equation of "heat conduction"

$$\frac{\partial^2 f}{\partial x^2} - \frac{\partial f}{\partial c} = 0 \qquad (c > 0).$$

Example 2. From Theorem 3 it follows that if $Mn=2$ is fixed, then only the following stable density functions can be the solution of the partial differential equation (3).

(a) *The Cauchy density function* $f(x, c)=\dfrac{1}{\pi c}\dfrac{1}{1+x^2/c^2}$. Here $\varphi(t)=e^{-c|t|}$, $A=1$, $\Omega(A)=0$, $B=0$. Thus $m=n=1$, $P=0$, and we have to take $M=2$,

$\varkappa=1$. The partial differential equation (3) becomes the Laplace partial differential equation

$$\frac{\partial^2 f}{\partial x^2} + \frac{\partial^2 f}{\partial c^2} = 0 \qquad (c > 0).$$

(b) *The density function of type* V *of Pearson*

$$f(x, c) = \begin{cases} \dfrac{c}{\sqrt{2\pi}} \dfrac{e^{-\frac{c^2}{2x}}}{x^{3/2}} & (x > 0) \qquad (c > 0). \\[2ex] 0 & (x \le 0) \end{cases}$$

Here $\varphi(t) = e^{-c|t|^{1/2}\{1 - i\,\mathrm{sgn}\,t\}}$, $A = 1/2$, $B = -1$, $\Omega(A) = 1$. Thus $m=1$, $n=2$, $P=-1$. We have to take $M=1$, $\varkappa=0$. The partial differential equation (3) becomes

$$2\frac{\partial f}{\partial x} - \frac{\partial^2 f}{\partial c^2} = 0 \qquad (c > 0).$$

(c) *The density function*

$$f(x, c) = \begin{cases} \dfrac{c}{\sqrt{2\pi}} \dfrac{e^{\frac{c^2}{2x}}}{(-x)^{3/2}} & (x < 0) \qquad (c > 0) \\[2ex] 0 & (x \ge 0) \end{cases}$$

(this is the mirror image of that in (b)). Here $\varphi(t) = e^{-c|t|^{1/2}\{1 + i\,\mathrm{sgn}\,t\}}$ $A = 1/2$, $B = 1$, $\Omega(A) = 1$. Thus $m=1$, $n=2$, $P=1$. We have to take $M=1$, $\varkappa=1$. The partial differential equation (3) becomes

$$2\frac{\partial f}{\partial x} + \frac{\partial^2 f}{\partial c^2} = 0 \qquad (c > 0).$$

(d) *The density function*

$$f(x, c) = \frac{1}{2\sqrt{3}\,\pi} \frac{e^{\frac{x^3}{2c^2}}}{x} W_{-\frac{1}{2}, \frac{1}{6}}\left[\frac{2x^3}{27c^2}\right] \qquad (c > 0).$$

Here $W_{\nu,\mu}(x)$ is the Whittaker function (cf. Zolotarev 1954). Here $\varphi(t) = e^{-c|t|^{3/2}\{1 + i\,\mathrm{sgn}\,t\}}$, $A = 3/2$, $B = -1$, $\Omega(A) = -1$. Thus $m=3$, $n=2$, $P=1$. We have to take $M=1$, $\varkappa=2$. The partial differential equation (3) becomes

$$2\frac{\partial^3 f}{\partial x^3} - \frac{\partial^2 f}{\partial c^2} = 0 \qquad (c > 0).$$

6*

(e) *The density function*

$$f(x, c) = -\frac{1}{2\sqrt{3}\,\pi} \frac{e^{-\frac{x^3}{2c^2}}}{x} W_{-\frac{1}{2},\,\frac{1}{6}}\left[-\frac{2x^3}{27c^2}\right] \qquad (c > 0)$$

(this is the mirror image of that in (d)). Here $\varphi(t) = e^{-c\,|t|^{3/2}\{1 - i\,\operatorname{sgn} t\}}$, $A = 3/2$, $B = 1$, $\Omega(A) = 1$. Thus $m=3$, $n=2$, $P=-1$. We have to take $M=1$, $\varkappa=1$. The partial differential equation becomes

$$2\frac{\partial^3 f}{\partial x^3} + \frac{\partial^2 f}{\partial c^2} = 0 \qquad (c > 0).$$

4. *Some expansions for stable density functions and related results.*

The partial differential equations in Section 3 are useful for obtaining expansions for stable density functions, which are based on

Theorem 4. Let $s_A(x, B, c)$ be a stable density function of characteristic function

$$\varphi(t) = e^{-c\,|t|^A\{1 + iB\,\operatorname{sgn} t\cdot\Omega(A)\}}$$

where $c > 0$, $0 < A \leq 2$, $|B| \leq 1$ $\Omega(A) = \tan\dfrac{\pi A}{2}$ $(A \neq 1)$, $\Omega(1) = 0$. Then

$s_A(x, B, c-y)$ *where $y < c$ is an analytic function of y if $y < \dfrac{c}{\sqrt{1 + B^2 \tan^2(\pi A/2)}}$,*

for any x (Medgyessy 1956).

By the theorem we have

$$(4) \quad s_A(x, B, c-y) = \sum_{n=0}^{\infty} \frac{(-y)^n}{n!} \frac{\partial^n}{\partial c^n} s_A(x, B, c) \quad \left(y < \frac{c}{\sqrt{1 + B^2 \tan^2(\pi A/2)}}\right).$$

(For negative y-values the expansion is always valid.)

For the remainder term when taking a finite number of terms in this expansion and its generalizations see Medgyessy (1956), (1961*a* pp. 204—207).

Let $A = 2$; then $s_2(x, B, c) = \dfrac{e^{-\frac{x^2}{4c}}}{\sqrt{4\pi c}}$. By Example (1) in Section 3, $s_2(x, B, c)$ satisfies the partial differential equation

$$\frac{\partial s_2}{\partial c} - \frac{\partial^2 s_2}{\partial x^2} = 0 \qquad (c > 0, \ -\infty < x < \infty).$$

We have

Theorem 5.

(5)
$$s_2(x, B, c-y) = \frac{e^{-\frac{x^2}{4(c-y)}}}{\sqrt{4\pi(c-y)}} = \sum_{n=0}^{\infty} \frac{(-y)^n}{n!} \frac{\partial^{2n}}{\partial x^{2n}} s_2(x, B, c) =$$

$$\sum_{n=0}^{\infty} \frac{(-y)^n}{n!} \frac{\partial^{2n}}{\partial x^{2n}} \left(\frac{e^{-\frac{x^2}{4c}}}{\sqrt{4\pi c}} \right) \qquad (y < c)$$

(Medgyessy 1961*a* pp. 93—101, 213—216).

The proof is based on (4) and the differentiability infinitely many times of the stable density functions. It can be found in Medgyessy (1961*a* pp. 93—101, 213—216) where the theorem is stated for certain analytic functions instead of $s_2(x, B, c)$ and by using a more general operator than $\frac{\partial^2}{\partial x^2}$, which includes this latter one. Our result can also be obtained immediately and holds also by writing $x-y$ instead of x (cf. Theorem 1 in Ch. V, § 3).

The investigation of the partial differential equation standing in the background of Theorem 5 shows that there is no stable density function $s_A(x, B, c)$ other than the *normal* one, which satisfies a partial differential equation of type $\frac{\partial s_A}{\partial c} - A_1 \frac{\partial^r s_A}{\partial x^r} = 0$ $(c>0, -\infty<x<\infty, A_1$ is a constant) and for which there holds, consequently, an expansion similar to that figuring in Theorem 5.

Theorem 6. If $s_A(x, B, c)$ satisfies the partial differential equation

$$\frac{\partial^2 s_A}{\partial c^2} - A_2 \frac{\partial^m s_A}{\partial x^m} \qquad (c>0, \quad -\infty < x < \infty)$$

where A_2 is a constant, $m=1, 2, \ldots$ then

(6)
$$\frac{1}{2}[s_A(x, B, c-y) + s_A(x, B, c+y)] = G(x, y) =$$

$$\sum_{n=0}^{\infty} \frac{y^{2n}}{(2n)!} A_2^n \frac{\partial^{mn}}{\partial x^{mn}} s_A(x, B, c) \quad \left(0 < y < \frac{c}{\sqrt{1+B^2 \tan^2(\pi A/2)}} \right)$$

(Medgyessy 1961*a* pp. 213—216).

The proof based on the same idea as the one of Theorem 5 and stated for certain analytic functions instead of $s_A(x, B, c)$ and for a more general type of operator, comprising also the operator $\frac{\partial^m}{\partial x^m}$ of our theorem can be found in Medgyessy's book (1961*a* pp. 213—216).

Example 1. Let $A=1$, $B=0$ and consequently $s_1(x, 0, c) = \dfrac{1}{\pi c}\dfrac{1}{1+x^2/c^2}$. By

(a) in Example 2 of Section 3, $\dfrac{\partial^2 s_1}{\partial c^2}+\dfrac{\partial^2 s_1}{\partial x^2}=0$ $(c>0)$. Consequently, (6) gives

$$\frac{1}{2}[s_1(x, 0, c-y)+s_1(x, 0, c+y)] =$$

$$\frac{1}{2}\left[\frac{1}{\pi(c-y)}\frac{1}{1+x^2/(c-y)^2} + \frac{1}{\pi(c+y)}\frac{1}{1+x^2/(c+y)^2}\right]=$$

$$\sum_{n=0}^{\infty}\frac{y^{2n}}{(2n)!}(-1)^n\frac{\partial^{2n}}{\partial x^{2n}}s_1(x, 0, c) = \sum_{n=0}^{\infty}\frac{y^{2n}}{(2n)!}(-1)^n\frac{\partial^{2n}}{\partial x^{2n}}\left(\frac{1}{\pi c}\frac{1}{1+x^2/c^2}\right).$$

This holds also on writing $x-y$ instead of x.

Example 2. Let $A=1/2$, $B=-1$ and

$$s_{1/2}(x, -1, c) = \begin{cases} \dfrac{c}{\sqrt{2\pi}}\dfrac{e^{-\frac{c^2}{2x}}}{x^{3/2}} & (x>0) \\[2ex] 0 & (x\leq 0) \end{cases} \quad (c>0).$$

By (b) in Example 2 of Section 3, $\dfrac{\partial^2 s_{1/2}}{\partial c^2}-2\dfrac{\partial s_{1/2}}{\partial x}=0$ $(c>0)$. Thus (6) gives:

$$\frac{1}{2}[s_{1/2}(x, -1, c-y)+s_{1/2}(x, -1, c+y)] =$$

$$\frac{1}{2}\left[\frac{c-y}{\sqrt{2\pi}}\frac{e^{-\frac{(c-y)^2}{2x}}}{x^{3/2}} + \frac{c+y}{\sqrt{2\pi}}\frac{e^{-\frac{(c+y)^2}{2x}}}{x^{3/2}}\right]=$$

$$\sum_{n=0}^{\infty}\frac{y^{2n}}{(2n)!}2^n\frac{\partial^n}{\partial x^n}s_{1/2}(x, -1, c) = \sum_{n=0}^{\infty}\frac{y^{2n}}{(2n)!}2^n\frac{\partial^n}{\partial x^n}\left(\frac{c}{\sqrt{2\pi}}\frac{e^{-\frac{c^2}{2x}}}{x^{3/2}}\right) \quad \left(y<\frac{c}{\sqrt{2}}\right)$$

if $x>0$ and equals 0, if $x\leq 0$ (Medgyessy 1961a pp. 110—121, 213—216).

5. *Partial integro-differential equations for stable density functions.*

Let $s_A(x, B, c)$ be as above. While in Section 3 A was rational, here *irrational* values of A will be allowed.

Theorem 7. For $0<A<1$ and $1<A<2$, $|B|\leq 1$, $c>0$, $s_A(x, B, c)$ satisfies one of the partial integro-differential equations

$$\frac{\partial s_A}{\partial c} = \frac{1}{2\Gamma(1-A)\cos(A\pi/2)} \int_{-\infty}^{\infty} \frac{B-\text{sgn}(x-y)}{|x-y|^A} \frac{\partial s_A}{\partial y} dy \quad (0 < A < 1);$$

$$\frac{\partial s_A}{\partial c} = \frac{1}{2\Gamma(2-A)\cos(A\pi/2)} \int_{-\infty}^{\infty} \frac{B\,\text{sgn}(x-y)-1}{|x-y|^{A-1}} \frac{\partial^2 s_A}{\partial y^2} dy \quad (1 < A < 2)$$

(Medgyessy 1958).

The proof based on a remark of Feller (1952) is given in the works of Medgyessy (1958), (1961a pp. 199—203).

6. A partial differential equation for the Gamma density function.

Let

$$f(x, c) = \begin{cases} \dfrac{x^{\gamma-1}e^{-\frac{x}{c}}}{c^{\gamma}\Gamma(\gamma)} & (x > 0) \\ 0 & (x \leq 0) \end{cases} \quad (\gamma > 0, \ c > 0)$$

be a Gamma density function of order γ.

Theorem 8. The Gamma density function of order γ, $f(x, c)$, satisfies the partial differential equation

$$\frac{\partial f}{\partial c} = x\frac{\partial^2 f}{\partial x^2} - (\gamma-2)\frac{\partial f}{\partial x} \quad (c > 0)$$

(Medgyessy 1961a pp. 139—143).

For the proof and the relevant supplementary statements see Medgyessy (1961a) pp. 139—142.

7. An expansion for the Gamma density function.

The partial differential equation in Section 6 can be utilized to obtain an expansion for the Gamma density function.

Theorem 9. If $f(x, c)$ is the above Gamma density function of order γ, we have

$$f(x, c-y) = \sum_{n=0}^{\infty} C_m(x, y)\frac{\partial^m}{\partial x^m}f(x, c) \quad (x > 0, \quad 0 < y < c)$$

where

$$C_m(x, y) = \sum_{r=m-[m/2]}^{m} \frac{y^r}{r!} E_r^{(m+1-r)} x^{m-r}$$

and the constants $E_r^{(v)}$ are defined by the recurrence relations

$$E_r^{(v)} = -E_{r-1}^{(v-1)} + (\gamma - 2v)E_{r-1}^{(v)} + v(\gamma - v - 1)E_{r-1}^{(v+1)}$$

$$(v = 1, 2, ..., r+1; \quad r = 0, 1, 2, ...),$$

$E_r^0 \equiv 0, \quad E_r^{(r+k)} \equiv 0 \quad (k > 1), \quad E_0^{(1)} = 1$ (Medgyessy 1961a pp. 140—141).

The proof is given in Medgyessy's book (1961a pp. 140—141).

The explicit form of $C_m(x, y)$ is, for $m=0, 1, ..., 5$ as follows:

$$C_0(x, y) = 1;$$

$$C_1(x, y) = (\gamma - 2)y;$$

$$C_2(x, y) = -xy + (\gamma^2 - 5\gamma + 12)\frac{y^2}{2!};$$

$$C_3(x, y) = -(\gamma - 3)xy^2 + (\gamma^3 - 7\gamma^2 + 22\gamma - 24)\frac{y^3}{3!};$$

$$C_4(x, y) = \frac{x^2 y^2}{2!} - (\gamma^2 - 5\gamma + 12)\frac{xy^3}{2!} + (\gamma^4 - 12\gamma^3 + 57\gamma^2 - 134\gamma + 120)\frac{y^4}{4!};$$

$$C_5(x, y) = (\gamma - 4)\frac{x^2 y^3}{2!} - (\gamma_3 - 10\gamma^2 + 40\gamma - 60)\frac{xy^4}{3!} +$$

$$(\gamma^5 - 18\gamma^4 + 129\gamma^3 - 488\gamma^2 + 948\gamma - 720)\frac{y^5}{5!};$$

8. *The number of changes of sign of the functions represented by a convolution transform with Pólya frequency function kernel.*

A property of the Pólya frequency functions defined in Ch. II, § 1 suggested

Theorem 10. If $\Lambda(x)$ is a Pólya frequency function and $f(x)$ a bounded function, then the number of the changes of sign of the convolution transform $g(x) = \int_{-\infty}^{\infty} \Lambda(x-t)f(t)\,dt$ is less than or equal to the number of changes of sign of $f(x)$.

For the proof see Schoenberg (1951), (1953).

Supplements and problems to Ch. II, § 8

1. In connection with Theorem 1 we notice that if $\varphi_1(t)$ and $\varphi_2(t)$ belong to distribution functions that are *not* infinitely divisible then the question: when will $\dfrac{\varphi_1(t)}{\varphi_2(t)}$ be a characteristic function can be investigated only by means of the few relevant criteria; they can be found e.g. in the works of Pólya (1949); Girault (1955); Linnik (1960); Lukács (1964).

2. Certain stable density functions can be identified, by means of a special transformation, with the solution of a boundary value problem connected with the Laplace equation (Zolotarev 1956; Ibragimov, Linnik 1965 p. 57). For stable density functions not only partial but also ordinary differential equations hold; see Linnik (1954); Zolotarev (1954); Ibragimov, Linnik (1965) p. 54.

3. Beside the expansions enumerated in Section 4 other expansions of stable density functions are known; see Pollard (1946); Bergström (1952), (1953); Ibragimov, Linnik (1965 pp. 60—63).

4. Theorem 4 can also be extended easily to complex values of y (Medgyessy 1950).

5. In connection with Theorem 6 we notice that a closer investigation of the partial differential equation (3), being the basis of that theorem, shows that *beside the Cauchy density function and the density function of type V of Pearson, there are still three other stable density functions* (see Section 3, Example 2, (c)—(e)) for which the expansion (6) in the theorem is valid. Also the general partial differential equation (3) valid for the stable density functions could be combined with the expansion (4) in Theorem 4, but the result can hardly be surveyed. The proof of this can be found in Medgyessy (1956), (1961a pp. 110—121).

6. Theorems 3—6 in § 8 are valid also for the *convolution* of the stable density functions figuring in them with another density function which is independent of the parameter c.

7. An extension of the results of Theorem 7 is given by Ibragimov, Linnik (1956 pp. 54—56).

8. It is interesting that the stable density functions characterized by the parameter values $A=1$, $B \neq 0$ *lie outside* the sphere of effectiveness of both Theorem 3 and Theorem 7. It is an unsolved **problem** whether they satisfy some differential — or integro-differential — equation at all (they satisfy an integral equation; see Ch. III, § 1, 1.1.1).

III. DECOMPOSITION OF SUPERPOSITIONS, I. DECOMPOSITION OF SUPERPOSITIONS OF DENSITY FUNCTIONS

§ 1. FIRST TYPE OF SUPERPOSITION. THE $\mathscr{A}$-METHOD

This chapter deals with Problem 1 of Ch. II, § 1; i.e. with the different methods of the (numerical) decomposition of a superposition of density functions

$$k(x) = \sum_{k=1}^{N} p_k f(x, \alpha_k, \beta_k),$$

especially in that case which is of fundamental importance from the practical viewpoint, when **the superposition $k(x)$ to be decomposed belongs to the type**

$$k(x) = \sum_{k=1}^{N} p_k f(x - \alpha_k, \beta_k) \qquad (\Lambda_1 < \beta_1 \leqq \beta_2 \leqq \ldots \leqq \beta_N \leqq \Lambda_2)$$

where $f(x, \beta_k)$ is a strictly unimodal (A$_1$) (or (A$_2$) or (A$_3$)) density function for all values of β_k (cf. Ch. II, § 4) and where **the abscissae of the peaks of the graphs of $f(x - \alpha_k, \beta_k)$ are all different** (see Fig. 1).

The present paragraph will deal with a method of decomposition of $k(x)$ called the $\mathscr{A}$-method. Its basic idea was originated by Sen (1922); Doetsch (1928), (1936) and Medgyessy (1954b), (1961a) Chapter II (see also *Supplements and problems to* Ch. III, § 1).

Essentially, the method is as follows:

Let the so-called **test function**

$$b(y) = \sum_{k=1}^{N} p_k g(y, \alpha_k, \beta_k)$$

correspond to the superposition $k(x)$ represented by its graph or by some ordinate values of that. For this test function let the following *conditions* hold:

I. $g(y, \alpha_k, \beta_k)$ is a unimodal transform of increased narrowness of $f(x - \alpha_k, \beta_k)$ $(k = 1, \ldots, N)$ (cf. Ch. II, § 6) and the implied well-defined analytical relation between $g(y, \alpha_k, \beta_k)$ and $f(x - \alpha_k, \beta_k)$ is independent of α_k, β_k.

II. The distance between the abscissae of the peaks of $g(y, \alpha_k, \beta_k)$ and $g(y, \alpha_l, \beta_l)$ $(k \neq l)$ is greater than or equal to the distance between the abscissae of the peaks of $f(x - \alpha_k, \beta_k)$ and $f(x - \alpha_l, \beta_l)$, respectively (see Fig. 1).

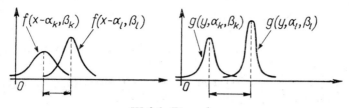

III. § 1. Figure 1

III. $b(y)$ can be constructed by the help of $k(x)$ only, without the knowledge of N, p_k, α_k, β_k and utilizing, eventually, the analytical relation between $g(y, \alpha_k, \beta_k)$ and $f(x - \alpha_k, \beta_k)$.

Evidently $b(y)$ is the superposition of the component density functions $g(y, \alpha_k, \beta_k)$.

In consequence of the character of $g(y, \alpha_k, \beta_k)$ we have that if we were able to plot the graph of $b(y)$ then, *probably*, the graphs of the different components of it would present themselves *more separated* in this than in the graph of $k(x)$ (see Fig. 2; in this the abscissae of the peaks remain unmoved). If the separation is sufficiently great the graphs of the components will appear almost without disturbing each other. Then from the graph of $b(y)$ the number N of the components — and, eventually, also the approximate value of certain parameters — can be determined.

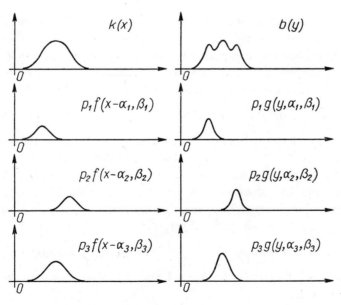

III. § 1. Figure 2

Let us consider the following steps:

(A) Determination of the density function $g(y, \alpha_k, \beta_k)$ with the help of $f(x-\alpha_k, \beta_k)$.

(B) Determination, with the help of (A), of the test function $b(y)$, with the aid of $k(x)$, serving as the basis for the decomposition method.

(C) On the basis of (B) and of the "*measured*" values of $k(x)$, the elaboration of a numerical method for the construction of an *approximation* to the test function $b(y)$.

For the graph $\mathscr{G}_1$ of this approximation to $b(y)$ the statements concerning the evaluation of the graph of the "exact" test function are approximately true.

Definition 1. The construction of the graph $\mathscr{G}_1$ of the test function *approximation*, followed by the determination of the number and of the abscissae etc. of the peaks of $\mathscr{G}_1$ is called the $\mathscr{A}$-**method**, yielding the (numerical) decomposition of the superposition $k(x)$.

In the following test functions $b(y)$ will be determined for different types of superpositions $k(x)$. Naturally, the results of Ch. II, § 4 will be used.

Example. An interesting application of the $\mathscr{A}$-method (in the general case) can be seen in the following, which is due to Židkov, Ščedrin, Rambidi, Egorova (1968).

In gas electronography when characterizing the geometrical configuration of molecules in the gaseous phase, the important so-called radial distribution function $f(r)$ $(r>0)$, being a superposition of certain *well-separated* unimodal density functions whose peak places may reveal some required physical data, is to be determined from a certain measured function $g(r)$ connected with $f(r)$ by the relation $g(r)=\int\limits_0^\infty K(r, \varrho)f(\varrho)d\varrho$ where $K(r, \varrho)$ is a given kernel. Here $g(r)$ is also a superposition of unimodal density function whose peaks have the same abscissae as those of $f(r)$. The graphs of the components of $f(r)$ are *narrower* than those of the components of $g(r)$. It is found that the graph of $f(r)$ is appropriate for determining the requested peak abscissae while the graph of $g(r)$ is not. We see that determining $f(r)$ by means of the graph of $g(r)$ is equivalent to the (numerical) decomposition of the superposition $g(r)$, by means of the inversion of the integral transformation expressed by $g(r)=\int\limits_0^\infty K(r, \varrho)f(\varrho)d\varrho$. Here $f(r)$ behaves as a test function. Conditions I and II in Ch. III, § 1 are, as can easily be seen, fulfilled.

The authors mentioned calculated $f(r)$ numerically from the measured values of $g(r)$ using the regularization method of A. N. Tihonov (see Ch. V, § 1, Section 4). In Fig. 3, taken from the above-mentioned article of Židkov *et al.* the solid line shows the function $g(r)$ obtained from measurements on P_4O_{10}. The dashed line shows an approximation of the corresponding function $f(r)$, obtained by means of the regularization method. 50 measured points of $g(r)$ were used. The curve shows 8 peaks over those due to approximation errors. The abscissae of the peaks could be identified easily with certain physical data which one would expect to have.

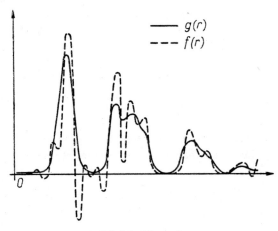

$$\text{———} \quad g(r)$$
$$\text{- - -} \quad f(r)$$

III.§ 1. Figure 3

1. First particular case of the $\mathscr{A}$-method: The method of formant changing

A very important type of the superposition

$$k(x) = \sum_{k=1}^{N} p_k f(x - \alpha_k, \beta_k) \qquad (\Lambda_1 < \beta_1 \leqq \beta_2 \leqq \dots \leqq \beta_N < \Lambda_2)$$

is that one in which β_k is a (Λ_1, Λ_2) (or (Λ_2, Λ_1)) *monotone formant* of $\mathscr{G}f(x - \alpha_k, \beta_k)$. In the case of such a superposition let us consider the function

$$b(y) = \sum p_k f(y - \alpha_k, \theta(\beta_k))$$

where $\theta(x)$ is a strictly monotone function not depending on β_k as a parameter and such that $\theta(\beta_k)$ is one of the possible values of β_k, and for it there holds: (1) if β_k is a (Λ_1, Λ_2) monotone formant of $\mathscr{G}f(x - \alpha_k, \beta_k)$, then $\Lambda_1 \leqq \beta_k < \theta(\beta_k) < \Lambda_2$; (2) if β_k is a (Λ_2, Λ_1) monotone formant of $\mathscr{G}f(x - \alpha_k, \beta_k)$, then $\Lambda_1 < \theta(\beta_k) < \beta_k \leqq \Lambda_2$.

As $f(x - \alpha_k, \beta_k)$ was strictly unimodal, $f(x - \alpha_k, \theta(\beta_k))$ will be also. From the foregoing it follows that $\mathscr{G}f(y - \alpha_k, \theta(\beta_k)) < \mathscr{G}f(x - \alpha_k, \beta_k)$. *Let us suppose* that a well-defined analytical relation, independent of α_k, β_k holds between $f(y-\alpha_k, \theta(\beta_k))$ and $f(x-\alpha_k, \beta_k)$. Then Condition I in Ch. III, § 1 is satisfied for $f(y-\alpha_k, \theta(\beta_k))$; i.e. $f(y-\alpha_k, \theta(\beta_k))$ can be considered as the unimodal transform of increased narrowness of $f(x-\alpha_k, \beta_k)$ of the *same* type. If the abscissa of the peak of $\mathcal{G}f(x, \beta_k)$ is *independent* of β_k, Condition II in Ch. III, § 1 is also satisfied for $f(y-\alpha_k, \theta(\beta_k))$; in the converse case *let us suppose* that this condition is fulfilled. Finally *let us suppose* that $b(y)$ can be constructed with the help of $k(x)$, without any knowledge of N, p_k, α_k, β_k, utilizing eventually the relation between $f(y-\alpha_k, \theta(\beta_k))$ and $f(x-\alpha_k, \beta_k)$. Then Condition III in Ch. III, § 1 is also fulfilled. Thus $b(y)$ can be taken as the test function corresponding to $k(x)$ and by carrying out the steps (A), (B) and (C) in Ch. III, § 1, the $\mathcal{A}$-method can be applied.

The nearer $\theta(\beta_k)$ lies to Λ_2 or Λ_1, the more peaked and separated are the graphs of the components of $b(y)$ and thus, the better a decomposition can be expected.

In the case of the present type of superposition $k(x)$ the $\mathcal{A}$-method will also, for a trivial reason, be called the **method of formant changing**. Its idea appeared first in the works of Medgyessy (1954b), (1961a pp. 121—127).

In the following we shall give examples to illustrate the present type of superposition.

1.1. The simple diminishing of the formant

The most frequent type of the superposition

$$k(x) = \sum_{k=1}^{N} p_k f(x-\alpha_k, \beta_k)$$

dealt with in Ch. III, § 1, Section 1 is that in which $0 < \beta_1 \leqq \beta_2 \leqq \ldots \leqq \beta_N < \Lambda_2$, β_k is a $(\Lambda_2, 0)$ monotone formant of $\mathcal{G}f(x, \beta_k)$ and $\theta(\beta_k) = \beta_k - \lambda$ where λ is a parameter and $0 < \lambda < \beta_1$; this latter condition is *crucial*. This $\theta(\beta_k)$ satisfies the restrictions prescribed for $\theta(\beta_k)$ in Ch. III, § 1, Section 1. Here $\mathcal{G}f(y-\alpha_k, \beta_k-\lambda) \prec \mathcal{G}f(x-\alpha_k, \beta_k)$. *Let us suppose* that a well-defined analytical relation independent of α_k, β_k holds between $f(y-\alpha_k, \beta_k-\lambda)$ and $f(x-\alpha_k, \beta_k)$. Then Condition I in Ch. III, § 1 is satisfied for $f(y-\alpha_k, \beta_k-\lambda)$ i.e. $f(y-\alpha_k, \beta_k-\lambda)$ can be considered as the unimodal transform of increased narrowness of $f(x-\alpha_k, \beta_k)$, of the *same* type. If the abscissa of the peak of $\mathcal{G}f(x, \beta_k)$ is *independent* of β_k, Condition II in Ch. III, § 1 is also satisfied for $f(y-\alpha_k, \beta_k-\lambda)$; in the converse case *let us suppose* that this condition is fulfilled. Finally *let us suppose* that $b(y)$ can be constructed by the help of $k(x)$ without the knowledge of N, p_k, α_k, β_k, utilizing eventually, the relation between $f(y-\alpha_k, \beta_k-\lambda)$ and $f(x-\alpha_k, \beta_k)$. Then Condition III in Ch. II, § 1 is fulfilled. Writing now $b(y, \lambda)$ instead of $b(y)$,

$$b(y, \lambda) = \sum_{k=1}^{N} p_k f(y-\alpha_k, \beta_k-\lambda) \qquad (0 < \lambda < \beta_1)$$

can be taken as the test function corresponded to the superposition

$$k(x) = \sum_{k=1}^{N} p_k f(x-\alpha_k, \beta_k)$$

and, carrying out the steps (A), (B) and (C) in Ch. III, § 1 the 𝒜-method can be applied.

In the present case the 𝒜-method is called the **method of the (simple) diminishing of the formant**. (This, as a special case of the formant changing, appears earlier than the latter one: Sen 1922; Doetsch 1928, 1936; Medgyessy 1953.)

The nearer λ lies to β_1, the more peaked the components of $b(y)$ are (moreover, the component belonging to α_1 tends to $+\infty$ at $x=\alpha_1$ and to 0 otherwise), and the better a decomposition can be expected to be. In the case of a numerical decomposition we do not know β_1; λ must, however, be less than β_1. Therefore, it is advisable to carry out the numerical procedure for a whole set of λ-values increasing monotonely from 0 on and then to compare the results. While the λ used is less than β_1, the different components will appear more and more separated in the graphs of the results and the result obtained with the value of λ the nearest to β_1 will show

the best separation of the components of the test function. Because of the character of numerical methods it is clear that if we take a λ-value equal to or greater than β_1 the increasing separation will not necessarily fail; methodological examples show that for $\lambda=\beta_1$ the component with β_1, though it remains unimodal at the first mode, *does not* tend to a "Dirac delta function"

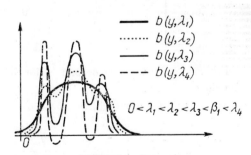

$$0 < \lambda_1 < \lambda_2 < \lambda_3 < \beta_1 < \lambda_4$$

III.§ 1. 1.1. Figure 1

and negative values of $b(y, \lambda)$ at points where the $k(x)$ is at about 0 influence neither the significance of its peaks at other places, nor the utility of the whole procedure. At any rate a greater oscillation into negative ordinate values of $b(y, \lambda)$ points to the rejecting of the value of λ which has been used. It has also to be taken into account that little peaks appear in consequence of the numerical methods during the decomposition procedure. Hence we can also see how to evade the disturbing effect of that λ has to be less than β_1 (cf. Fig. 1).

A very frequent relevant type of superposition to which the decomposition method of the diminishing of the formant can be applied is that in which

$$f(x - \alpha_k, \beta_k) = \frac{1}{\beta_k^\omega} f\left(\frac{x - \alpha_k}{\beta_k^\omega}\right) \qquad (\omega > 0; \quad \beta_k > 0),$$

so that

$$k(x) = \sum_{k=1}^{N} p_k \frac{1}{\beta_k^\omega} f\left(\frac{x - \alpha_k}{\beta_k^\omega}\right) \qquad (\omega > 0; \quad 0 < \beta_1 \leq \beta_2 \leq \ldots \leq \beta_N < \Lambda_2).$$

Here, by Example 1 in Ch. II, §4 $\mathscr{G} \dfrac{1}{\beta_k^\omega} f\left(\dfrac{x - \alpha_k}{\beta_k^\omega}\right)$ possesses β_k, which acts as a scale parameter, as a $(\Lambda_2, 0)$ monotone formant — and so on. In the case of this superposition, the test function, which is the basis of the decomposition, is

$$b(y, \lambda) = \sum_{k=1}^{N} p_k \frac{1}{(\beta_k - \lambda)^\omega} f\left(\frac{x - \alpha_k}{(\beta_k - \lambda)^\omega}\right) \qquad (0 < \lambda < \beta_1).$$

The construction of $b(y, \lambda)$ by means of $k(x)$ is to be determined in every concrete case.

Example. **Decomposition of a superposition of Gamma density functions of the same order.** Let us consider the superposition of Gamma density functions

$$k(x) = \begin{cases} \displaystyle\sum_{k=1}^{N} p_k \frac{x^{\gamma-1} e^{-\frac{x}{\beta_k}}}{\beta_k^\gamma \Gamma(\gamma)} & (x > 0) \\ 0 & (x \leq 0) \end{cases} \qquad (0 < \beta_1 < \beta_2 < \ldots < \beta_N \leq \Lambda_2).$$

For $r = 1$ we meet the **decomposition of a superposition of exponential density functions.**

This superposition belongs to the above type and we have as test function

$$b(y, \lambda) = \begin{cases} \displaystyle\sum_{k=1}^{N} p_k \frac{x^{\gamma-1} e^{-\frac{x}{\beta_k - \lambda}}}{(\beta_k - \lambda)^\gamma \Gamma(\gamma)} & (x > 0) \\ 0 & (x \leq 0) \end{cases} \qquad (0 < \lambda < \beta_1).$$

Since the components are strictly unimodal, Conditions I, II and III in Ch. III, § 1 are fulfilled, as may be proved easily, taking into account that the components of $b(y, \lambda)$ satisfy, as functions of y and λ, partial differential equations derived easily from the type given in Theorem 8 of Ch. II, § 8, Section 6 for the components of $k(x)$. Concretely $b(x, \lambda)$, as a function of y and λ, satisfies the partial differential equation

(1) $$(\gamma-2)\frac{\partial b}{\partial y} - y\frac{\partial^2 b}{\partial y^2} - \frac{\partial b}{\partial \lambda} = 0 \qquad (y > 0,\ 0 < \lambda < \beta_1).$$

Applying Theorem 9 in Ch. II, § 8, Section 7 to the components of $b(y, \lambda)$ it can easily be shown that the test function is given by

(2) $$b(y, \lambda) = \sum_{m=0}^{\infty} C_m(y, \lambda)\frac{d^m}{dy^m}k(y) \qquad (y > 0,\ 0 < \lambda < \beta_1)$$

where $C_m(y, \lambda)$ is the polynomial referred to in Theorem 9.

Unfortunately, the expansion (2) is of little value in practice: to differentiate numerically is an *incorrect* problem. It is solved by the regularization method described in Ch. V, § 1, Section 4 only for very particular cases. At most three terms in (2) can be taken without introducing major errors if a numerical problem is under examination.

Theoretically also the solution of the partial differential equation (1) gives $b(y, \lambda)$ with the initial condition $b(y, 0)=k(y)$; however, this is again an *incorrect* problem. In practice, regularization methods cited in Ch. V, § 1, Section 4 can eventually be utilized in solving (1).

Other ways for the decomposition of a superposition of Gamma density functions of the same order, at least in case of a superposition of exponential density functions, will be given in Ch. III, § 2.

1.1.1. A particular case

A special investigation is devoted to the superposition

$$k(x) = \sum_{k=1}^{N} p_k f(x-\alpha_k, \beta_k) \qquad (0 < \beta_1 \leqq \beta_2 \leqq \ldots \leqq \beta_N \leqq \Lambda_2)$$

being a special case of that in Ch. III, § 1, Subsection 1.1, in which the characteristic function $\varphi(t, \beta_k)$ of $f(x, \beta_k)$ has the form

(1) $$\mathbf{F}[f(x, \beta_k); t] = \varphi(t, \beta_k) = \Phi(t)^{\beta_k}$$

were β_k is a $(\Lambda_2, 0)$ monotone formant of the graph of the density function $f(x, \beta_k)$ and $\Phi(t)$ is a characteristic function. (It appears first in the work of Medgyessy 1954c; cf. also Medgyessy 1961a pp. 72—80.)

For the test function

$$b(y, \lambda) = \sum_{k=1}^{N} p_k f(x-\alpha_k, \beta_k - \lambda)$$

introduced above, it follows from the conditions enumerated in Ch. III, § 1, Subsection 1.1 that a well-defined analytical relation, independent of α_k, β_k, exists between $f(y-\alpha_k, \beta_k-\lambda)$ and $f(x-\alpha_k, \beta_k)$; for

$$\mathbf{F}[f(y-\alpha_k, \beta_k-\lambda); t] = e^{i\alpha_k t} \Phi(t)^{\beta_k-\lambda} =$$

$$e^{i\alpha_k t} \frac{\Phi(t)^{\beta_k}}{\Phi(t)^\lambda} = \frac{\mathbf{F}[f(x-\alpha_k, \beta_k); t]}{\Phi(t)^\lambda} \qquad (0 < \lambda < \beta_1).$$

Thus Condition I in Ch. III, § 1 is fulfilled. Condition II in Ch. III, § 1 is fulfilled, further Condition III in Ch. III, § 1 is also fulfilled as, in consequence of the last equality,

$$(2) \qquad \mathbf{F}[b(y, \lambda); t] = \frac{1}{\Phi(t)^\lambda} \mathbf{F}[k(x); t],$$

that is

$$b(y, \lambda) = \mathbf{F}^{-1}\left[\frac{1}{\Phi(t)^\lambda} \mathbf{F}[k(x); t]; y\right].$$

Then $b(y, \lambda)$ can be taken as the test function corresponding to the superposition $k(x)$. From the preceding we have, since

$$\mathbf{F}[f(x, \lambda); t] = \Phi(t)^\lambda,$$

$$(3) \qquad k(x) = \int_{-\infty}^{\infty} f(x-y, \lambda) b(y, \lambda) dy,$$

i.e. $b(y, \lambda)$ is the solution of a Fredholm integral equation of the Ist kind of convolution type. This can be treated numerically.

In case of the numerical decomposition of $k(x)$, one of the convenient methods of Chapter V (e.g. the regularization method) can be taken into account here — as we are dealing with an *incorrect* problem.

Example 1. **Decomposition of a superposition of Gamma density functions of different order.**

Let

$$k(x) = \begin{cases} \sum_{k=1}^{N} p_k \dfrac{x^{\beta_k-1} e^{-\frac{x}{B}}}{B^{\beta_k} \Gamma(\beta_k)} & (x > 0) \\ 0 & (x \le 0) \end{cases} \qquad (0 < \beta_1 < \beta_2 < ... < \beta_N \le \Lambda_2)$$

($B>0$ is given). Here $\alpha_k=0$ $(k=1, ..., N)$. The abscissae of the peaks of the graphs of the components are different. The characteristic function of the kth component is

$$\varphi(t, \beta_k) = \left(\frac{1}{1-iBt}\right)^{\beta_k},$$

i.e. it is of the type (1) and β_k is a $(\Lambda_2, 0)$ monotone formant of the graph of the kth component — as was shown in Example 2 of Ch. II, § 4. The corresponding test function is

$$b(y, \lambda) = \begin{cases} \sum_{k=1}^{N} \dfrac{x^{\beta_k - \lambda - 1} e^{-\frac{x}{B}}}{B^{\beta_k - \lambda} \Gamma(\beta_k - \lambda)} & (x > 0) \\ 0 & (x \leqq 0) \end{cases} \qquad (0 < \lambda < \beta_1).$$

It can easily be seen that the distances between the abscissae of the peaks of the graphs of the components of $b(y, \lambda)$ are the same as in the case of the components of $k(x)$; the other conditions of the applicability for the method of the diminishing of the formant are also fulfilled. Now we can make use of the above results; finally, $b(y, \lambda)$ will be the solution of the Volterra integral equation of the Ist kind of convolution type

$$k(x) = \int_0^x \frac{(x-y)^{\lambda-1} e^{-\frac{x-y}{B}}}{B^{\lambda} \Gamma(\lambda)} b(y, \lambda) \, dy.$$

The numerical treatment and the resolution of this will be ignored here; it presents an *incorrect* problem. It is needed only to be remarked that several papers are concerned with the solution of such an integral equation — e.g. by means of the regularization method described in Ch. V, § 1, Section 4 (e.g. Arsenin, Ivanov 1968; Schmaedeke 1968).

Example 2. **Decomposition of a superposition of stable density functions (excluding the normal ones).**

Let $s_{AB}(x)$ be a *stable* density function whose characteristic function is of the form

(4) $$\varphi_{AB}(t) = e^{-|t|^A \{1 + iB \operatorname{sgn} t \cdot \Omega(A)\}}$$

where $0 < A < 2$, $|B| \leqq 1$, $\Omega(A) = \tan \dfrac{\pi A}{2}$ $(A \neq 1)$, $\Omega(1) = 0$ $(A, B$ are given).

As the characteristic function of the members of the family of stable density functions $s_A(x, B, c)$ is

$$\varphi(t) = e^{-c|t|^A \{1 + iB \operatorname{sgn} t \cdot \omega(t, A)\}}$$

$$\left[c > 0, \ 0 \leqq A \leqq 2, \ |B| \leqq 1, \ \omega(t, A) = \tan \frac{\pi A}{2} \ (A \neq 1), \ \omega(t, 1) = \frac{2}{\pi} \log |t| \right]$$

(Gnedenko, Kolmogorov 1954 § 34; see also Ch. II, § 1), $s_{AB}(x) \equiv s_A(x, B, 1)$ cannot be identical with *any* stable density function by an appropriate choice of A and B because of the form of $\Omega(A)$. Evidently $\dfrac{1}{\beta_k^{1/A}} s_{AB} \left(\dfrac{x - \alpha_k}{\beta_k^{1/A}} \right)$ is also a stable density function (writing down its characteristic function will convince us of this).

Given A and B, let us consider the problem of decomposing the superposition of stable density functions

$$(5) \qquad k(x) = \sum_{k=1}^{N} p_k \frac{1}{\beta_k^{1/A}} s_{AB}\left(\frac{x-\alpha_k}{\beta_k^{1/A}}\right)$$

$$(\alpha_i \neq \alpha_j \quad (i \neq j); \quad 0 < \beta_1 \leq \beta_2 \leq \dots \leq \beta_N \leq \Lambda_2).$$

This superposition belongs to the type mentioned at the end of Ch. III, § 1, Subsection 1.1. Thus β_k is a $(\Lambda_2, 0)$ monotone formant of the graph of its kth component. It appears first in Medgyessy (1954b). The characteristic function of $s_{AB}\left(\dfrac{x}{\beta_k^{1/A}}\right)$ is, by (4), $e^{-\beta_k |t|^A \{1+iB \operatorname{sgn} t \cdot \Omega(A)\}}$. The corresponding test function is

$$b(y, \lambda) = \sum_{k=1}^{N} p_k \frac{1}{(\beta_k - \lambda)^{1/A}} s_{AB}\left(\frac{x-\alpha_k}{(\beta_k - \lambda)^{1/A}}\right).$$

The stable density functions are strictly unimodal (Ibragimov, Černin 1959; Ibragimov, Linnik 1965 p. 81). Thus

$$\frac{1}{(\beta_k - \lambda)^{1/A}} s_{AB}\left(\frac{x-\alpha_k}{(\beta_k - \lambda)^{1/A}}\right) \qquad (k = 1, \dots N)$$

are strictly unimodal, too, and the abscissae of the peaks of the graphs of these functions are also all different. The conditions for the applicability of the method of formant changing are fulfilled here in case of a (0) *symmetrical* $s_{AB}(x)$ as Condition II in Ch. III, § 1 is here always satisfied. In other cases investigated the fulfilment of Condition II in Ch. III, § 1 has to be controlled. If it is fulfilled, then the difference of the abscissae of the peaks of the components of the test function $b(y, \lambda)$ does not diminish. With all conditions fulfilled the results of the present section can be applied and, finally, we shall have for the test function the integral equation

$$(6) \qquad k(x) = \int_{-\infty}^{\infty} s_{AB}(x-y, \lambda)\, b\,(y, \lambda)\, dy$$

i.e. $b(y, \lambda)$ is the solution of a Fredholm integral equation of the Ist kind of convolution type.

In the case of a numerical decomposition of $k(x)$ we will be faced with the numerical solution of this integral equation. This is delicate, the solution of such an integral equation being, as it is commonly known, an *incorrect* problem. The numerical methods that can be applied in this case are summarized in § 1 and § 2 of Chapter V. In a concrete case we shall take one of them.

Particular cases:

(a) **Decomposition of a superposition of Cauchy density functions.** This super-position is obtained from (4) and (5) by putting $A=1$; then we have

$$k(x) = \sum_{k=1}^{N} p_k \frac{1}{\pi \beta_k} \frac{1}{1+(x-\alpha_k)^2/\beta_k^2}$$

$$(\alpha_i \neq \alpha_j \quad (i \neq j); \quad 0 < \beta_1 \leq \beta_2 \leq \dots \leq \beta_N \leq \Lambda_2).$$

The test function

$$b(\gamma, \lambda) = \sum_{k=1}^{N} p_k \frac{1}{\pi(\beta_k-\lambda)} \frac{1}{1+(x-\alpha_k)^2/(\beta_k-\lambda)^2} \qquad (0 < \lambda < \beta_1)$$

will satisfy the integral equation

$$k(x) = \int_{-\infty}^{\infty} \frac{1}{\pi\lambda} \frac{1}{1+(x-\gamma)^2/\lambda^2} b(\gamma, \lambda)\, dy.$$

Condition II in Ch. III, §1 is, evidently, fulfilled.

Let us give a "methodological" *numerical example*. In Fig. 1a the solid line shows the graph of the superposition of two Cauchy density functions

$$k(x) = \frac{1}{1+(x-0.5)^2} + \frac{1}{1+(x+0.5)^2}.$$

It is unimodal. Let us construct, numerically, by means of the "measured" ordinate values of the graph (with a random second decimal figure distributed uniformly), approximations $\hat{b}(y, \lambda)$ of the test functions $b(y, \lambda)$ belonging to $\lambda=0.454\,(=5/11)$ (dotted line) and $\lambda=0.636\,(=7/11)$ (thin line), the computations having been performed by the aid of the method described in Ch. V, §2 based on data at the points $x_k=k/11\,(k=-60, \dots, 0, \dots, 60)$ (see Fig. 1a). Already the dotted graph reveals the two components, showing (at least) two separated peaks. Of course, secondary peaks also appear at points where $b(y, \lambda)$ is in fact small. For comparison the graph of the exact $b(y, \lambda)$ is also given (dashed line) for $\lambda=0.636\,(=7/11)$ (Medgyessy 1971b, 1972a).

Figure 1b presents the graphs of an approximation to $F[k(x); t]e^{-\lambda|t|}$ (dotted line for $\lambda=0.454\,(=5/11)$, thin line for $\lambda=0.636\,(=7/11)$) which can also be used in the numerical procedure instead of those to $F[k(x); t]$ (cf. Ch. V, §2), at appropriate scales for the ordinate values and t respectively and trapezoidal formula parameters. The places of "cut off" are marked by arrows. For the greater λ-value the function oscillates more intensively after the "cut off" because $e^{\lambda|t|}$ magnifies more the parts of the function originating from the errors of $k(x)$.

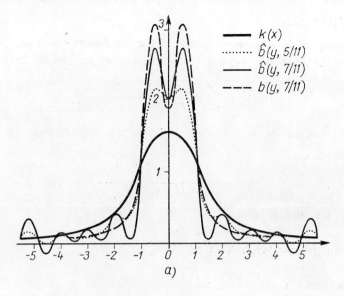

a)

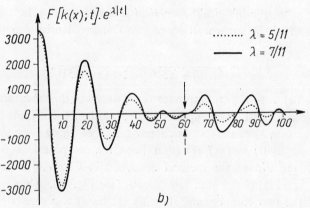

b)

III. § 1. 1.1.1. Figure 1

(b) **Decomposition of a superposition of density functions of type V of Pearson.**
This superposition is obtained from (4) and (5) by putting $A=1/2, B=-1$. Then,
with regard to the results of Ch. II, § 8, we have

$$k(x) = \begin{cases} \sum_{k=1}^{N} p_k \dfrac{\beta_k}{\sqrt{2\pi}} \dfrac{e^{-\frac{\beta_k^2}{x-\alpha_k}}}{(x-\alpha_k)^{3/2}} & (x > \min_{1 \leq k \leq N} \alpha_k) \\ 0 & (x \leq \min_{1 \leq k \leq N} \alpha_k) \end{cases}$$

$$(\alpha_i \neq \alpha_j \quad (i \neq j); \quad 0 < \beta_1 \leq \beta_2 \leq \ldots \leq \beta_N < \Lambda_2)$$

the components being defined as zero if the appropriate $x-\alpha_k$ becomes negative. The test function is

$$b(y, \lambda) = \begin{cases} \displaystyle\sum_{k=1}^{N} p_k \frac{\beta_k - \lambda}{\sqrt{2\pi}} \frac{e^{-\frac{(\beta_k-\lambda)^2}{x-\alpha_k}}}{(x-\alpha_k)^{3/2}} & (x > \min_{1\leq k\leq N} \alpha_k) \\ 0 & (x \geqq \min_{1\leq k\leq N} \alpha_k) \end{cases} \qquad (0 < \lambda < \beta_1)$$

(the components being defined to be zero if the appropriate $x-\alpha_k$ is negative). Here Condition I in Ch. III, § 1 is satisfied. However, Condition II in Ch. III, § 1 *is not always fulfilled* because the components of $k(x)$ are not (0) symmetrical and *the maxima of the ith and jth components depend on α_k, β_k and α_j, β_j respectively*, being at $x=\alpha_i+\dfrac{\beta_i^2}{3}$ and $\alpha_j+\dfrac{\beta_j^2}{3}$ respectively. Those of the test function $b(y, \lambda)$ are at $\alpha_i+\dfrac{(\beta_i-\lambda)^2}{3}$ and at $\alpha_j+\dfrac{(\beta_j-\lambda)^2}{3}$ respectively. Condition II in Ch. III, § 1 is equivalent to

$$\left| \left(\alpha_j + \frac{(\beta_j-\lambda)^2}{3}\right) - \left(\alpha_i + \frac{(\beta_i-\lambda)^2}{3}\right) \right| > \left| \left(\alpha_j + \frac{\beta_j^2}{3}\right) - \left(\alpha_i + \frac{\beta_i^2}{3}\right) \right|.$$

This does not hold for all values of the parameters. Thus the $\mathscr{A}$-method *cannot always be applied* to this example (cf. Medgyessy 1954c).

If the method is applicable, $b(y, \lambda)$ satisfies the Volterra integral equation of the Ist kind

$$k(x) = \int_{-\infty}^{x} \frac{\lambda}{\sqrt{2\pi}} \frac{e^{-\frac{\lambda^2}{(x-y)}}}{(x-y)^{3/2}} b(y, \lambda)\,dy.$$

No numerical example will be given here.

1.1.1.1. Decomposition of a superposition of normal density functions

Because of its importance, the case of the stable density functions corresponding to $A=2$, B arbitrary in (4) and (5) must be investigated in a separate subsection.

Here $\varphi_{AB}(t)=e^{-|t|^2}=e^{-t^2}$, hence $s_{AB}(x)=\dfrac{e^{-\frac{x^2}{4}}}{\sqrt{4\pi}}$, i.e., a normal density function (Gaussian function); the superposition $k(x)$ in (5) of Ch. III, § 1, Subsection 1.1.1 will take the form

(1)
$$k(x) = \sum_{k=1}^{N} p_k \frac{e^{-\frac{(x-\alpha_k)^2}{4\beta_k}}}{\sqrt{4\pi\beta_k}}$$

$$(\alpha_i \neq \alpha_j (i \neq j); \quad 0 < \beta_1 \leqq \beta_2 \leqq \ldots \leqq \beta_N < \Lambda_2)$$

of a **superposition of normal density functions.** Thus the **decomposition of a super-position of normal density functions** appears.

In accordance with the result of the preceding subsection the conditions of the application of the method of the simple diminishing of the formant are fulfilled (Sen 1922; Doetsch 1928, 1936; Medgyessy 1953). The corresponding test function is

$$(2) \qquad b(y, \lambda) = \sum_{k=1}^{N} p_k \frac{e^{-\frac{(x-\alpha_k)^2}{4(\beta_k - \lambda)}}}{\sqrt{4\pi(\beta_k - \lambda)}} \qquad (0 < \lambda < \beta_1).$$

This is determined by the integral equation

$$(3) \qquad k(x) = \int_{-\infty}^{\infty} \frac{e^{-\frac{(x-y)^2}{4\lambda}}}{\sqrt{4\pi\lambda}} b(y, \lambda) \, dy.$$

The numerical solution of this is, in general, an *incorrect* problem, but can be carried out by means of certain methods in Ch. V.

One of the methods in Ch. V, § 3 based on the exact solution of (3) has the form of a series of functions. A proof of this solution is given in Ch. V, § 3; the result is

$$(4) \qquad b(y, \lambda) = \sum_{v=0}^{\infty} \frac{(-\lambda)^v}{v!} k^{(2v)}(x).$$

There is a very important fact: the solution of our integral equation with the aid of the series is *not* an *incorrect* problem. Thus the numerical treatment of this series expansion, which is much easier than that of the integral equation, comes readily to prominence. For the numerical treatment, see Ch. V, § 3.; its essence is that from the "measured" equidistant values $\hat{k}(y+jh)$ ($j=0, \pm 1, \ldots$; h is given) of $k(x)$, the approximation of $b(y, \lambda)$ at the point y, $\hat{b}(y, \lambda)$, is obtained in the form

$$(5) \qquad \hat{b}(y, \lambda) = \sum_{j=-m}^{m} c_j^{(m)}(\lambda) \hat{k}(y+jh) \qquad (0 < \lambda < \beta_1).$$

The calculation of the constants $c_j^{(m)}(\lambda)$, *depending on λ only,* is given in Ch. V, § 3.

It is very important that (5) frequently provides a fairly good approximation also in the case of a small m, i.e. *we can also work with the knowledge of some section of the graph of $k(x)$ (local* character), in contrast to other procedures which make use of the measured values of $k(x)$ over a long (theoretically, infinite) interval (Medgyessy 1954b, 1954c, 1955b, 1956, 1961a pp. 93—101, 1966a; see also Ch. V, § 3).

Another numerical solution of the integral equation (3) may be provided by the method in Ch. V, § 2 (Medgyessy, Varga 1968). In the case of the superposition (1) this performs, in fact, the transformation (2) in Ch. III, § 1, Subsection 1.1.1, represented by

$$\text{(6)} \qquad \mathbf{F}[f(y, \lambda); t] \cdot \mathbf{F}\left[\frac{e^{-\frac{x^2}{4\lambda}}}{\sqrt{4\pi\lambda}}; t\right] = \mathbf{F}[k(x); t],$$

on the measured values $\hat{k}(x)$ (instead of $k(x)$), together with a certain filtering of the effects of the error. It is remarkable that the solution of (3) in this way is a *correct* problem. This method assumes in advance, of course, the knowledge of $\hat{k}(x)$-values in a large interval.

In fact, it is on the preceding relation between the Fourier transforms that a representation of the approximation of the test function $b(y, \lambda)$ can be established by means of a Fourier expansion and a subsequent Fourier synthesis (Doetsch 1928, 1936; Medgyessy 1953, 1954b, 1954c, 1955b, 1957, 1961a pp. 84—86); it is strongly related to the method in Medgyessy, Varga (1968). In the above-mentioned works, the method was as follows.

A sufficiently long interval, denoted by $(0, l)$ was chosen so that $k(x)$ and $b(y, \lambda)$ should be approximately zero outside this interval. Let

$$k_1(x) = \begin{cases} k(x) & (0 \le x \le l) \\ 0 & \text{otherwise} \end{cases},$$

$$b_1(y, \lambda) = \begin{cases} b(y, \lambda) & (0 \le y \le l) \\ 0 & \text{otherwise} \end{cases}$$

and let $k_1(x), b_1(y, \lambda)$ be extended to even functions in $(-l, l)$ and then, be repeated periodically. By Fourier cosine expansions we have

$$k_1(x) = \frac{g_0}{2} + \sum_{v=1}^{\infty} a_v \cos\frac{v\pi x}{l} \qquad (0 < x \le l)$$

and

$$b_1(y, \lambda) = \frac{b_0}{2} + \sum_{v=1}^{\infty} b_v \cos\frac{v\pi y}{l}$$

where

$$a_v = \frac{2}{l}\int_0^l k_1(x)\cos\frac{v\pi x}{l}\,dx, \; b_v = \frac{2}{l}\int_0^l b_1(y, \lambda)\cos\frac{v\pi y}{l}\,dy \quad (v = 0, 1, 2, \ldots).$$

If $(0, l)$ is sufficiently long, we have approximately

$$a_v \approx \frac{2}{l}\int_{-\infty}^{\infty} k_1(x)\cos\frac{v\pi x}{l}\,dx = \frac{2}{l}\,\mathrm{Re}\,\mathbf{F}\left[k_1(x); \frac{v\pi}{l}\right] \approx \frac{2}{l}\,\mathrm{Re}\,\mathbf{F}\left[k(x); \frac{v\pi}{l}\right]$$

and, similarly,

$$b_v \approx \frac{2}{l} \operatorname{Re} \mathbf{F}\left[b(y, \lambda); \frac{v\pi}{l}\right].$$

By (6)

$$\mathbf{F}[b(y, \lambda); t] = e^{\lambda t^2} \mathbf{F}[k(x); t]$$

that is

$$\mathbf{F}\left[b(y, \lambda); \frac{v\pi}{l}\right] = e^{\lambda\left(\frac{v\pi}{l}\right)^2} \mathbf{F}\left[k(x); \frac{v\pi}{l}\right],$$

and, finally,

$$b_v \approx e^{\lambda\left(\frac{v\pi}{l}\right)^2} a_v.$$

The determination of the Fourier cosine expansion coefficients a_v ($v=0, 1, \dots, M$) numerically or by a Fourier analyser gives an approximation of $b(y, \lambda)$ in $(0, l)$ by the Fourier synthesis (M is a given integer)

$$b(y, \lambda) \approx \frac{a_0}{2} + \sum_{v=1}^{M} e^{\lambda\left(\frac{v\pi}{l}\right)^2} a_v \cos\frac{v\pi}{l} y \qquad (0 < y \leq l).$$

For error analysis (including the errors in the determination of a_v) see Medgyessy (1954c, 1955b). It is mentioned here merely that the error bounds increase if $\lambda \uparrow \beta_1$.

A generalization of this procedure described by Medgyessy (1961a pp. 172—188, 81—84) might be omitted here.

We begin the presentation of numerical examples with *methodological examples,* of which the exact solutions are known. They may show the usefulness of the procedures.

Example 1. In Fig. 1 the solid line is the graph of the function

$$k(x) = \frac{e^{-\frac{(x-1)^2}{2}}}{\sqrt{2\pi}} + \frac{e^{-\frac{(x+1)^2}{2}}}{\sqrt{2\pi}}$$

i.e. $p_1=p_2=1/2$, $\alpha_1=1$, $\alpha_2=-1$, $\beta_1=\beta_2=1/2$. It has one *only* peak. The dotted line is the graph of an approximation $\hat{b}(y, \lambda)$ of the corresponding test function computed by means of the formula on the basis of the "measured" values $\hat{k}(x)$ of $k(x)$:

$$\hat{b}(y, \lambda) = \sum_{j=-m}^{m} c_j^{(m)}(\lambda)\, \hat{k}(y+jh)$$

taking $\lambda=3/8$, $m=3$. It shows the *two* components well. For the sake of comparison, the thin line shows the graph of the exact test function

$$b(y, 3/8) = 2\,\frac{e^{-2(x-1)^2}}{\sqrt{2\pi}} + 2\,\frac{e^{-2(x+1)^2}}{\sqrt{2\pi}}.$$

The maxima are already fairly well indicated by the dotted line (Medgyessy 1953, 1954c, 1961a pp. 90—101).

Example 2. In Fig. 2 the solid line shows the graph of the function

$$k(x) = 1000\,e^{-\frac{(x-5)^2}{4}} + 1000\,e^{-\frac{(x-7)^2}{2}} + 500\,e^{-\frac{(x-8.75)^2}{2}} + 1000\,e^{-\frac{(x-11.25)^2}{4}} +$$

$$500\,e^{-\frac{(x-13.25)^2}{2}} + 500\,e^{-\frac{(x-15.25)^2}{4}}$$

(Medgyessy, Varga 1968). It shows *two* peaks only. The dashed line is the graph

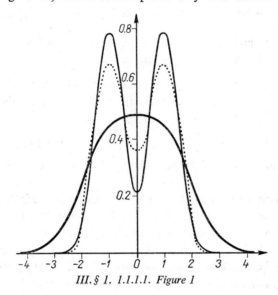

III. § 1. 1.1.1.1. Figure 1

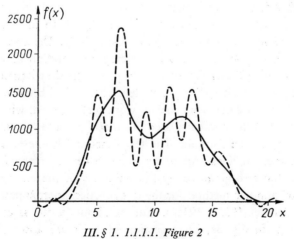

III. § 1. 1.1.1.1. Figure 2

of an approximation $\hat{b}(y, \lambda)$ to the corresponding test function $b(y, \lambda)$ with $\lambda = 3/8$, when the method described in Ch. V, § 2 was applied. The computation was based on the values $k(x_k)$ where $x_k = 0.25\,k$ ($k = 0, 1, \ldots, 80$), distorted with artificial random errors ξ_k (i.e. on $k(x_k) + \xi_k$) where the random variables ξ_k were independent and uniformly distributed with $\mathbf{D}^2 \xi_k = 1/12$. The approximate "spectrum" calculated with $k(x_k) + \xi_k$ was, according to the method, "cut off" at certain points (approximation of a "cut off filter"). The graph reveals *all* the *six* components and is appropriate to the determination of parameters, too. The error analysis as well as other considerations concerning the applicability of the numerical method used will not be dealt with; see Medgyessy, Varga (1968).

The following examples are concerned with decompositions of superpositions of normal density functions arising in practice.

Example 3. Let us consider the graph of Example 1 in Ch. I, § 1, a section of the intensity distribution of a Fe arc spectrum (solid line in Fig. 3 a, c and e). Considering its measured ordinate values as the values $\hat{k}(x)$, let us construct the test function approximation $\hat{b}(y, \lambda) = \sum_{j=-3}^{3} c_j^{(3)}(\lambda)\,\hat{k}\,(x + jk)$ at the values $\lambda = 0.45$, $h = 1$ and $m = 3$ from 60 data points. The result is shown by the dotted line in Fig. 3a.

As a comparison we present the graph of a section of the intensity distribution of a Fe arc spectrum (solid line in Fig. 3b) which has been photographed with the aid of a spectrograph of greater resolving power than that used for obtaining the intensity distribution graph of the same spectrum shown with a thick line in Fig. 3a. The dotted line shows $\hat{b}(y, \lambda)$, for comparison. The decomposition revealed almost all lines still hidden in the solid line graph.

The approximation of $b(y, \lambda)$ has also been constructed using the method of Ch. V, § 2 and it has been plotted with a dotted line both in Fig. 3c and d. The solid lines are the same as in Fig. 3a and b, respectively. The λ-value is the same as above; the other parameters also influencing the filtration of the noise effect have, of course, other values. According to the procedure the corresponding Fourier transform of $\hat{k}(x)$ was "cut off" appropriately when it had become near to 0 (application of a "cut off filter") in order to work further, then, with the "cut off" transform. The integrals were calculated from 60 equidistant values of $\hat{k}(x)$ by means of the trapezoidal formula. The test function also reveals the places of several hidden spectral lines. All peaks but one can be recognized except for the tails where the errors caused by the finiteness of the ending data appear.

The result of another method for the test function approximation is given by Fig. 3e (from Berencz 1955a, 1955b). The solid line is the same intensity distribution curve as in Fig. 3a, i.e. a section of a Fe arc absorption spectrum.

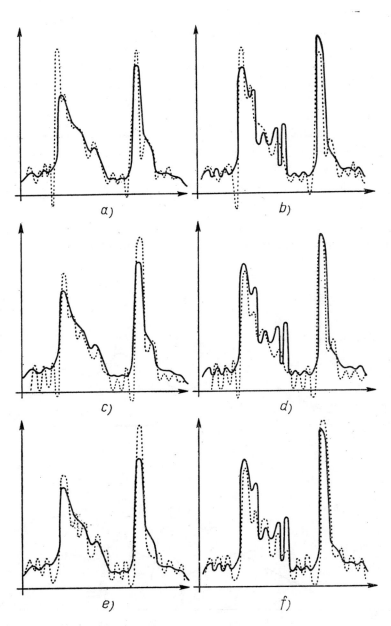

$a)$ $b)$ $c)$ $d)$ $e)$ $f)$

III. § 1. 1.1.1.1. Figure 3

The dotted line shows the test function $b(y, \lambda)$ obtained by Fourier analysis and Fourier synthesis at $\lambda = 0.45$. In Fig. 3f the solid line shows the intensity distribution curve photographed with a more powerful spectrograph, in order to compare the actual spectrum line places with those given by the test function (dotted line). The number and the abscissae of the hidden peaks can be read off fairly well from the graph of the test function. The necessary Fourier analysis was carried out with a Mader–Ott harmonic analyser, up to the 33rd cosine coefficient.

In this example the last method proved the best, because it utilized more starting "information" than the procedures which gave Fig. 3a—d based on 60 "measured" values only. The latter ones can, of course, also rely on more data. In any case, *the decomposition might serve as a substitute for the application of an apparatus of greater resolving power.*

Example 4. In Fig. 2 of Ch. I, § 1 the data of the protein fractions of human blood serum are to be determined (Medgyessy 1953). In Fig. 4a the solid line

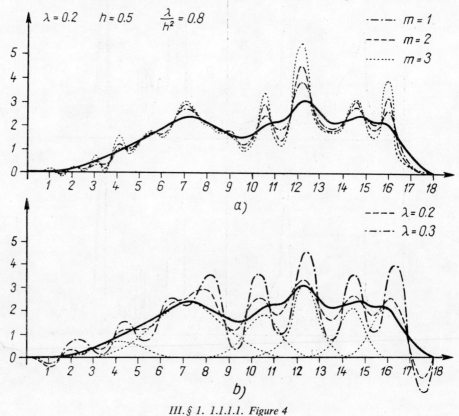

III.§ 1. 1.1.1.1. Figure 4

shows the graph yielded by Tiselius' apparatus during electrophoresis. The test function approximation $\hat{b}(y, \lambda)$ was constructed by means of the formula $\hat{b}(y, \lambda) = \sum\limits_{j=-m}^{m} c_j^{(m)}(\lambda)\hat{k}(y+jk)$ with $\lambda=0.2$, $h=0.5$ for several m values ($m=1$: „dots and dashes"; $m=2$: dashed line; $m=3$: dotted line). The graphs of the components are well separated (Medgyessy 1966a).

In Fig. 4b (Medgyessy 1953, 1954c) the solid line shows the same electrophoretic curve as in Fig. 4a. The dashed line shows the approximation of the test function $b(y, \lambda)$ obtained by Fourier analysis and Fourier synthesis for $\lambda=0.2$; „dots and dashes" show it for $\lambda=0.3$. We had $l=18$, $M=18$. The coefficients a_ν were determined with the help of a Mader–Ott harmonic analyser. The Fourier synthesis was carried out numerically at points of distance 0.5. The conjectured component Gaussian function curves are also plotted (dotted line) by means of the procedure described in Ch. III, § 3, (A), 3.

Example 5. In Fig. 5 taken from the paper of Dobozy, Volly (1970) the solid line shows a section of the graph of the intensity distribution $E(v^*)$ of the ultraviolet absorption spectrum of L-cystine $N/100$ in Na_2CO_3 (v^* is the wave number). $E(v^*)$ is assumed to be a superposition $k(x)$ of normal density functions in a certain interval. The test function approximation was constructed by means of $b^*(y, \lambda) = \sum\limits_{j=-m}^{m} c_j^{(m)}(\lambda)\hat{k}(x+jh)$ with $\lambda=2$ (dotted line) and $\lambda=8$ (thin line) (the left part of the relevant graph is omitted) and $k=0.5$, $m=2$. At places where the starting superposition is at about zero, $\hat{f}(y, \lambda)$ takes on negative values seemingly not influencing the other parts of the graphs of the test functions. On the basis of the dotted line one conjectures the presence of two components; this is supported by the shape of the thin line. The separation of the graphs of the components is observable as λ increases. As to the rest, spectroscopical considerations have justified the conjecture on two hidden lines in the present spectrum section.

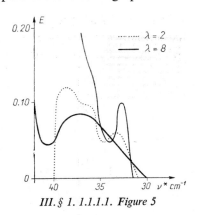

III. § 1. 1.1.1.1. Figure 5

Example 6. Szőke, Varga, Nagypál (1967) and Varga (1968) investigated a section of a fluorescence spectrum produced at a low temperature; they assumed the curve of intensity distribution to belong to a superposition $k(x)$ of normal density functions. This curve is shown by the solid line in Fig 6a, adopted from the mentioned paper of Varga (1968). The dotted line is the graph of the test function approximation calculated by means of the method described in Ch. V, § 2, while the "spectrum" of the $k(x)$ values, loaded with errors, is shown after a multiplication with $e^{\lambda t^2}$ in Fig. 6b. ↓ indicates the place of the "cut off". Six components are recognizable; their reality was supported by the experimental background. — We mention that after the determination of the number of components the parameters of the components of $k(x)$ were determined by Szőke, Varga, Nagypál (1967) by a fitting procedure, during a detailed treatment of the problem.

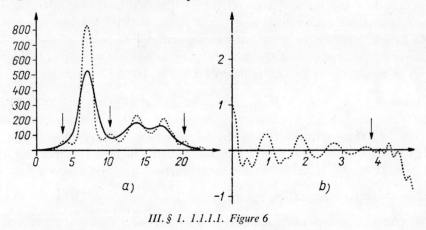

III.§ 1. 1.1.1.1. Figure 6

Example 7. Although newer methods have been developed in the meantime, the method of obtaining a test function based on Fourier analysis and Fourier synthesis, is also in use at present. For instance a successful spectroscopic

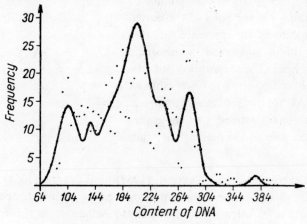

III.§ 1. 1.1.1.1. Figure 7

application of this method was published by Lehotai (1960). Gregor (1969) investigated the data from a measurement of the DNA content in the nuclei of liver cells of rats in control and experimental groups (i.e. a *sample*). From these, relative frequencies were calculated. After some smoothing he got from them still somewhat oscillating data to an approximation of a superposition of normal density functions, being assumed to be the background to the DNA content. In the case of the experimental animals they are shown by dots in Fig. 7. Nevertheless these data were already sufficient to carry out numerically the decomposition by the use of the described Fourier analysis and synthesis. From the test function obtained, shown by a solid line in Fig. 7, after some additional procedures several parameters of the initial superposition which have a biological meaning could be read off. It is worth mentioning that Gregor (1969) has also given, in Algol–60 language, an algorithm for this method of decomposition.

2. Second particular case of the $\mathscr{A}$-method

2.1. The first type of test function

Now we consider a newer type of the superposition

$$k(x) = \sum_{k=1}^{N} p_k f(x-\alpha_k, \beta_k) \qquad (A_1 < \beta_1 \leq \beta_2 \leq ... \leq \beta_N \leq A_2)$$

in which $f(x, \beta_k)$ is characterized by the fact that it is strictly unimodal and that if

(1) $$\mathbf{F}[f(x, \beta_k); t] = \varphi(t, \beta_k) \qquad (k = 1, ..., N)$$

then 1. $\varphi(t,0)=1$; 2. $\varphi(t, \beta_k)$ is the characteristic function of an infinitely divisible density function; and 3. $\dfrac{\varphi(t, \beta_k)}{\varphi(t, \lambda)}$ $(0\leq\lambda<\beta_1)$ is the characteristic function of some infinitely divisible, strictly unimodal density function $f^*(y, \beta_k, \lambda)$ where λ is a $(0, \beta_k)$ *monotone formant* of $\mathscr{G}f^*(y, \beta_k, \lambda)$.

Let us consider the function

$$b(y, \lambda) = \sum_{k=1}^{N} p_k f(y-\alpha_k, \beta_k, \lambda).$$

For this, Condition I of Ch. III, § 1 is fulfilled by the present assumptions since $f^*(y-\alpha_k, \beta_k, \lambda)$ is also strictly unimodal, and λ is a $(0, \beta_k)$ monotone formant of $\mathscr{G}f(y-\alpha_k, \beta_k)$, thus $\mathscr{G}f^*(y-\alpha_k, \beta_k, \lambda)\prec \mathscr{G}f(x-\alpha_k, \beta_k)$ i.e. $f^*(y-\alpha_k, \beta_k, \lambda)$ can be considered as the unimodal transform of increased narrowness of $f(x-\alpha_k, \beta_k)$. Also there holds the relation, independent of α_k, β_k

$$(2) \qquad \mathbf{F}[f^*(y-\alpha_k, \beta_k, \lambda); t] = e^{i\alpha_k t} \frac{\varphi(t, \beta_k)}{\varphi(t, \lambda)} = \frac{\mathbf{F}[f(x-\alpha_k, \beta_k); t]}{\varphi(t, \lambda)}$$

between $f^*(y-\alpha_k, \beta_k, \lambda)$ and $f(x-\alpha_k, \beta_k)$. If the abscissa of the peak of $f^*(y-\alpha_k, \beta_k, \lambda)$ is independent of β_k, then Condition II in Ch. III, § 1 is also satisfied; otherwise the fulfilment of Condition II in Ch. III, § 1 *is to be supposed* additionally. Condition III in Ch. III, § 1 is, however, fulfilled as, by (2)

$$\mathbf{F}[b(y, \lambda); t] = \frac{1}{\varphi(t, \lambda)} \mathbf{F}[k(x); t]$$

i.e.

$$b(y, \lambda) = \mathbf{F}^{-1}\left[\frac{1}{\varphi(t, \lambda)} \mathbf{F}[k(x); t]; y\right].$$

Thus $b(y, \lambda)$ can be taken for the test function corresponding to the superposition $k(x)$. Then, the steps (A), (B) and (C) in Ch. III, § 1 having been carried out, the $\mathscr{A}$-method can be applied here taking into account that, using $\mathbf{F}[\varphi(x, \lambda); t] = \varphi(t, \lambda)$,

$$(3) \qquad k(x) = \int\limits_{-\infty}^{\infty} f(x-y, \lambda) b(y, \lambda) \, dy,$$

that is $b(y, \lambda)$ is also here the solution of a Fredholm integral equation of the Ist kind and of convolution type (Medgyessy 1961a pp. 58—63).

The numerical decomposition of $k(x)$ also leads to an *incorrect* problem here; we can only make use of some method in Ch. V, § 1 (e.g. the regularization method).

The types of superposition in Ch. III, § 1, Subsection 1.1.1, are, in fact, particular cases of the present type; they could, however, be included in a simpler category.

The concrete form of $f^*(y, \beta_k, \lambda)$ can be given by fixing the type of $f(x, \beta_k)$.

Example. **Decomposition of a superposition of ch density functions.** Let

$$k(x) = \sum_{k=1}^{N} p_k \frac{1}{2\beta_k \operatorname{ch}[\pi(x-\alpha_k)/2\beta_k]}$$

$$(\alpha_i \neq \alpha_j \,(i \neq j); 0 < \beta_1 \leqq \beta_2 \leqq \ldots \beta_N \leqq \Lambda_2).$$

The components are strictly unimodal. Here

$$\mathbf{F}\left[\frac{1}{2\beta_k \operatorname{ch}(\pi x/2\beta_k)}; t\right] = \varphi(t, \beta_k) = \frac{1}{\operatorname{ch}\beta_k t},$$

$\varphi(t, 0) = 1$, $\varphi(t, \beta_k)$ is the characteristic function of an infinitely divisible density function (Medgyessy 1961a p. 64) and (cf. Ch. II, § 8, Section 1)

$$\frac{\varphi(t, \beta_k)}{\varphi(t, \lambda)} = \frac{\text{ch}\,\lambda t}{\text{ch}\,\beta_k t} \qquad (0 < \lambda < \beta_1)$$

is also the characteristic function of an infinitely divisible density function. Thus conditions (1)—(3) above are fulfilled. By Example 3 of Ch. II, § 4,

$$\mathbf{F}^{-1}\left[\frac{\text{ch}\,\lambda t}{\text{ch}\,\beta_k t}; y\right] = \frac{1}{\beta_k}\,\frac{\text{ch}\,\dfrac{\pi x}{2\beta_k} \cdot \cos\dfrac{\pi \lambda}{2\beta_k}}{\text{ch}\,\dfrac{\pi x}{\beta_k} + \cos\dfrac{\pi \lambda}{\beta_k}} = g^*(y, \beta_k, \lambda).$$

It can be proved easily that this is strictly (0) unimodal. The same example showed that λ is a $(0, \beta_k)$ monotone formant of $g^*(y, \beta_k, \lambda)$. Thus $g^*(y, \beta_k, \lambda)$ can be taken for the unimodal transform of increased narrowness of $\dfrac{1}{2\beta_k \,\text{ch}\,[\pi(x-\alpha_k)/2\beta_k]}$.

With the application of this $g^*(y, \beta_k, \lambda)$, the distances between the abscissae of the peaks of the components $g^*(y-\alpha_k, \beta_k, \lambda)$ of the test function are, evidently, the same as in the case of the components of $k(x)$. Thus the above results can, finally, be applied and the corresponding test function is

$$b(y, \lambda) = \sum_{k=1}^{N} p_k \frac{1}{\beta_k}\,\frac{\text{ch}\,\dfrac{\pi(x-\alpha_k)}{2\beta_k} \cdot \cos\dfrac{\pi \lambda}{2\beta_k}}{\text{ch}\,\dfrac{\pi(x-\alpha_k)}{\beta_k} + \cos\dfrac{\pi \lambda}{\beta_k}} \qquad (0 < \lambda < \beta_1)$$

which is the solution of the integral equation

$$k(x) = \int_{-\infty}^{\infty} \frac{1}{2\lambda\,\text{ch}\,[\pi(x-y)/2\lambda]}\,b(y, \lambda)\,dy,$$

as

$$\mathbf{F}^{-1}\left[\frac{1}{\text{ch}\,\lambda t}; x\right] = \frac{1}{2\lambda\,\text{ch}\,(\pi x/2\lambda)}.$$

The numerical solution is not dealt with here but it may be attempted by means of the methods described in Ch. V, § 1 (e.g. the regularization method).

8*

2.2. The second type of test function

Let us again consider the superposition

$$k(x) = \sum_{k=1}^{N} p_k f(x - \alpha_k, \beta_k) \qquad (\Lambda_1 < \beta_1 \leqq \beta_2 \leqq \ldots \leqq \beta_k < \Lambda_2)$$

already investigated in Ch. III, § 1, Section 2, where $f(x, \beta_k)$ was strictly unimodal, $\mathbf{F}[f(x, \beta_k); t] = \varphi(t, \beta_k)$ and $\dfrac{\varphi(t, \beta_k)}{\varphi(t, \lambda)}$ $(0 \leqq \lambda \leqq \beta_1)$ was the characteristic function of an infinitely divisible strictly unimodal density function $f^*(y, \beta_k, \lambda)$, and λ was a $(0; \beta_k)$ monotone formant of $\mathscr{G}f^*(y, \beta_k, \lambda)$. The test function $b(y, \lambda)$ of the decomposition of this superposition by the aid of the $\mathscr{A}$-method was given by

$$b(y, \lambda) = \mathbf{F}^{-1}\left[\frac{1}{\varphi(t, \lambda)} \mathbf{F}[k(x); t]; y\right]$$

i.e. $b(y, \lambda)$ was the solution of the integral equation

$$k(x) = \int_{-\infty}^{\infty} f(x - y, \lambda) b(y, \lambda) \, dy;$$

(cf. Ch. III, § 1, Subsection 2.1).

The test function introduced above had the disadvantage of being derived by means of the solution of a convolution Fredholm integral equation of the Ist kind; thus its explicit construction represents an *incorrect* problem.

However, a new method of decomposition can be found by means of the following idea (Medgyessy 1967a), prompted by the article of Kreisel (1949).

To decompose the above superposition $k(x)$ we define, in certain cases, *another test function*. However, this test function **can be constructed by means of a convolution transformation** from the superposition to be decomposed; thus the problem is *correct* and its numerical treatment will not be delicate, or sensitive with regard to measurement errors.

Considering the superposition

$$k(x) = \sum_{k=1}^{N} p_k f(x - \alpha_k, \beta_k) \qquad (0 < \beta_1 \leqq \beta_2 \leqq \ldots \leqq \beta_N < \Lambda_2),$$

we investigate merely that case when $f(x, \beta_k)$ is a (0) *unimodal* $(\mathbf{A_1})$ *density function* (cf. Ch. II, § 4) whose expectation exists. Again let us have for $\mathbf{F}[f(x, \beta_k); t] = \varphi(t, \beta_k)$ $(k = 1, \ldots, N)$ 1. $\varphi(t, 0) = 1$; 2. $\varphi(t, \beta_k)$ is the characteristic function of an infinitely divisible density function; evidently this latter one is now (0) symmetrical and also (0) unimodal; 3. $\dfrac{\varphi(t, \beta_k)}{\varphi(t, \lambda)}$ $(0 \leqq \lambda < \beta_1)$ is the characteristic function of an infinitely divisible, strictly unimodal density function $f^*(y, \beta_k, \lambda)$ being (0) symmetrical and, consequently, (0) unimodal. From the above statements it follows that its expectation exists.

For a test function let us now take the function

$$b_1(y, \lambda) = \sum_{k=1}^{N} p_k h(y - \alpha_k, \beta_k, \lambda)$$

in which

$$h(y, \beta_k, \lambda) = \int_{-\infty}^{\infty} f^*(x, \beta_k, \lambda) A(y - x, \varepsilon) \, dy$$

where $A(x, \varepsilon)$ $(0 < \varepsilon < T)$ is some (0) symmetrical Pólya frequency function (cf. Ch. II, § 1) for which $\int_{-\infty}^{-\xi} A(x, \varepsilon) dx < \omega A(-\xi, \varepsilon)$ $(\omega > 0, \; \xi > 0$ are arbitrary) holds if $\xi > \xi(\omega) > 0$, where $\xi(\omega)$ is a threshold number depending on ω and ε is a $(T, 0)$ monotone formant of $\mathscr{G}A(x, \varepsilon)$. Since the Pólya frequency functions are strongly unimodal (Ch. II, § 1, Theorem 8), $h(y, \beta_k, \lambda)$ is, in virtue of Ch. II, § 1, Theorem 4, (0) symmetrical and (0) unimodal. By Theorem 1 in the *Supplements and problems to* Ch. II, § 4, λ is a $(0, \beta_k)$ monotone formant of $h(y, \beta_k, \lambda)$, too; for this is equivalent to asserting that if $\lambda_i < \lambda_j < \beta_k$ then $\mathscr{G}h(y, \beta_k, \lambda_j) < \mathscr{G}h(y, \beta_k, \lambda_i)$ which follows already from the theorem since, by assumption $\mathscr{G}f^*(y, \beta_k, \lambda_j) < \mathscr{G}f^*(y, \beta_k, \lambda_i)$. In other words if $\lambda \downarrow \beta_k$ then the graph of $h(y, \beta_k, \lambda)$ becomes narrower and narrower. Hence it follows that the kth component of $b_1(y, \lambda)$ is (α_k) symmetrical, (α_k) unimodal and that λ is its $(0, \beta_k)$ monotone formant; i.e. λ is a $(0, \beta_1)$ monotone formant of all components. For $b_1(y, \lambda)$ Condition I in Ch. III, § 1 is satisfied as by the preceding, $h(y - \alpha_k, \beta_k, \lambda)$ can be considered to be the unimodal transform of increased narrowness of $f(x - \alpha_k, \beta_k)$, and between $h(y - \alpha_k, \beta_k, \lambda)$ and $f(x - \alpha_k, \beta_k)$ there holds the relation, independent of α_k, β_k:

$$\mathbf{F}[h(y, p_k, \lambda); t] = \frac{\varphi(t, \alpha)}{\varphi(t, \lambda)} \alpha(t, \varepsilon)$$

where

$$\alpha(t, \varepsilon) = \mathbf{F}[A(x, \varepsilon); t].$$

Condition II of Ch. III, § 1 is also satisfied, as is Condition III in Ch. III, § 1, since by the preceding equality

$$\mathbf{F}[b_1(y, \lambda); t] = \frac{\alpha(t, \varepsilon)}{\varphi(t, \lambda)} \mathbf{F}[k(x); t],$$

that is

$$b_1(y, \lambda) = \mathbf{F}^{-1}\left[\frac{\alpha(t, \varepsilon)}{\varphi(t, \lambda)} \mathbf{F}[k(x); t]; y\right].$$

Thus $b(y, \lambda)$ can be taken as the test function corresponding to $k(x)$.

Now let us examine the nature of $A(x, \varepsilon)$. *Let $A(x, \varepsilon)$ and $\alpha(t, \varepsilon)$ respectively be such that* $\int\limits_{-\infty}^{\infty} \left| \dfrac{\alpha(t, \varepsilon)}{\varphi(t, \lambda)} \right| dt < \infty$; *then* $\dfrac{\alpha(t, \varepsilon)}{\varphi(t, \lambda)}$ *can be considered as the Fourier trans-form of a function $M(x, \lambda, \varepsilon)$,* $\mathbf{F}[M(x, \lambda, \varepsilon); t] = \dfrac{\alpha(t, \varepsilon)}{\varphi(t, \lambda)}$, $M(x, 0, \varepsilon) = A(x, \varepsilon)$, ($M(x, \lambda, \varepsilon)$ is, naturally, *not* a density function). In this case, however,

$$h(y, \beta_k, \lambda) = \int\limits_{-\infty}^{\infty} M(y-x, \lambda, \varepsilon) f(x, \beta_k) \, dx,$$

and consequently the test function is in this case

$$b_1(y, \lambda) = \int\limits_{-\infty}^{\infty} M(y-x, \lambda, \varepsilon) k(x) \, dx \qquad (0 < \lambda < \beta_1),$$

whence the relation between $b_1(y, \lambda)$ and $k(x)$ can be seen.

The test function $b_1(y, \lambda)$ can serve the decomposition of the superposition $k(x)$ of density functions; from the graph of this, N and, eventually, α_k, β_k can be determined. *The test function $b_1(y, \lambda)$ can be represented in the form of a convolution transformation* applied to the superposition $k(x)$; this also allows a numerical treatment and the $\mathscr{A}$-method can be applied.

In any case the following facts must not be forgotten as can be easily seen from the foregoing: in the case of $\lambda = 0$

$$b_1(y, 0) = \int\limits_{-\infty}^{\infty} A(y-x, \varepsilon) k(x) \, dx$$

is *not* equal to $k(y)$; it equals a superposition consisting of the components of $k(x)$ transformed by a convolution transformation of the above type. It can be expected that in the case of a small ε these components will differ only slightly from the components of $k(x)$ because $A(x, \varepsilon)$ "filters out" a very short section of $k(x)$; the latter are, however, less separated than the former. This is still not too interesting. If λ increases, the components of $b_1(y, \lambda)$ — of which λ is a monotone formant — become narrower and even in the case of $\lambda = \beta_1$ the component with parameter β_1 does not go over to a "Dirac delta function": it goes over to $A(x-\alpha_k, \varepsilon)$, i.e. to a "peak" function becoming more and more peaked when ε decreases. This, however, distorts little the total picture of the test functions. The other components will have only a slightly broader graph as a result of this. Thus by this procedure we get a test function by means of a convolution transformation applied to $k(x)$, the evaluation of which is already a *correct* problem and will be — if a sufficiently small ε has been taken — almost as appropriate for determining unknown parameters — if $\lambda \uparrow \beta_1$ — as the test function $b(y, \lambda)$ in Ch. III, § 1, Section 2. This slight disadvantage is, however, counterbalanced in that we have to calculate numerically a convolution integral and not to solve an integral equation as in Ch. III, § 1, Section 2, for the representation of the test function; thus it can be expected that the result will be less charged with errors, in spite of the "measurement" errors, than in the procedure in Ch. III, § 1, Section 2. *Therefore this method acquires a particular importance.*

If $\varphi(t, \beta_k) = e^{-\beta_k |t|^A}$ $(0 < A < 2)$ i.e. it is the characteristic function of a certain (0) symmetrical stable density function, then $\alpha(t, \varepsilon) = e^{-\varepsilon t^2}$ satisfies the conditions as $A(x, \varepsilon) = \dfrac{e^{-\frac{x^2}{4\varepsilon}}}{\sqrt{4\pi\varepsilon}}$ and $\displaystyle\int_{-\infty}^{-\xi} A(x, \varepsilon)\, dx < \omega A(-\xi, \varepsilon)$ $(\omega > 0, \quad \xi > 0$ and is sufficiently large) because by a well-known inequality,

$$\int_{-\infty}^{-\xi} \frac{e^{-\frac{x^2}{4\varepsilon}}}{\sqrt{4\pi\varepsilon}}\, dx < \sqrt{\frac{\varepsilon}{\pi}}\, \frac{e^{-\frac{\xi^2}{4\varepsilon}}}{\xi} < \omega\, \frac{e^{-\frac{\xi^2}{4\varepsilon}}}{\sqrt{4\pi\varepsilon}}, \quad \text{if } \xi > \frac{2\varepsilon}{\omega}.$$

In the case of $\varphi(t, \beta_k) = e^{-\beta_k t^2}$, $\alpha(t, c) = e^{-\varepsilon t^2}$ is not appropriate.

In certain cases $M(x, \lambda, \varepsilon)$ can be given only in the form of a series; this is, however, no disadvantage in the numerical computation of $b_1(y, \lambda)$.

Example 1. **Decomposition of a superposition of ch density functions.** Let again

$$k(x) = \sum_{k=1}^{N} p_k \frac{1}{2\beta_k \operatorname{ch}[\pi(x - \alpha_k)/2\beta_k]}$$
$$(\alpha_i \neq \alpha_j (i \neq j); \; 0 < \beta_1 \leq \beta_2 \leq \ldots \leq \beta_N < \Lambda_2).$$

Here $\varphi(t, \beta_k) = \dfrac{1}{\operatorname{ch}\beta_k t}$. Let $\alpha(t, \varepsilon) = e^{-\varepsilon t^2}$ $(0 < \varepsilon < T)$; this satisfies all conditions imposed on $\alpha(t, \varepsilon)$ and

$$\int_{-\infty}^{\infty} \left| \frac{\alpha(t, \varepsilon)}{\varphi(t, \lambda)} \right| dt = \int_{-\infty}^{\infty} \operatorname{ch}\lambda t \cdot e^{-\varepsilon t^2}\, dt < \infty \qquad (0 \leq \lambda < \beta_1);$$

further

$$M(x, \lambda, \varepsilon) = \frac{1}{2\pi} \int_{-\infty}^{\infty} e^{-ixt} \operatorname{ch}\lambda t \cdot e^{-\varepsilon t^2}\, dt = e^{\frac{\lambda^2}{4\varepsilon}}\, \frac{e^{-\frac{x^2}{4\varepsilon}}}{\sqrt{4\pi\varepsilon}} \cos\frac{\lambda x}{2\varepsilon}.$$

According to the foregoing, the test function is

$$b_1(y, \lambda) = e^{\frac{\lambda^2}{4\varepsilon}} \int_{-\infty}^{\infty} \frac{e^{-\frac{(y-x)^2}{4\varepsilon}}}{\sqrt{4\pi\varepsilon}} \cos\frac{\lambda(y-x)}{2\varepsilon} \cdot k(x)\, dx.$$

λ is a $(0, \beta_1)$ monotone formant of the graphs of the components of the test function (Medgyessy 1967a) (we recall that, as we have seen in Ch. III, § 1, Section 2, λ was a monotone formant of the graph of the test function $b(y, \lambda) = \mathbf{F}^{-1}[\mathrm{ch}\,\lambda t \cdot \mathbf{F}[k(x); t]; \; y])$.

By virtue of the relation

$$b_1(y, \lambda) = \int_{-\infty}^{\infty} M(y-x, \lambda, \varepsilon)\, k(x)\, dx$$

a numerical approximation $b_1^*(y, \lambda)$ to $b_1(y, \lambda)$ can also be given in the present case, in a very manageable form by means of the method in Ch. V, § 4, Section 2,

$$b_1^*(y, \lambda) = \sum_{j=-m}^{m} W_j^{(m)}(\lambda)\, \hat{k}(x+jh)$$

(writing $W_j^{(m)}(\lambda)$ instead of $W_j^{(m)}$) where $\hat{k}(x)$ is the "measured" value of $k(x)$, h is given and

$$W_j^{(m)}(\lambda) = \sum_{k=0}^{m} B_{j,k}^{(m)} (-1)^k \frac{m_{2k}}{h^{2k}}$$

(m_{2k} is the $2k$th moment of $M(x, \lambda, \varepsilon)$ and the $B_{j,k}^{(m)}$ are the numbers introduced, and collected in a table in Ch. V, § 3). *In our case m_{2k} can also be calculated by means of the Fourier transform of $M(x, \lambda, \varepsilon)$.* Some values of it, calculated by the use of this remark, are as follows:

$$m_0 = 1$$
$$m_2 = -\lambda^2 + 2\varepsilon,$$
$$m_4 = -\lambda^4 + 12\varepsilon^2 - 12\lambda^2\varepsilon,$$
$$m_6 = -\lambda^6 + 30\varepsilon\lambda^4 - 180\varepsilon^2\lambda^2 + 120\varepsilon^3.$$

Thus after the choice of λ, ε, h, $W_j^{(m)}$ i.e. the concrete form of the approximation, $b^*(y, \lambda)$ can already be put down. The error of this approximation is no dealt with.

The convolution integral yielding the test function $b_1(y, \lambda)$ can also be computed numerically by means of the method described in Ch. V, § 4, Section 1, supposing that sufficiently many $k(x)$ values are at one's disposal.

Let us consider as a *methodological example* the superposition of ch density functions

$$k(x) = \frac{1}{\mathrm{ch}\,[\pi(x-0.5)/2]} + \frac{1}{\mathrm{ch}\,[\pi(x+0.5)/2]}.$$

Its graph is unimodal; it is shown by the solid line in Fig. 1. In the way described above we have constructed the approximation $b_1^*(y, \lambda)$ to its test function with $\lambda=3/4$, $\varepsilon=1/64$, $m=2$, $h=0.3$; it is shown by the dotted line. The values of $\hat{k}(x)$ have been obtained from those of $k(x)$, computed from a table to only 3 decimal figures. The graph of the exact test function

$$b_1(y, \lambda) = 2\,\frac{\operatorname{ch}[\pi(x-0.5)/2]\cdot\cos[0.75\pi/2]}{\operatorname{ch}\pi(x-0.5)+\cos 0.75\pi} + 2\,\frac{\operatorname{ch}[\pi(x+0.5)/2]\cdot\cos[0.75\pi/2]}{\operatorname{ch}\pi(x+0.5)+\cos 0.75\pi}$$

is shown by the thin line for comparison. The two, originally hidden, components present themselves quite well (Medgyessy 1972b).

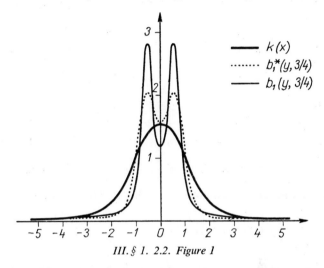

III.§ 1. 2.2. Figure 1

Example 2. **Decomposition of a superposition of Cauchy density functions.** Although this problem has appeared already in dealing with the simple diminishing of the formant, we also construct the new type test function of this, as the superposition of Cauchy density functions belongs to the present type of superposition too. Thus let

$$k(x) = \sum_{k=1}^{N} p_k\,\frac{1}{\pi\beta_k}\,\frac{1}{1+(x-\alpha_k)^2/\beta_k^2}$$

$$\left(\alpha_i \neq \alpha_j\ (i\neq j);\ 0 < \beta_1 \leqq \beta_2 \leqq \dots \leqq \beta_N < \Lambda_2\right).$$

Here $\varphi(t, \beta_k)=e^{-\beta_k|t|}$. Let us put again $\alpha(t, \varepsilon)=e^{-\varepsilon t^2}$. $(0<\varepsilon<T)$; this is appropriate from the viewpoint of the method; e.g.

$$\int_{-\infty}^{\infty}\left|\frac{\alpha(t, \varepsilon)}{\varphi(t, \lambda)}\right|dt = \int_{-\infty}^{\infty} e^{\lambda|t|}e^{-\varepsilon t^2}\,dt < \infty.$$

Further

$$M(x, \lambda, \varepsilon) = \frac{1}{2\pi} \int_{-\infty}^{\infty} e^{-ixt} e^{\lambda |t|} e^{-\varepsilon t^2} dt.$$

By a simple modification

$$M(x, \lambda, \varepsilon) = \frac{1}{\pi} \int_{-\infty}^{\infty} e^{-ixt} \operatorname{ch} \lambda t \cdot e^{-\varepsilon t^2} dt - \frac{1}{2\pi} \int_{-\infty}^{\infty} e^{-ixt} e^{-\lambda |t|} e^{-\varepsilon t^2} dt.$$

The first integral already occurred in Example 1, the second is the convolution of a Cauchy density function and of a normal density function. Explicitly:

$$M(x, \lambda, \varepsilon) = \frac{1}{\sqrt{\pi\varepsilon}} e^{\frac{\lambda^2}{4\varepsilon}} e^{-\frac{x^2}{4\varepsilon}} \cos\frac{\lambda x}{2\varepsilon} - \frac{1}{2\sqrt{\pi\varepsilon}\,\pi} \int_{-\infty}^{\infty} \frac{\left(\frac{\lambda}{2\sqrt{\varepsilon}}\right) e^{-y^2}}{\left(\frac{\lambda}{2\sqrt{\varepsilon}}\right)^2 + \left(\frac{x}{2\sqrt{\varepsilon}} - y\right)^2}\, dy.$$

The second integral is equal to

$$\pi \operatorname{Re} \frac{i}{\pi} \int_{-\infty}^{\infty} \frac{e^{-t}}{z-t}\, dt = \pi \operatorname{Re} e^{-z^2} \left(1 + \frac{2i}{\sqrt{\pi}} \int_{0}^{z} e^{u^2}\, du\right)$$

at $z = \pm \frac{x}{2\sqrt{\varepsilon}} + i\frac{\lambda}{2\sqrt{\varepsilon}}$ (Faddeeva, Terent'ev 1954 a p. 7). This real part denoted, for $z = \xi + i\eta$, by $u(\xi, \eta)$ is tabulated in the book referred to. Finally, we have

$$M(x, \lambda, \varepsilon) = \frac{1}{2\sqrt{\pi\varepsilon}} \left[2 e^{\left(\frac{\lambda}{2\sqrt{\varepsilon}}\right)^2} e^{-\left(\frac{x}{2\sqrt{\varepsilon}}\right)^2} \cos 2\left(\frac{\lambda}{2\sqrt{\varepsilon}}\right)\left(\frac{x}{2\sqrt{\varepsilon}}\right) - u\left(\frac{x}{2\sqrt{\varepsilon}}, \frac{\lambda}{2\sqrt{\varepsilon}}\right)\right].$$

The convolution transform representation of the test function $b_1(y, \lambda)$, by the *immediate* use of $M(x, \lambda, \varepsilon)$, can be computed *numerically* only by the method described in Ch. V, § 4, Section 1, because in our case the moments of $M(x, \lambda, \varepsilon)$ do not exist, as can be seen from its Fourier transform. Clearly one should have sufficiently many measured $k(x)$ values at one's disposal. Also special mathematical instruments can be used in this work (cf. Ch. V, § 4, Section 1). In general $M(x, \lambda, \varepsilon)$ is to be tabulated.

The method described in point 5 of the *Supplements and problems to* Ch. V, § 2 can also be applied here to obtain the approximation of the test function, by performing numerically the steps corresponding to the relation

$$b_1(y, \lambda) = \mathbf{F}^{-1}\left[e^{\lambda |t|} e^{-\varepsilon t^2} \mathbf{F}[k(x); t]; y\right].$$

This way often yields a better result than the method in Ch. V, § 4, Section 1, for example if $M(x, \lambda, \varepsilon)$ varies "suddenly", the latter one gives a fairly good result only if built on values of $k(x)$ and $M(x, \lambda, \varepsilon)$ belonging to x-points, located in a very dense manner, but this procedure may give rise to difficulties.

To give a methodological *numerical example* let us consider again the superposition of two Cauchy density functions, investigated in Example 2, (a) of Ch. III, § 1, Subsection 1.1.1. We had

$$k(x) = \frac{1}{1+(x-0.5)^2} + \frac{1}{1+(x+0.5)^2}.$$

Its graph plotted with a solid line in Fig. 2 is unimodal. To this we have constructed numerically, by means of the method described in Ch. V, § 2 the ap-

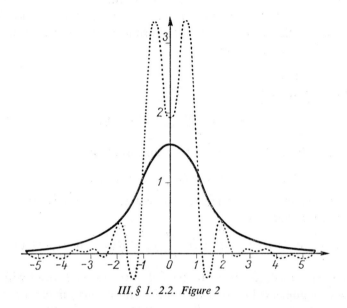

III. § 1. 2.2. Figure 2

proximation of the test function, with $\lambda=3/4$, $\varepsilon=1.64$ shown by the dotted line (the details of the numerical procedure may be omitted). The two peaks of the test function approximation reveal the two components (Medgyessy 1972*b*).

3. Decomposition of superpositions of the first type by means of the transformation of the superposition

In some cases a superposition of density functions

$$k(x) = \sum_{k=1}^{N} p_k f(x-\alpha_k, \beta_k) \qquad (\Lambda_1 < \beta_1 \leq \beta_2 \leq \ldots \leq \beta_N < \Lambda_2)$$

cannot be decomposed by the methods we have described. However, by means of an appropriate linear transformation it can be transformed into a superposition of density functions $k^*(y)$ that can already be decomposed by one of the above methods — or sometimes it gives a possibility of introducing a perfectly new decomposition method. Here examples of such transformations will be given for such cases when the transformation can be realized numerically and it does not much influence the numerical treatment of the decomposition. The ideas recognizable in the examples may, of course, be applied in other cases also (see Medgyessy 1961a, pp. 143—152).

Theoretically it is possible sometimes to unify the present transformation and the calculation of the test function belonging to $k^*(y)$ into a transformation acting on $k(x)$. However, this leads to formidable formulae and so such calculations will not be dealt with in what follows.

Particular cases:

(a) There are superpositions

$$k(x) = \sum_{k=1}^{N} p_k \, f(x - \alpha_k, \beta_k)$$

that can be transformed by means of the transformation

$$k(M(y)) \cdot M'(y) = k^*(y),$$

where $M(y)$ is strictly monotone and differentiable in the domain of definition of y, into such a superposition

$$k^*(y) = k(M(y)) \cdot M'(y) = \sum_{k=1}^{N} p_k \, f(M(y) - \alpha_k, \beta_k) \cdot M'(y)$$

which can already be decomposed by the aid of the methods dealt with hitherto. In the case of a numerical decomposition this transformation can be carried out simply by a scale changing etc. (Medgyessy 1961a pp. 148—149).

Example. **Decomposition of a superposition of lognormal density functions.** Let

$$k(x) = \begin{cases} \displaystyle\sum_{k=1}^{N} p_k \dfrac{e^{-\frac{(\log x - \alpha_k)^2}{4\beta_k}}}{x\sqrt{4\pi\beta_k}} & (x > 0) \\[4mm] 0 & (x \le 0) \end{cases} \qquad \left(\begin{matrix} \alpha_i \ne \alpha_j \quad (i \ne j); \\ 0 < \beta_1 \le \beta_2 \le \ldots \le \beta_N < \Lambda_2 \end{matrix} \right).$$

Then under the transformation

$$k^*(y) = k(e^y)e^y$$

this goes into the superposition of normal density functions

$$k^*(y) = \sum_{k=1}^{N} p_k \, \frac{e^{-\frac{(y-\alpha_k)^2}{4\beta_k}}}{\sqrt{4\pi\beta_k}}$$

which can be decomposed by the aid of the methods in Ch. III, § 1, Subsection 1.1.1.1.

We shall not deal with any concrete example here.

(b) Let us consider the superposition of density functions

$$k(x) = \sum_{k=1}^{N} p_k v(x, \beta_k) \qquad (\Lambda_1 < \beta_1 \leqq \beta_2 \leqq \ldots \leqq \beta_N < \Lambda_2)$$

where $v(x, \beta_k)$ are (0) unimodal density functions. We suppose that $k(x)$ cannot be decomposed by any method above. It has been proved (Girault 1955; see also the formulae of Theorem 2 in Ch. II, § 1) that the function $v^*(y, \beta_k)$ defined by the relations

$$v^*(y+0, \beta_k) = v(y, \beta_k) - y v'(y+0, \beta_k) \qquad (y > 0)$$
$$v^*(y-0, \beta_k) = v(y, \beta_k) + y v'(y-0, \beta_k) \qquad (y < 0)$$

is again a density function; let us suppose that it is unimodal. Let us consider the transformation defined by

$$k^*(y+0) = k(y) - y k'(y+0) \qquad (y > 0)$$
$$k^*(y-0) = k(y) + y k'(y-0) \qquad (y < 0).$$

It then follows that

$$k^*(y) = \sum_{k=1}^{N} p_k v^*(y, \beta_k)$$

is a superposition of unimodal density functions. In several cases $k^*(y)$ can already be decomposed by some of the existing methods, which also yields the unknown parameters of $k(x)$ (Medgyessy 1961a pp. 145—146).

The present transformation can also be carried out numerically (if formulae for the derivative are not delicate). No concrete example will be dealt with here.

(c) **Decomposition of a superposition of Beta density functions.** Let

$$k(x) = \begin{cases} \sum\limits_{k=1}^{N} p_k \dfrac{x^{\beta_k-1}(1-x)^{q-\beta_k-1}}{B(\beta_k, q-\beta_k)} & (0 < x \leqq 1) \\ 0 & \text{otherwise} \end{cases} \qquad (0 < \beta_1 < \beta_2 < \ldots < \beta_N < \Lambda_2)$$

where $q - \beta_k > 0$ $(k=1, \ldots, N)$ and q is given (Medgyessy 1961a pp. 149—152). This superposition cannot be decomposed by any of the above methods.

By the transformation

$$k^*(y) = \int_0^1 \frac{y^{q-1} e^{-\frac{y}{x}}}{x^q \Gamma(q)} k(x) \, dx$$

$k(x)$ goes over in the superposition of Gamma density functions of different order

$$k^*(y) = \begin{cases} \sum_{k=1}^{N} p_k \dfrac{y^{\beta_k-1}e^{-y}}{\Gamma(\beta_k)} & (y > 0) \\ 0 & (y \leqq 0) \end{cases}$$

since

$$\int_0^1 \frac{y^{q-1}e^{-\frac{y}{x}}}{x^q \Gamma(q)} \cdot \frac{x^{\beta_k-1}(1-x)^{q-\beta_k-1}}{B(\beta_k, q-\beta_k)}\, dx = \frac{y^{\beta_k-1}e^{-y}}{\Gamma(\beta_k)}$$

which can be decomposed by the aid of the method of formant changing, as discussed in Example 1 in Ch. III, § 1, Subsection 1.1.1.

This and similar transformations are dealt with in detail in Medgyessy (1961a, pp. 149—152). These transformations can also be carried out numerically, but no example will be dealt with here.

Supplements and problems to Ch. III, § 1

1. The basic idea of the $\mathscr{A}$-method can be found in Nikhilranjan Sen's work (1922) although his article had a different objective; Doetsch (1926), (1928) gave the exact elaboration of the basic idea while Medgyessy (1954b), (1961a Ch. II), (1971a), (1971b) provided its development.

2. Obviously a decomposition procedure can be applied also to decide whether a density function is a *superposition of density functions* or not (cf. Pearson 1894).

3. If in a decomposition procedure the number N of the components has already been determined, then the determination of the other parameters is equivalent to the *solution of a certain system of nonlinear algebraic equations*, or to a problem of fitting; both will be mentioned in Ch. III, § 3.

4. If, during the decomposition procedure, the graph of some component changes into a very sharp peak, then it can be subtracted graphically from the original graph. After the elimination of a component in this way, the decomposition procedure *can be iterated* on what remains. The present book does not in general deal with such iterations.

5. The $\mathscr{A}$-method and the other methods to be treated later can be applied, evidently, also in the case when N is *known*. Then they serve as special procedures for the determination of the unknown parameters of the original superposition

(Békési 1967). E.g. the earliest publication dealing with a decomposition method (Sen 1922) used the method exclusively in this sense.

6. Clearly the method in Ch. III, § 1, Subsection 1.1.1 can be applied also in such cases when only *the graph of the characteristic function* of the given superposition $k(x)$ is known. Such cases often occur in physics, meteorology etc.

7. Evidently the method in Ch. III, § 1, Subsection 1.1.1 — and the second special case of the $\mathscr{A}$-method also — *can be generalized* easily to cases where

$$\varphi(t, \beta_k) = \Phi_1(t) \Phi(t)^{\beta_k},$$

$\Phi_1(t)$ is some function not depending on β_k, and the conditions have been modified conveniently. This occurs e.g. in such cases where the components of the superposition to be decomposed are *convolutions* corresponding to the above relation (e.g. convolutions of normal density functions and some other density functions not depending on β_k). In the latter case — and in several other cases, too — the different methods of representation of test functions can be applied immediately (also numerically); cf. *Supplements and problems to* Ch. II, § 8. Such a described case occurs e.g. in astrophysics where the intensity distribution curve of certain spectral lines is supposed to be the graph of the convolution of a normal density function and a Cauchy density function with (partly) given parameters (the so-called **Voigt function**). Then a spectrum intensity curve is the graph of a *superposition of Voigt functions*. The reconstruction of a distorted Voigt function graph from a measured curve ("**Entschmierung**", "**Entzerrung**") as well as a **decomposition of a superposition of Voigt functions** have often appeared in astrophysics; see e.g. Elste (1953); Larson, Kenneth (1967).

The case when $\varphi(t, \beta_k)$ is the characteristic function of a Čebyšev–Hermite expansion evidently belongs to this sphere.

8. For the superpositions of certain stable density functions the test function $b(y, \lambda)$ can be determined not only from the integral equation (6) in Ch. III, § 1, Subsection 1.1.1, but from one of the following *partial integro-differential equations* to be deduced in a simple way from Theorem 7 in Ch. II, § 8:

$$\frac{\partial b}{\partial \lambda} = -\frac{1}{2\,\Gamma(1-A)\cos(\pi A/2)} \int_{-\infty}^{\infty} \frac{B - \operatorname{sgn}(y-\xi)}{|y-\xi|^A} \frac{\partial b}{\partial \xi}\, d\xi \quad (0 < A < 1),$$

$$\frac{\partial b}{\partial \lambda} = -\frac{1}{2\,\Gamma(2-A)\cos(\pi A/2)} \int_{-\infty}^{\infty} \frac{B \operatorname{sgn}(y-\xi)-1}{|y-\xi|^{A-1}} \frac{\partial^2 b}{\partial \xi^2}\, d\xi \quad (1 < A < 2)$$

where $0 \leqq \lambda < \beta_1$ (Medgyessy 1958, 1961a pp. 101—104, 199—203). Numerical problems of their application, questions of correctness etc. will not, however, be investigated here. Furthermore a test function could be found in an easier way.

9. In the case of the decomposition of a superposition of Cauchy density functions

$$k(x) = \sum_{k=1}^{N} p_k \frac{1}{\pi \beta_k} \frac{1}{1+(x-\alpha_k)^2/\beta_k^2}$$

with the test function

$$b(y, \lambda) = \sum_{k=1}^{N} p_k \frac{1}{\pi(\beta_k - \lambda)} \frac{1}{1+(x-\alpha_k)^2/(\beta_k - \lambda^2)}$$

we can give, beside the solution of an integral equation, another representation of the test function. We have

$$\mathbf{F}[b(y, \lambda); t] = \sum_{k=1}^{N} p_k e^{i\alpha_k t} e^{-\beta_k |t|} e^{\lambda |t|} = \mathbf{F}[k(x); t] e^{\lambda |t|} =$$

$$\sum_{k=1}^{N} p_k 2 \left[e^{i\alpha_k t} e^{-\beta_k |t|} \left(\frac{e^{-\lambda |t|} + e^{\lambda |t|}}{2} \right) - \frac{e^{i\alpha_k t} e^{-(\beta_k + \lambda)|t|}}{2} \right] =$$

$$2 \operatorname{ch} \lambda t \cdot \mathbf{F}[k(x); t] - e^{-\lambda |t|} \mathbf{F}[k(x); t] =$$

$$2 \sum_{v=0}^{\infty} \frac{\lambda^{2v} t^{2v}}{(2v)!} \mathbf{F}[k(x); t] - e^{-\lambda |t|} \mathbf{F}]k(x); t].$$

In our case

$$(-1)^v k^{(2v)}(x) = \frac{1}{2\pi} \int_{-\infty}^{\infty} e^{-ixt} t^{2v} \mathbf{F}[k(x); t] dt.$$

Thus

$$b(y, \lambda) = 2 \sum_{v=0}^{\infty} \frac{(-1)^v \lambda^{2v}}{(2v)!} k^{(2v)}(x) - \int_{-\infty}^{\infty} \frac{1}{\pi\lambda} \frac{1}{1+(x-y)^2/\lambda^2} k(y) dy,$$

i.e. the test function $b(y, \lambda)$ is obtained by means of a series of the even derivatives of the original superposition $k(x)$ and a convolution transformation of $k(x)$. The numerical computation of the value of this series is, however, an *incorrect* problem not to be handled at present even by the regularization method of A. N. Tihonov (cf. Ch. V, § 1, Section 4). (As to the convolution transformation it represents a correct problem.) Thus, this result has little practical value.

10. The special case of the $\mathscr{A}$-method — the simple diminishing of the formant — treated above is *not* restricted, in fact, to superpositions of *density functions*. It can also be carried out on a superposition of unimodal *(non-density)* functions

$$k^*(x) = \sum_{k=1}^{N} p_k f^*(x - \alpha_k, \beta_k) \qquad (\Lambda_1 < \beta_1 \leq \beta_2 \leq \ldots \leq \beta_N < \Lambda_2)$$

where $\int\limits_{-\infty}^{\infty} f^*(x, \beta_k)\, dx = \infty$ but $\mathbf{F}[f^*(x, \beta_k); t]$ exists and belongs to the type $\Phi_1(t)\, \Phi(t)^{\beta_k}$, the conditions and the technique being the same as above. E.g. let

$$f^*(x, \beta_k) = \sqrt{\frac{\beta_k + \sqrt{\beta_k^2 + x^2}}{2\pi(\beta_k^2 + x^2)}} \qquad (0 < \beta_1 < \beta_2 < ... < \beta_N < \Lambda_2)$$

(Medgyessy 1954c).

This is *not* a density function; however,

$$\mathbf{F}[f^*(x, \beta_k); t] = \frac{e^{-\beta_k |t|}}{\sqrt{|t|}}$$

and this belongs to the type we have mentioned. Since β_k regulates the "narrowness" of the graph of the "component" $f^*(x, \beta_k)$, we have a full analogy with the "regular" superpositions and can apply the same "decomposition" technique as described for them. We do not continue these investigations here because it is improbable that such superpositions will be met in experimental science.

11. **The decomposition of a superposition of normal density functions** is one of the most frequent decomposition problems. Nowadays it is a classical one. In the present form it was first dealt with about 50 years ago by Nikhilranjan Sen (1922) and — already in an exact framework — by Doetsch (1928); later it was investigated by Doetsch (1936) and Medgyessy (1953). G. Doetsch called the procedure of decomposition "*Gauss-Analysis*" (in German) (Doetsch 1936 p. 312). A great number of investigations have been devoted, also since then, to this problem (see e.g. Tricomi 1938; Medgyessy 1954b, 1955c; Berencz 1955a, 1955b; Medgyessy 1956, 1957a, 1961a pp. 80—101, 1966a; Varga 1966a; Bhattacharya 1967; Medgyessy, Varga 1968; Tricomi 1968; Varga 1968; Gregor 1969) from both a theoretical and a numerical viewpoint.

The test function as the solution of a Fredholm integral equation of the Ist kind, appeared first in Doetsch (1928), (1936).

12. The integral equation $k(x) = \int\limits_{-\infty}^{\infty} \dfrac{e^{-\frac{(x-y)^2}{4\lambda}}}{\sqrt{4\pi\lambda}}\, b(y, \lambda)\, dy$ arises, independently from its decomposition background, in a great number of contexts. Even its *exact* solution has a vast literature. Its *numerical* solution, often appearing under problems called "**Entschmierung**" (i.e. the reconstruction of the "true" shape of the intensity distribution curve of a spectral line from the broadened "measured" ("verschmiert") one) in the experimental literature — is, as it is commonly known, delicate because it represents an *incorrect* problem. Many works have been devoted to the exact and numerical solution: Eddington (1913); Dyson (1926); Doetsch (1928); Schulz (1934); Doetsch (1936); Trumpler (1951);

Trumpler, Weaver (1953) pp. 101—108; Pollard (1953); Medgyessy (1953), (1954b); Bracewell (1955a); Medgyessy (1954c), (1955b), (1957), (1961a) pp. 80—93; Kurth (1965); — or, in case of a *non-Gaussian* kernel function: van Cittert (1931); Burger, van Cittert (1932), (1933); van de Hulst (1941); Kremer (1941); Righini (1941); van de Hulst (1946a), (1946b); Stokes (1948); Keating, Warren (1952); Elste (1953); Trumpler, Weaver (1953) Chapter 1.5; Bracewell (1955b); Unsöld (1955) pp. 252—265; Flynn, Seymour (1960); Medgyessy (1961a) pp. 172—188; Larson, Kenneth (1967); Armstrong (1967). However, the early methods could not be successful as they ignored the fact of *incorrectness*. At present (3) can be treated by the newer methods — e.g. by the *regularization method* — surveyed in Chapter V.

13. It is surprising that although the *"Entschmierung"* methods that make a curve *"narrower"* can also be applied without any alteration to decompose a superposition of normal density function — *this particular property has not been recognized by the scientists considering (since 1922) decomposition problems, although the "Entschmierung" also appeared earlier* (cf. Eddington 1913).

14. The form $b(y, \lambda) = \sum_{v=0}^{\infty} \frac{(-1)^v}{v!} k^{(2v)}(y)$ of the test function used in decomposing superpositions of normal density functions already appeared earlier (Eddington 1913; Pollard 1953; Medgyessy 1953; Bracewell 1955a). Its approximation $\hat{b}(y, \lambda)$ was also obtained earlier by substituting numerical approximations for the derivatives of a section of the series of $b(y, \lambda)$; errors were not investigated (Medgyessy 1953, 1954c, 1961a pp. 99—101). Formally the result is the same as shown in Ch. V, § 3 — as is to be expected.

15. It should be remarked, too, that $\hat{b}(y, \lambda)$ in Ch. III, § 1, Subsection 1.1.1.1 applied to a single Gaussian function with given β_1 gives no "Dirac delta function" even at $\lambda = \beta_1$. Numerical experimentation shows that $\hat{b}(y, \beta_1)$ is a function for which $\hat{b}(0, \beta_1) - \hat{k}(0) \geqq 0$ and increases with m; further $\mathscr{G}\hat{b}(y, \lambda)$ has a chief peak at $x=0$ and several smaller secondary peaks at points x where $\hat{k}(x) \approx 0$. The whole picture is similar to the graph of a narrower and higher Gaussian function graph. This follows from the numerical procedure; it does not, however, destroy the effectiveness of the present method. The estimation of the value of β_1 from the value of λ will merely be more difficult because the image of that function approximation does not become unsurveyable at $\lambda = \beta_1$ — as it occurs in the principles of application of the $\mathscr{A}$-method.

16. In Ch. II, § 8, Section 3, $f(x, c) = \dfrac{e^{-\frac{(x-\alpha_k)^2}{4c}}}{\sqrt{4\pi c}}$ as a stable density function

satisfied a partial differential equation. Hence it follows that $b(y, \lambda)$ in (2) of Ch. III, § 1, Subsection 1.1.1.1, namely

$$b(y, \lambda) = \sum_{k=1}^{N} p_k \frac{e^{-\frac{(y-\alpha_k)^2}{4(\beta_k-\lambda)}}}{\sqrt{4\pi(\beta_k - \lambda)}},$$

satisfies the partial differential equation

$$\frac{\partial b}{\partial \lambda} + \frac{\partial^2 b}{\partial y^2} = 0 \qquad (0 \leq \lambda < \beta_1, \ -\infty < y < \infty)$$

where $b(x, 0) = \sum_{k=1}^{N} p_k \dfrac{e^{-\frac{(x-\alpha_k)^2}{4\beta_k}}}{\sqrt{4\pi\beta_k}} = k(x)$ (cf. (1) in Ch. III, § 1, Subsection 1.1.1.1).

Besides, Ch. II, § 8, Section 4 showed that $f(x, c-\lambda)$ could be developed into the function series

$$f(y, c-\lambda) = \sum_{n=0}^{\infty} \frac{(-\lambda)^n}{n!} f^{(2n)}(y).$$

Consequently

$$b(y, \lambda) = \sum_{n=0}^{\infty} \frac{(-\lambda)^n}{n!} k^{(2n)}(y),$$

in full agreement with (4) in the same subsection (Medgyessy 1954b, 1955b, 1961a pp. 93—101). This way of obtaining the expansion of $b(y, \lambda)$ is, however, rather special; therefore we have worked with the result of Ch. V, § 3.

17. The statement that the kth component $f(y-\alpha_k, \beta_k-\lambda)$ of $b(y, \lambda)$ and, consequently, $b(y, \lambda)$ itself, satisfies the partial differential equation $\dfrac{\partial f}{\partial \lambda} + \dfrac{\partial^2 f}{\partial y^2} = 0$ above is obvious if we realize that this component,

$$f(y-\alpha_k, \beta_k-\lambda) = \frac{e^{-\frac{(x-\alpha_k)^2}{4(\beta_k-\lambda)}}}{\sqrt{4\pi(\beta_k-\lambda)}}$$

is the unique solution of *Kolmogorov's first and second equations* in the theory of diffusion processes in a special case (here the other equation is $\dfrac{\partial f}{\partial \beta_k} - \dfrac{\partial^2 f}{\partial \alpha_k^2} = 0$, the equation of heat conduction) (Gnedenko 1954 pp. 291—292).

18. According to Ch. V, § 3 the sum (5) for $\hat{b}(y, \lambda)$ in Ch. III, § 1, Subsection 1.1.1.1 would give the exact value of $b(y, \lambda)$, except for measurement errors, if $k(x)$ were substituted by an approximating polynomial, as the approximation is derived in accordance with the general principles of

9*

numerical analysis. Varga (1966a) modified the method, by claiming that if

$$k(x) = \frac{e^{-\frac{x^2}{4\beta}}}{\sqrt{4\pi\beta}} \text{ then } \sum_{j=-m}^{m} a_j k(x+jh) \text{ should give the best least squares approxi-}$$

mation to $\dfrac{e^{-\frac{x^2}{4\sigma\beta}}}{\sqrt{4\pi\sigma\beta}}$ where $\sigma < 1$ and β are given. From this condition the coeffi-

cients a_j can be calculated. Thus the linear operation represented by $\sum\limits_{j=-m}^{m} a_j k(x+jh)$
makes the graph of a Gaussian function of parameter β narrower; it introduces,
however, *not* $\beta - \lambda$ but $\sigma\beta$ as a new parameter. The application of this operation
to a superposition of normal density functions, with certain β-value will, evidently,
make that component the narrowest, to the parameter β_1 of which β is the nearest;
the other components will be influenced by another character. Carrying out this
procedure on the "measured" superposition data $\hat{k}(x)$ with monotonely increasing
β values, one can expect, even taking into account the approximations, in the
case of a small σ that if β is near to β_1, then the graph of the component
containing β_1 will emerge strongly from the whole picture and the other ones
will also become narrower, i.e. the basic idea of the $\mathscr{A}$-method can be applied.
During the procedure, the lack of the value of β_1 appears in the same manner
as in the simple diminishing of the formant. The cited article gives a table of
the a_j values for several parameter values, and a methodological example as well.

The detailed exact investigation of the effectiveness of the method is an unsolved **problem**.

Example. In Fig. 1 the solid line is the graph of the function

$$k(x) = e^{-\frac{x^2}{2}} + e^{-\frac{(x-2.5)^2}{4}} + e^{-\frac{(x-5)^2}{2}} + 0.5\,e^{-\frac{(x-7.25)^2}{2}} + 0.5\,e^{-\frac{(x-9.25)^2}{2}}$$

(see Varga 1966 a). It shows *three* peaks only. The dashed line is the graph of an

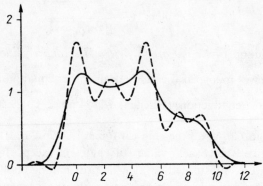

Supplements and problems to Ch. III, § 1. Figure 1

approximation to the corresponding test function $b(y, \lambda)$ with an appropriate λ-value, using the approach described in Varga (1966a) above. It reveals *all the five* components; the positions of the peaks give the exact maximum places fairly well, too.

19. The numerical solution of the integral equation (3) in Ch. III, § 1, Subsection 1.1.1.1 yielding $b(y, \lambda)$ by means of $k(x)$ may also be tried by means of the regularization method of A. N. Tihonov considered in Ch. V, § 1, Section 4. Using its simplest form, the approximation to $b(y, x)$ is yielded by that function $z(y)$ which, with its values z_j taken at $y_j = a + jh - \dfrac{h}{2} \left(j = 0, 1, ..., N; \ h = \dfrac{b-a}{N} \right)$, where (a, b) is a conveniently large interval and N is sufficiently large, minimizes the functional (written as a function of z_j)

$$\hat{M}_1^{\alpha}[z, u] = \sum_{i=1}^{M} \left(\sum_{j=1}^{N} K_{i-j} z_j h - k_i \right)^2 h_1 + \alpha \sum_{j=1}^{N} \left\{ \frac{K_0^{(j)}}{h} (z_{j+1} - z_j)^2 + K_1^{(j)} z_j^2 h \right\}.$$

Here K_{i-j} is defined by the quadrature formula, written for an appropriate function $f(y)$ as $\displaystyle\int_{-\infty}^{\infty} \frac{e^{-\frac{(x-y)^2}{4\lambda}}}{\sqrt{4\pi\lambda}} z(y)\, dy = \sum_{j=1}^{N} K_{i-j} z_j h + O(h^\gamma) \ (\gamma > 0), \ k_i = k(x_i)$,

$x_i = c + ih_1 - \dfrac{h_1}{2} \left(i = 0, 1, ..., M; \ h_1 = \dfrac{d-c}{M} \right)$, where (c, d) is a conveniently large interval and M is sufficiently large, $K_\nu^{(j)} > 0$ $(\nu = 0, 1)$ are appropriately chosen positive constants (e.g. $K_\nu^{(j)} = 1$ is suitable), and finally, α is a parameter varied in order to get the best approximation of $b(y, \lambda)$. Naturally, the result will depend on the choice of a, b, c, d, M, N, too. However, we do not touch upon this here.

20. The method described for the computation of the test function approximation $\hat{b}(y, \lambda)$ by means of Fourier expansion and Fourier synthesis also gives, evidently, a general method for the numerical solution of convolution Fredholm integral equations of the Ist type (Schulz 1934; Stokes 1948; Medgyessy 1961a pp. 81—84, 172—188). Instruments used in Fourier synthesis have a vast literature (e.g. Serebrennikov 1948). Also Medgyessy (1957b) constructed an apparatus for this purpose.

21. **Historical remarks.** In the first form of the decomposition of normal density functions, a section of the intensity distribution in a spectrum was given with the supposed form $k(x) = \displaystyle\sum_{k=1}^{N} p_k \frac{e^{-\frac{(x-\alpha_k)^2}{4\beta}}}{\sqrt{4\pi\beta}}$ at a temperature proportional to β and with N given. The distribution was wanted at a *lower* temperature in which the

peaks of the lines were *narrower*, and the places of the spectrum lines became more observable. This was the main objective of Sen (1922) and the starting point of Doetsch (1928). The solution was based on the *formal* statement that $k(x)$, being a function of x and β, described the temperature distribution of an infinite rod at time β and satisfied the partial differential equation of heat con-

duction $\dfrac{\partial k}{\partial \beta} = \dfrac{\partial^2 k}{\partial x^2}$ $(\beta > 0)$ (cf. Ch. II, § 8, Section 3). $b(y, \lambda) = \displaystyle\sum_{k=1}^{N} p_k \dfrac{e^{-\dfrac{(x-\alpha_k)^2}{4(\beta-\lambda)}}}{\sqrt{4\pi(\beta-\lambda)}}$

was *formally* identical with the temperature distribution in an infinite rod at time $\beta - \lambda$, *prior* to time β provided the temperature distribution at time β was $k(x)$.

To put it differently, the backward temperature distribution had to be calculated from the subsequent one. The A. N. Tihonov integral, yielding the *forward* temperature distribution in the present example, was not applicable to calculating the backward temperature distribution to solve the converse problem; nevertheless it is equivalent to an expansion containing the derivatives of $k(x)$ — at least, formally. Sen (1922) *extended* this expansion to *former* time points — without looking into the validity of this — and obtained an expansion $b(y, \lambda) = \displaystyle\sum_{n=0}^{\infty} \dfrac{(-\lambda)^n}{n!} k^{(2n)}(x)$ formally identical with the one occurring above. In practice, Sen (1922) approximated the spectrum intensity curve with an interpolatory polynomial based on the "measured" values of $k(x)$, and applied the expansion to this.

In Fig. 2 taken from Sen (1922) the solid line shows the initial distribution graph; the other ones show $b(y, \lambda)$ at different increasing λ-values (i.e. at former time points) (I, II). The abscissae of the peaks (i.e. spectrum lines) are marked.

Originally Sen (1922) wanted only to make the graphs of the intensity distribution of the two components in the H_α spectrum line narrower, and *did not* realize that the method was independent of the number of components, though he carried out, *formally*, a decomposition.

It was essentially with this background that Doetsch (1928) treated the decomposition problem, and rendered the procedure of Sen (1922) exact, with the analogy of the heat conduction equation. On the one hand, he proved that $b(y, \lambda)$ satisfied the corresponding

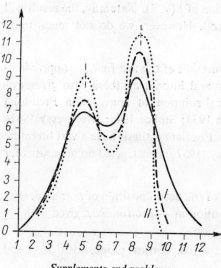

Supplements and problems
to Ch. III, § 1. Figure 2

integral equation with the aid of an earlier technique of handling the
heat conduction problem (Doetsch 1926); *formally,* the results of the latter
work can be connected with decomposition problems. On the other hand he
gave the solution of that particular integral equation in an explicit form by
means of Fourier integrals, approximating this with the aid of Fourier synthesis
in which the coefficients depended on the Fourier coefficients, calculated in a
finite interval of $k(x)$ — and on λ. G. Doetsch was the *first* to realize that the
method is applicable *to separate* components whose graphs remained hidden
in the intensity distribution curve; i.e. the basic idea of the 𝒜-method was his.

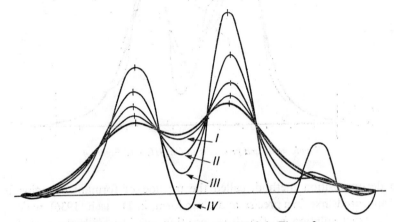

Supplements and problems to Ch. III, § 1. Figure 3

In Fig. 3 taken from Doetsch (1928) the solid line shows the intensity distribution
curve of the H_α line, considered as the graph of a superposition of normal density
functions. This curve was handled by Sen (1922). The thin lines I, II, III and
IV show the graphs of the corresponding test functions for increasing λ-values,
constructed by the mentioned procedure, with 25 Fourier coefficients, calculated
instrumentally by means of a Mader–Ott analyser. The separation of the components is easily visible. Line IV indicates that the λ-value applied here was
too great, i.e. the λ used in obtaining III was already near to β_1. *This was the first
(numerical) decomposition performed with an exact method.*

Figure 4, taken also from Doetsch (1928), shows the decomposition of the superposition of normal density functions whose graph (solid line) is the intensity
distribution curve of the H_α line, determined by a more exact measurement method.
It had long been suspected that there was a *third* line between those two revealed
by the two high peaks. To investigate this, Doetsch (1928) let the superposition
be decomposed by the mentioned method, with increasing λ-values. The thin lines
I, II, III show the corresponding results. Line III reveals well the presence of a

third (rather hidden) component curve, i.e. the presence of a third spectroscopic line; thus the investigation described *succeeded.*

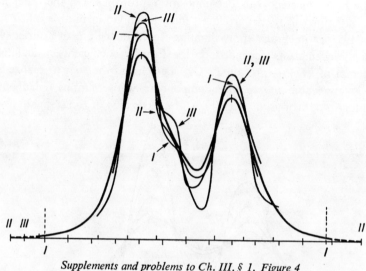

Supplements and problems to Ch. III, § 1. Figure 4

The decomposition method, called the method of formant changing in this work, appeared first, in a precise and clear form, in Doetsch (1936), which paper gave the exact solution of the decomposition of $k(x)$ with the test function $b(y, \lambda)$, by means of the solution of the corresponding integral equation, in case of *different* β-values. The test function was provided in the form of repeated Fourier transforms, as (2) in Ch. III, § 1, Subsection 1.1.1; its approximation was also obtained here by the Fourier expansion of a section of the "measured" starting superposition $\hat{k}(x)$, and subsequent Fourier synthesis, as described already.

Medgyessy (1954c), (1955b) was the first to investigate the *errors* in the case of this numerical determination of the test function approximation.

22. The idea described above, that in case of *identical* β_k-values the test function can be obtained as a *backward* solution of a certain equation of heat conduction, can be exploited immediately in order to obtain, at least formally, an approximation of the test function, as H. Freudenthal remarked in 1953. Namely let us consider the partial differential equation $\dfrac{\partial k^*}{\partial t} = \dfrac{\partial^2 k^*}{\partial x^2}$ satisfied by the superposition $k^*(x, t) = \sum\limits_{k=1}^{N} p_k \dfrac{e^{-\frac{(x-\alpha_k)^2}{4t}}}{\sqrt{4\pi t}}$ and transform it into a finite difference form used in numerical analysis, with mesh distances Δx, Δt. Then, starting with

the "initial" values $k^*(x_i, t)$ we can construct $k^*(x_i, t-\varDelta t)$, $(x_i, t-2\varDelta t)$ etc., giving the test function approximations.

The same result could have been obtained by starting from the equation $\frac{\partial b}{\partial \lambda} = -\frac{\partial^2 b}{\partial y^2}$ which holds for a test function $b(y, \lambda) = k^*(y, t-\lambda)$ $(0<\lambda<t)$. Moreover, the latter equation holds also for a test function $b(y, \lambda)$ with *non-equal* β_k-values; thus its numerical treatment yields a method of decomposition of a general superposition of normal density functions (Medgyessy 1953, 1955c, 1971b, 1972b).

However, it has not been decided whether the problem of solving the above-mentioned backward equation is *incorrect* or not. Thus the practical application of the above ideas is an unsolved **problem**. This question is not investigated here; reference is made solely to works concerning incorrect problems in the field of partial differential equations (Lattès, Lions 1967; Čudov 1967; Arcangeli 1968).

Example. In Fig. 5 the solid line is the graph, plotted with certain errors, of the superposition

$$k(x) = \frac{e^{-\frac{(x-1)^2}{2}}}{\sqrt{2\pi}} + \frac{e^{-\frac{(x+1)^2}{2}}}{\sqrt{2\pi}}$$

already investigated in Example 1 in Ch. III, § 1, Subsection 1.1.1.1. We have applied the present method to decompose this superposition with the simplest (central) finite difference formulae for the partial derivatives in the equation $\frac{\partial b}{\partial \lambda} = -\frac{\partial^2 b}{\partial y^2}$ which holds for the test function $b(y, \lambda)$ belonging to $k(x)$. The

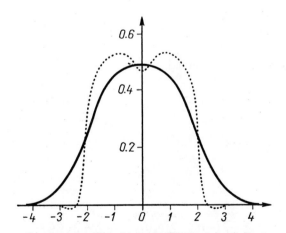

Supplements and problems to Ch. III, § 1. **Figure 5**

dotted line shows the (rather modest) result at certain mesh distances and λ-values. The two components have become separated.

23. Theoretically in the case of a superposition $k(x)$ of normal density functions the test function $b(y, \lambda)$ can also be obtained by utilizing the Čebyšev–Hermite expansion of it as well as of the starting superposition $k(x)$ as follows (Medgyessy 1953, 1954c, 1961a pp. 87—93).

The H. Cramér condition of the expansion of a function $f(x)$ $(-\infty < x < \infty)$ in such a series (Cramér 1925, 1946 pp. 226—227) ($f(x)$ is of bounded variation in any finite interval, $\int\limits_{-\infty}^{\infty} |f(x)| e^{\frac{x^2}{4}}\, dx < \infty$) is valid for both $k(x), b(y, \lambda)$ if $0 < \beta_k < 1$ $(k = 1, \ldots, N)$, which can always be attained by a scale changing; thus

$$k(x) = \sum_{n=0}^{\infty} \frac{K_n}{n!}\, \varphi^{(n)}(x),$$

$$b(y, \lambda) = \sum_{n=0}^{\infty} \frac{B_n}{n!}\, \varphi^{(n)}(x)$$

where ($He_n(x)$ is the nth Hermite polynomial generated with $e^{-\frac{x^2}{2}}$)

$$K_n = (-1)^n \int\limits_{-\infty}^{\infty} He_n(x)\, k(x)\, dx, \quad B_n = (-1)^n \int\limits_{-\infty}^{\infty} He_n(y)\, b(y, \lambda)\, dy,$$

$$\varphi(x) = \frac{e^{-\frac{x^2}{2}}}{\sqrt{2\pi}}, \qquad \varphi^{(n)}(x) = (-1)^n He_n(x)\, \varphi(x).$$

We have

$$\mathbf{F}[k(x); t] = e^{-\lambda t^2} \mathbf{F}[b(y, \lambda); t].$$

Since

$$\int\limits_{-\infty}^{\infty} \varphi^{(n)}(x)\, e^{itx}\, dx = e^{-\frac{t^2}{2}} (-it)^n,$$

and the term-by-term integration is justified, we have

$$e^{-\frac{t^2}{2}} \sum_{n=0}^{\infty} \frac{K_n}{n!} (-it)^n = e^{-\lambda t^2} e^{-\frac{t^2}{2}} \sum_{n=0}^{\infty} \frac{B_n}{n!} (-it)^n.$$

Since $\dfrac{1}{e^{-\lambda t^2}}$ is an integral function having an expansion, say, of form

$$\frac{1}{e^{-\lambda t^2}} = \sum_{n=0}^{\infty} b_n (it)^n,$$

we have

$$\sum_{n=0}^{\infty} \frac{K_n}{n!} (-it)^n \sum_{n=0}^{\infty} b_n (it)^n = \sum_{n=0}^{\infty} \frac{B_n}{n!} (-it)^n,$$

whence

$$B_n = n! \sum_{r=0}^{n} (-1)^r b_r \frac{K_{n-r}}{(n-r)!}$$

or, inserting the values of b_n $\left(b_{2r} = \frac{\lambda^r}{r!}, \ b_{2r+1} = 0, \ r = 0, 1, \ldots\right)$

$$B_n = n! \sum_{r=0}^{[n/2]} \frac{\lambda^r K_{n-2r}}{r!(n-2r)!},$$

and

$$b(y, \lambda) = \sum_{n=0}^{\infty} \left(\sum_{r=0}^{[n/2]} \frac{\lambda^r K_{n-2r}}{r!(n-2r)!} \right) \varphi^{(n)}(x).$$

As B_n increases if λ does so, the expansion converges slower if λ increases.

The procedure can obviously be generalized to other superpositions $k(x)$ and test functions $b(y, \lambda)$ related by $\mathbf{F}[k(x); t] = \Phi(t)\mathbf{F}[b(y, \lambda); t]$ that is, to the solution of a Fredholm integral equation of the Ist kind of convolution type where $\Phi(t)$ is an appropriate characteristic function for which $\dfrac{1}{\Phi(t)}$ is integral (Medgyessy 1961a pp. 181—188; in a more general form: Runge 1914).

In practice this method is complicated; moreover it is an unsolved **problem** as to whether the determination of $b(y, \lambda)$, i.e. the solution of the corresponding equation etc. by this method, is a *correct* problem. First in some appropriate finite interval $(0, l)$, the coefficients K_n should be computed approximately, and then, having computed B_n, a partial sum of the series of $b(y, \lambda)$ should be constructed. All these steps would bring in errors that could be estimated only partly (Medgyessy 1961a pp. 87—93). Thus we do not continue these somewhat out-of-date investigations further.

24. The condition in Ch. III, § 1, Section 2, i.e. that the *quotient of the two characteristic functions* belonging to infinitely divisible density functions is again a characteristic function of an infinitely divisible density function, can be stated, equivalently, in form of a condition imposed on the "spectral function" in the Lévy–Hinčin canonical representation of the corresponding characteristic functions (cf. Ch. 1, § 8, Section 1).

25. *Clearly the second particular case of the $\mathscr{A}$-method* can be applied in such cases when only *the characteristic function* of the superposition to be decomposed is known. Such cases occur in physics, meteorology, etc. (Emslie, King 1953).

26. In the case of the *decomposition of a superposition of* ch *density functions* we have for the test function $b(y, \lambda)$ defined by the integral equation

$$k(x) = \int_{-\infty}^{\infty} \frac{1}{2\lambda \operatorname{ch} [\pi(x-y)/2\lambda]} \, b(y, \lambda) \, dy \qquad (0 < \lambda < \beta_1)$$

also

$$b(y, \lambda) = \sum_{v=0}^{\infty} \frac{\lambda^v}{(2v)!} \, k^{(2v)}(x),$$

by Theorem 1 in Ch. V, § 3. In practice, however, this is of little value because the numerical differentiation, the only possible tool when working with this expansion, is an *incorrect* problem (cf. Ch. V, § 1); only the case of the first (or second) derivative has been treated e.g. by means of A. N. Tihonov's regularization method up to the present. Thus this question is not dealt with further. The use of the above expansion from the viewpoint of numerical analysis is, furthermore, an uninvestigated **problem**.

27. A primitive version of *the method described in* Ch. III, § 1, Subsection 2.2 can already be found in Medgyessy (1961a p. 181). In the test function $b(y, \lambda) =$
$\mathbf{F}^{-1} \left[\dfrac{1}{\varphi(t, \lambda)} \mathbf{F}[k(x); t]; y \right]$ of the method described in Ch. III, § 1, Subsection 2.1
only the curve of the function $k(x) + \xi(x)$ is at our disposal instead of the curve of $k(x)$ because the superposition is not exactly of the supposed form; in addition, in the case of a "measured" function an error, "noise", can also be superposed on it; this is collected in $\xi(x)$. In the Fourier transform — the "spectrum" — of $k(x) + \xi(x)$, the spectrum of $k(x)$ vanishes outside a certain interval related to the spectrum of $\xi(x)$, while that of $\xi(x)$ oscillates in a rather broad band. The spectrum of $\xi(x)$ interferes strongly in the further calculations and destroys the result. Now let us consider the test function $b_1(y, \lambda) = \mathbf{F}^{-1} \left[\dfrac{\alpha(t, \varepsilon)}{\varphi(t, \lambda)} \mathbf{F}[b(x); t]; y \right]$ occurring
in the method in Ch. III, § 1, Subsection 2.2. Here the multiplication by $\alpha(t, \varepsilon)$ diminishes the "superfluous" spectrum section that we have referred to while it does not affect much the main section due to $k(x)$ if ε is chosen appropriately. In the end, the application of $\alpha(t, \varepsilon)$ results in a "noise-filtering" effect; however, it influences little the exact final result. Thus our method for giving an appropriate test function exactly also filters out the "noise". This "noise filtering" character also remains in its numerical applications where it is even more essential. In Medgyessy (1967a) the basic ideas of a detailed investigation of the "noise filtering" can also be found, together with a concretization of the nature of $\xi(x)$; varying ε can result in an optimal situation, too. Thus the multiplication by $\alpha(t, \varepsilon)$ represents a certain "filter" in the sense used in electronics, in which respect it is *related* to the method described in Ch. V, § 2; however, the object of

the procedure described in Ch. V, § 2 is the "noise filtering" exclusively, while the aim of the present one is to construct an appropriate new test function — and the "noise filtering" appears as a useful companion phenomenon. The present method ensures the necessary *unimodality* of the test function components, which objective does not appear in Ch. V, § 2, where it can be shown that it cannot be fulfilled at all.

28. In case of the method described in Ch. III, § 1, Subsection 2.2 the question whether one can find, if $\varphi(t, \beta_k) = e^{-\beta_k t^2}$, an appropriate $\alpha(t, \varepsilon)$ at all, further, that if one does find such a one, what is its analytical form — presents an unsolved **problem**.

29. In Example 1 in Ch. III, § 1, Subsection 2.2 ch $\lambda t \cdot e^{-\varepsilon t^2}$ can be considered as the "characteristic function" of a **density function** which can take also **negative values**, its integral over $(-\infty, \infty)$ being equal to one. Let us define the variance of this "density function" as accepted in probability theory; then we shall get a **negative "variance"**. The test function $b_1(y, \lambda)$ occurring in Ch. III, § 1, Subsection 2.2 can be regarded as the convolution of $k(x)$ and the present "density function" and this convolution *diminishes* the true variances — we may say, monotone formants, — i.e. it *makes* the curve of $k(x)$ *narrower* (Medgyessy 1971*b*, 1972*b*). We have introduced this "density function" and negative "variance" respectively as a technical tool; the further exploitation of this idea in decomposition problems is still an unsolved **problem**.

Analogous — and more elaborated — considerations concerning discrete distributions will be presented in Ch. IV, § 1, Subsection 1.1.1 and Ch. IV, § 1, Section 2.

30. In connection with the *decomposition of a superposition of lognormal density functions* we mention that the distribution of personal incomes in a given population can be described by such a superposition (Fréchet 1939, 1945; Aitchison, Brown 1957). The same decomposition occurs also in geology or biology (see e.g. Harding (1949); for the graphical handling see Ch. III, § 3, (A), Section 5).

31. We realize that the transformation that carries the superposition of *Beta density functions* into a second one which can then be decomposed (cf. Ch. III, § 1, Section 3) represents, formally, a *mixing*. The mixing appears as a convenient linear transformation in other cases also. The kernel in the transformation need *not* be, however, a *density function* in general.

Other relevant examples, although of totally theoretical character, can be found in the literature (Medgyessy 1961*a* pp. 146—148).

Determining the appropriate linear transformation is equivalent, in these cases, to the **determination of the unknown kernel of an integral transform**, where the kernel must not contain unknown parameters (Medgyessy 1961*a* p. 156). Except for trial and error procedures, the general treatment of this interesting unsolved **problem** is not known at present.

32. The $\mathscr{A}$-*method* seems to be appropriate for a **generalization** of the following kind. Let us consider again a superposition

$$k(x) = \sum_{k=1}^{N} p_k f(x - \alpha_k, \beta_k) \qquad (\varLambda_1 < \beta_1 \leq \beta_2 \leq ... \leq \beta_N < \varLambda_2)$$

of unimodal density functions $f(x - \alpha_k, \beta_k)$ and suppose that

$$f(x, \beta_k) = \int_{-\infty}^{\infty} K_\lambda(x - y) \, g(y, \beta_k) \, dy$$

where $K_\lambda(x)$, being a density function, depends on the parameter λ $(\varLambda_1 \leq \lambda \leq \varLambda_2)$ but does not depend on β_k, that $g(y, \beta_k)$ is unimodal and that the variances of $f(x, \beta_k)$, $K_\lambda(x)$, $g(y, \beta_k)$ denoted by $\mathbf{D}^2 f(x, \beta_k)$, $\mathbf{D}^2 K_\lambda(x)$, $\mathbf{D}^2 g(y, \beta_k)$ all exist and characterize, in some sense to be defined in future investigations, the "narrowness" of the relevant density function graphs. In other words we suppose that $f(x, \beta_k)$ is *factorizable* into two factors, one of them being *unimodal* (as to the concepts see Linnik 1960; Lukács 1964); the possibility of such a factorization should be investigated separately for it presents an unsolved **problem**. — If $\mathbf{D}^2 K_\lambda(x)$ increases as $\lambda \uparrow \varLambda_2$, then

$$b(y) = \sum_{k=1}^{N} p_k g(y - \alpha_k, \beta_k)$$

can be used as a test function because in consequence of $\mathbf{D}^2 f(x, \beta_k) = \mathbf{D}^2 K_\lambda(x) + \mathbf{D}^2 g(x, \beta_k)$ the graphs of their components become "narrower" when λ increases. Obviously

$$k(x) = \int_{-\infty}^{\infty} K_\lambda(x - y) b(y) \, dy,$$

from which $b(y)$ can be determined, also numerically, in form of an expression

$$b(y) = \sum_{j=-m}^{m} C_j^{(m)}(\lambda) \, k(y + jh)$$

sometimes *without the exact knowledge of the analytical form of $b(y)$*: the knowledge of the characteristic function should be sufficient.

Several types of superposition belong to this sphere; however, a detailed treatment of all this is, at present, an unsolved **problem**.

33. In connection with Point 7 above it can be noticed that sometimes also the *decomposition of certain superpositions of multimodal density functions* can

be done by means of the ideas described in Point 7 for decomposing super-positions of unimodal density functions. Let us consider for instance (Medgyessy 1961a p. 72) a superposition $k(x)$ whose components possess characteristic functions belonging to the type $\psi_k(t) = \varphi(t)^{\beta_k} e^{i\alpha_k t} \cos \gamma t$ (γ is given, and $\mathbf{F}^{-1}[\varphi(t)^{\beta_k}; x]$ is (0) unimodal). At a convenient choice of the parameters we reach that $\mathbf{F}^{-1}[\psi_k(t); x]$ is bimodal, with modes $x_1 = \alpha_k + \gamma$ and $x_2 = \alpha_k - \gamma$. If the generalization of the method of formant changing described in Point 7 above can be applied to $k(x)$ then, by virtue of this, $\mathbf{F}[k(x); t]$ is to be multiplied by $\dfrac{1}{\varphi(t)^{\lambda} \cos \gamma t}$ $(0 < \lambda < \beta_1)$. The characteristic functions of the components of the test function to be used will have the form $\dfrac{\varphi(t)^{\beta_k} \cos \gamma t}{\varphi(t)^{\lambda} \cos \gamma t} = \varphi(t)^{\beta_k - \lambda}$. The components of the test function will already be unimodal and the decomposition procedure can be continued in the same way as in case of any test function considered in the present work.

§ 2. SECOND TYPE OF SUPERPOSITION. THE $\mathscr{B}$-METHOD

Up to the present we have dealt with the decomposition of superpositions in which the abscissae of the peaks of the (unimodal) components are *different*. The $\mathscr{A}$-method itself, as a method of decomposition, has been built totally on this fact.

Now let us consider the **decomposition of superpositions belonging to the type**

$$k(x) = \sum_{k=1}^{N} p_k f(x, \beta_k) \qquad (\Lambda_1 < \beta_1 \leqq \beta_2 \leqq \ldots \leqq \beta_N < \Lambda_2)$$

where the components are members of a family of strictly unimodal (A$_1$) (or (A$_2$) or (A$_3$)) density functions depending only on a *single* parameter and where **the abscissae of the peaks of the graphs of** $f(x, \beta_k)$ **are the same for any** k (see Fig. 1).

A simple example for this is given by the superposition

$$f(x) = \sum_{k=1}^{N} p_k \frac{e^{-\frac{x^2}{4\beta_k}}}{\sqrt{4\pi\beta_k}} \qquad (0 < \beta_1 < \beta_k < \ldots < \beta_N < \Lambda_2)$$

of (0) unimodal normal density functions or by the *classical example* of a super-position of exponential density functions

$$f(x) = \begin{cases} \sum_{k=1}^{N} p_k \dfrac{e^{-\frac{x}{\beta_k}}}{\beta_k} & (x > 0) \\ \\ 0 & (x \leqq 0) \end{cases} \qquad (0 < \beta_1 < \beta_2 < \ldots < \beta_N < \Lambda_2)$$

whose decomposition has interested several authors—unfortunately, not with a satisfactory result hitherto.

In principle the present superposition $k(x)$ cannot be treated by the $\mathscr{A}$-method owing to the type of the superposition to be decomposed: there is no distance between the peaks of the components. *Formally*, the $\mathscr{A}$-method can, however, be performed often on such a superposition.

Now we give a new method of decomposition for such superpositions in the following: it is called $\mathscr{B}$-*method*. It consists of a transformation of the superposition, recalling the type introduced in Ch. III, § 1, Section 3.

Let the function

$$b(y) = \sum_{k=1}^{N} p_k \, g\big(y, \omega(\beta_k), \tau\big),$$

called again a **test function**, correspond to the superposition $k(x)$ represented by its graph or by some ordinate values of that. For this test function let the following conditions hold:

I. $\omega(x)$ is a given strictly monotone function, not containing β_k and $g\big(y, \omega(\beta_k), \tau\big)$ $(0 < \tau < T)$ is an $\big(\omega(\beta_k)\big)$ *unimodal density function* (or is proportional to such a one), in which the mode $\omega(\beta_k) \neq 0$, and may depend on τ, too, and whose graph becomes, in some sense, "narrower" and its peak height increases as $\tau \downarrow 0$ (or $\tau \uparrow T$); finally, which is in a well-defined analytical relation, independent of α_k, β_k, with $f(x, \beta_k)$ $(k = 1, \ldots, N)$.

II. For *any* τ the distance between the abscissae of the peaks of $g\big(y, \omega(\beta_k), \tau\big)$ and $g\big(y, \omega(\beta_l), \tau\big)$, i.e. $|\omega(\beta_k) - \omega(\beta_l)|$, is not less than some quantity $D(\beta_k, \beta_l)$ depending *only* on β_k and β_l (and being independent of τ) (see Fig. 1).

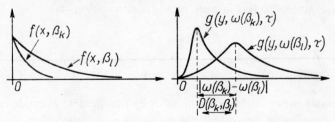

III. § 2. *Figure 1*

III. $\beta(y)$ can be constructed by the help of $k(x)$ only, without the knowledge of N, p_k, β_k and utilizing, eventually, the analytical relation between $g(y, \omega(\beta_k), \tau)$ and $f(x, \beta_k)$.

Evidently $b(y)$ is a superposition of density functions.

The peak height increases and, in addition, "narrowing" in I occurs if e.g. τ is a $(T, 0)$ (or $(0, T)$) monotone formant of $\mathscr{G}g(y, \omega(\beta_k), \tau)$. If τ is sufficiently near to 0 (or T), the kth component of this function will possess a graph with a sharp peak at the point $\omega(\beta_k)$. Further, there will be a certain minimum distance between the abscissae of the peaks of the components of $b(y)$ independent of τ (cf. Condition II). Thus the graph of $b(y)$ will, *probably*, show more and more *separated* peaks as $\tau \downarrow 0$ (or $\tau \uparrow T$). At a strong separation they present themselves almost one by one, the abscissae of the peaks being near to the $\omega(\beta_k)$; thus we can deduce the number N of the components — and, eventually, the value of $\omega(\beta_k)$ (i.e. β_k) — from them (see Fig. 2).

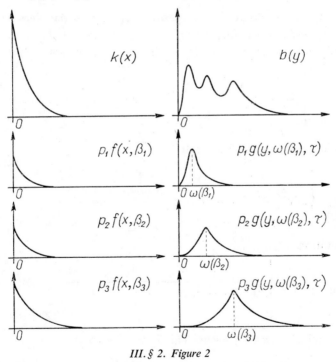

III. § 2. Figure 2

Let us consider the following steps:

(A) Determination of the described functions $g(y, \omega(\beta_k), \tau)$ with the help of $f(x, \beta_k)$.

(B) Determination, with the help of (A), of the test function $b(y)$ with the aid of $k(x)$, serving as the basis for the decomposition method.

(C) On the basis of (B) and of the *"measured"* values of $k(x)$, the elaboration of a numerical method for the construction of an *approximation* to the test function $b(y)$.

For the graph $\mathscr{G}_1$ of this approximation to $b(y)$, the statements concerning the evaluation of the graph of the "exact" test function remain approximately valid.

Definition 1. The construction of the graph $\mathscr{G}_1$ of the test function *approxima⁻ tion,* followed by the determination of the number and of the abscissae etc· of the peaks of $\mathscr{G}_1$ is called the $\mathscr{B}$-**method**, yielding the (numerical) decomposition of the superposition $k(x)$ (Medgyessy 1971b, 1972b).

In the following we shall find test functions $b(y)$ for certain types of superposition $k(x)$.

Example. **Decomposition of a superposition of exponential density functions.** An interesting type of the function $g(y, \omega(\beta_k), \tau)$ defined above was given by P. Medgyessy for the case of a superposition of exponential density functions

$$k(x) = \begin{cases} \displaystyle\sum_{k=1}^{N} p_k \, \frac{e^{-\frac{x}{\beta_k}}}{\beta_k} & (x > 0) \\ 0 & (x \le 0) \end{cases} \qquad (0 < \beta_1 < \beta_2 < \ldots < \beta_N < \Lambda_2).$$

Let us consider the following function:

$$b(y) = \begin{cases} \displaystyle\sum_{k=1}^{N} p_k \, \frac{\sigma-1}{B(1/\tau+1,\ 1/\tau(\sigma-1))} \, \frac{\left(e^{-\frac{y}{\beta_k}} - e^{-\frac{\sigma y}{\beta_k}}\right)^{\frac{1}{\tau}}}{\beta_k} & (y > 0) \\ 0 & (y \le 0) \end{cases}$$

with the notations

$$g(y, \omega(\beta_k), \tau) = \begin{cases} \displaystyle \frac{\sigma-1}{B(1/\tau+1,\ 1/\tau(\sigma-1))} \, \frac{\left(e^{-\frac{y}{\beta_k}} - e^{-\frac{\sigma y}{\beta_k}}\right)^{\frac{1}{\tau}}}{\beta_k} & (y > 0) \\ 0 & (y \le 0) \end{cases}$$

where $\dfrac{1}{\tau}$ is an integer, $\sigma > 1$ is given. (This is obtained by a certain transformation of a Beta density function.) It is shown easily by a more detailed discussion that this is a unimodal density function with mode $\omega(\beta_k) = \beta_k \dfrac{\log \sigma}{\sigma-1}$ whose graph becomes "narrower" and whose peak height increases if $\dfrac{1}{\tau}$ increases. *If one accepts this concept of "narrowness"* then Condition I in Ch. III, § 2 is fulfilled. Condition II in Ch. III, § 2 is fulfilled since the distance between the modes of the test function components having indices k and l is equal to $(\beta_l-\beta_k)\dfrac{\log\sigma}{\sigma-1}$.

Expanding $\dfrac{\left(e^{-\frac{y}{\beta_k}}-e^{-\frac{\sigma y}{\beta_k}}\right)^{\frac{1}{\tau}}}{\beta_k}$, we have

$$\frac{\left(e^{-\frac{y}{\beta_k}}-e^{-\frac{\sigma y}{\beta_k}}\right)^{\frac{1}{\tau}}}{\beta_k} = \sum_{v=0}^{1/\tau}\binom{1/\tau}{v}(-1)^v\frac{e^{-\frac{1}{\beta_k}\left[v(\sigma-1)+\frac{1}{\tau}\right]y}}{\beta_k}.$$

Consequently

$$b(y) = \begin{cases} \dfrac{\sigma-1}{B(1/\tau+1,\ 1/\tau(\sigma-1))}\sum\limits_{v=0}^{1/\tau}\binom{1/\tau}{v}(-1)^v\,k\left(\left[v(\sigma-1)+\frac{1}{\tau}\right]y\right) & (y>0) \\ 0 & (y\leqq 0) \end{cases}$$

i.e. its value at a fixed point y is constructed by the aid of its values at the points $\dfrac{1}{\tau}y$, $\left[(\sigma-1)+\dfrac{1}{\tau}\right]y$, $\left[2(\sigma-1)+\dfrac{1}{\tau}\right]y$, ..., $\dfrac{\sigma}{\tau}y$.

Thus $b(y)$ can be taken as the test function corresponding to the superposition $k(x)$.

In practice this method can be applied if $k(x)$ is known in a sufficiently long interval. However, if a section of the test function near to the origin is important and $k(x)$ is known for greater x-values only, our procedure may be of some use. The shape of the graph of the test function is regulated by both $\dfrac{1}{\tau}$ and σ; thus it should be constructed for several values of them, and these should be evaluated in their aggregate. At any rate if σ increases, the distance between the modes of the test function components diminishes; this should be counterbalanced by taking $\dfrac{1}{\tau}$ greater. As the latter requires $k(x)$-values from a longer interval, a σ-value near to 1 and a relatively small $\dfrac{1}{\tau}$ would suffice in many cases.

It should be indicated that $b(y)$ can be computed *exactly* (not only approximately) from the equidistant values of $k(x)$, which leads, in practice, to the **least inexact procedure**; one gets the approximate values of $b(y)$ by inserting the "measured" values of $k(x)$ into the formula for $b(y)$. This procedure is a *correct* problem. This is a very important feature of the present method that could not have been reached by any of the methods dealt with up to now in the present book; for, previously some complicated analytical expression was approximated by means of a linear expression of the measured data.

Thus the $\mathscr{B}$-method can, finally, be applied here.

A numerical error analysis is also very easy and thus has been omitted.

Let us now consider, as a *methodological example* the decomposition of the superposition of exponential density functions

$$k(x) = \begin{cases} \dfrac{1}{\sqrt{2\pi}}\,e^{-2x} + \dfrac{1}{\sqrt{2\pi}}\,e^{-8x} & (x > 0) \\ 0 & (x \leq 0). \end{cases}$$

Its graph is shown in Fig. 3a; it has been plotted by the help of a table, retaining 4 decimal figures. With these curve data we have constructed, in the way described above, approximations $b^*(y)$ to the test function $b(y)$ belonging to $k(x)$, at

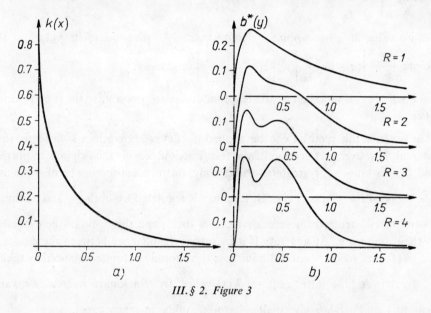

III.§ 2. Figure 3

$\sigma = 1.25$, $\dfrac{1}{\tau} = R = 1, 2, 3$ and 4. They are shown by the sub-figures in Fig. 3b.

Although the truncation errors influence the result, the curve of $b^*(y)$ presents, as R increases, more and more definitely two peaks pointing, by this, to the two — originally hidden — components of $k(x)$.

1. A particular case of the $\mathscr{B}$-method

In the superposition

$$k(x) = \sum_{k=1}^{N} p_k f(x, \beta_k) \qquad (\Lambda_1 < \beta_1 < \beta_2 < \ldots < \beta_N < \Lambda_2)$$

let $f(x, \beta_k)$ be a strictly (0) unimodal, and (A$_1$) (or (A$_2$)) density function.

With the help of the idea of the preceding example, let us construct the function $R_\sigma(y)$ ($\sigma > 1$) with a (0) unimodal *density function* $\omega(x)$ as follows:

$$R_\sigma(y) = \left(\frac{\sigma}{\sigma-1}\right)[\omega(y) - \omega(\sigma y)].$$

The right-hand side is a *density function*. In many cases it is strictly unimodal in $(0, \infty)$ with its mode at some $x_m \neq 0$. (This step was suggested by the form of simple curves of radioactive decay belonging to the type $f(x) = C(e^{-\lambda_1 x} - e^{-\lambda_2 x})$ and *non*-(0) unimodal.)

Let us suppose that *formally*, the steps of the $\mathscr{A}$-method can be applied to $k(x) = \sum_{k=1}^{N} p_k f(x, \beta_k)$ and the corresponding test function is

$$b_1(y) = \sum_{k=1}^{N} p_k h(y, \beta_k)$$

in which $h(y, \beta_k)$ is also (0) unimodal and its graph is, in some fixed sense "narrower" than that of $f(x, \beta_k)$. Let us consider the function

$$g^*(y, \omega(\beta_k), \tau) = \frac{\sigma}{\sigma-1}[h(y, \beta_k) - h(\sigma y, \beta_k)],$$

the parameter τ (essentially, a monotone formant) being involved in $h(y, \beta_k)$ in some way. Further *let us suppose* that the function $g^*(y, \omega(\beta_k), \tau)$ is unimodal with mode at $x = \omega(\beta_k)$ and that its curve becomes, in some sense, narrower as $\tau \downarrow 0$ (or $\tau \uparrow T$). Then it is seen that for the function $g^*(y, \omega(\beta_k), \tau)$ Condition I in Ch. III, § 2 is fulfilled partly because of the preceding and, partly, because analytical relations independent of β_k hold, on the one hand, between $g^*(y, \omega(\beta_k), \tau)$ and $h(y, \beta_k)$ and, on the other hand, between $h(y, \beta_k)$ and $f(x, \beta_k)$. *Let us suppose* that Condition II in Ch. III, § 2 is also fulfilled for $g^*(y, \omega(\beta_k), \tau)$. Condition III in Ch. III, § 2 is fulfilled for $b(y, \tau)$ because of the analytical relation holding between $g^*(y, \omega(\beta_k), \tau)$ and $f(x, \beta_k)$. Finally $b(y, \tau)$ can be taken as the test function corresponding to $k(x)$ and after the steps (A), (B) and (C) in Ch. III, § 2, the $\mathscr{B}$-method can be applied.

Then, essentially, since the function $R_\sigma(y)$ **is a non-(0) unimodal density function created from a (0) unimodal density function**, we have constructed, by calculating the test function

$$b(y) = \sum_{k=1}^{N} p_k \, g^*(y, \omega(\beta_k), \tau)$$

with the help of $k(x)$, a particular case of the $\mathscr{B}$-method and *reduced the problem to another case that has already been investigated.*

The concrete type of the $\mathscr{A}$-method figuring above is, in several cases, identical with some type occurring in our discussion of the $\mathscr{A}$-method.

The conditions of applicability of the $\mathscr{B}$-method are fulfilled e.g. in the case where 1. $f(x, \beta_k) = \dfrac{1}{\beta_k^{\Omega}} f^*\left(\dfrac{x}{\beta_k^{\Omega}}\right)$ ($\Omega > 0$, Ω is given), $f^*(x)$ is (0) unimodal, $\dfrac{\sigma}{\sigma - 1} [f^*(x) - f^*(\sigma x)]$ is (m_σ) unimodal in $(0, \infty)$ and 2. the simple diminishing of the formant (in Ch. III, § 1, Subsection 1.1.) is adopted as the $\mathscr{A}$-method (there λ stood for τ); it can be applied in the case of components $\dfrac{1}{\beta_k^{\Omega}} f^*\left(\dfrac{x}{\beta_k^{\Omega}}\right)$ ($k = 1, 2 \ldots N$) when the graph of the kth component has β_k as a $(\Lambda_2, 0)$ monotone formant. The property of being (0) unimodal holds e.g. if $f^*(x)$ is a (0) symmetrical stable density function (cf. Ch. III, § 1, Subsection 1.1.1 — Ch. III, § 1, Subsection 1.1.1.1).

If 1. holds then Condition II in Ch. III, § 2 is satisfied because

$$g^*(y, \omega(\beta_k), \tau) = \frac{\sigma}{\sigma - 1}\left[\frac{1}{(\beta_k - \tau)^{\Omega}} f^*\left(\frac{y}{(\beta_k - \tau)^{\Omega}}\right) - \frac{1}{(\beta_k - \tau)^{\Omega}} f^*\left(\frac{\sigma y}{(\beta_k - \tau)^{\Omega}}\right)\right]$$

and this function is $((\beta_k - \tau)^{\Omega} m_\sigma)$ unimodal, i.e. $\omega(\beta_k) = (\beta_k - \tau)^{\Omega} m_\sigma$, its peak height increases as $\tau \downarrow \beta_k$ and the distance between the modes of $g^*(y, \omega(\beta_k), \tau)$ and $g^*(y, \omega(\beta_l), \tau)$ is $m_\sigma[(\beta_l - \tau)^{\Omega} - (\beta_k - \tau)^{\Omega}]$ $(k \neq l)$ which is greater than the distance between the modes of $\dfrac{1}{\beta_l^{\Omega}} f^*\left(\dfrac{x}{\beta_l^{\Omega}}\right) - \dfrac{1}{\beta_l^{\Omega}} f^*\left(\dfrac{\sigma x}{\beta_l^{\Omega}}\right)$ and $\dfrac{1}{\beta_k^{\Omega}} f^*\left(\dfrac{x}{\beta_k^{\Omega}}\right) - \dfrac{1}{\beta_k^{\Omega}} f^*\left(\dfrac{\sigma x}{\beta_k^{\Omega}}\right)$. Hence our assertions follow.

In practice, first the method of the diminishing of the formant is applied in the way already described to the superposition $k(x)$ and then the function $R_\sigma(y)$ is constructed e.g. by graphical procedures (compression, subtraction, multiplication). Thus the present test function $b(y, \tau)$ is obtained; the notation $b(y, \tau)$ is obvious and justified.

In the case of $f(x, \beta_k) = \dfrac{1}{\beta_k^{\Omega}} f^*\left(\dfrac{x}{\beta_k^{\Omega}}\right)$ the steps to be taken are sketched in Fig. 1.

It is difficult to give σ a priori. In general the curve approximating to the test

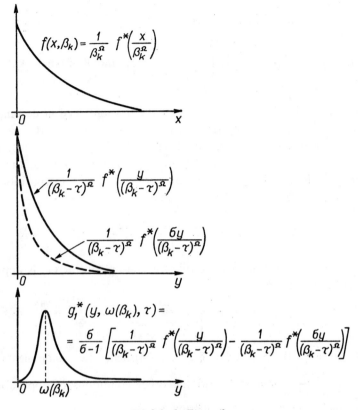

$$f(x,\beta_k) = \frac{1}{\beta_k^\Omega}\, f^*\!\left(\frac{x}{\beta_k^\Omega}\right)$$

$$\frac{1}{(\beta_k-\tau)^\Omega}\, f^*\!\left(\frac{y}{(\beta_k-\tau)^\Omega}\right)$$

$$\frac{1}{(\beta_k-\tau)^\Omega}\, f^*\!\left(\frac{6y}{(\beta_k-\tau)^\Omega}\right)$$

$$g_1^*\,(y,\,\omega(\beta_k),\,\tau) =$$
$$= \frac{6}{6-1}\left[\frac{1}{(\beta_k-\tau)^\Omega}\, f^*\!\left(\frac{y}{(\beta_k-\tau)^\Omega}\right) - \frac{1}{(\beta_k-\tau)^\Omega}\, f^*\!\left(\frac{6y}{(\beta_k-\tau)^\Omega}\right)\right]$$

III.§ 2. 1. Figure 1

function is constructed with several σ-values in order to obtain the decomposition by comparing these figures; this points to the optimal σ-value.

All this is shown best by a concrete example which also shows the basic idea very well.

Example. **Decompositions of a superposition of** (0) **symmetrical normal density functions.** Let us take

$$k(x) = \sum_{k=1}^{N} p_k \frac{e^{-\frac{x^2}{4\beta_k}}}{\sqrt{4\pi\beta_k}} \qquad (0 < \beta_1 < \beta_2 < \ldots < \beta_N < \Lambda_2).$$

The components satisfy the above assumptions because they are (0) symmetrical stable density functions and

$$R_\sigma^*(x) = \frac{\sigma}{\sigma-1}\left(\frac{e^{-\frac{x^2}{4}}}{\sqrt{4\pi}} - \frac{e^{-\frac{\sigma^2 x^2}{4}}}{\sqrt{4\pi}}\right) \qquad (\sigma > 1)$$

is $\left(2\sqrt{\dfrac{\log \sigma^2}{\sigma^2-1}}\right)$ unimodal in $(0,\infty)$ as $m_\sigma = 2\sqrt{\dfrac{\log \sigma^2}{\sigma^2-1}}$, the kth component be-

longs to the type $\dfrac{1}{\beta_k^{1/2}}\,\varphi\left(\dfrac{x}{\beta_k^{1/2}}\right)$, $\left(\varphi(x)=\dfrac{e^{-\frac{x^2}{4}}}{\sqrt{4\pi}}\right)$, and β_k is the $(\varLambda_2, 0)$ mono-

tone formant of its graph.

The method of the simple diminishing of the formant used in decomposing a superposition of normal density functions carries β_k into $\beta_k-\tau$ $(0<\tau<\beta_1)$ term by term and produces a narrowing of the graphs of the components. When it is applied to $k(x)$, as described in Ch. III, § 1, Subsection 1.1.1, the relevant numerical procedure will give an approximation to the superposition of (0) unimodal normal density functions

$$k_1(x) = \sum_{k=1}^{N} p_k \frac{e^{-\frac{x^2}{4(\beta_k-\tau)}}}{\sqrt{4\pi(\beta_k-\tau)}}.$$

With the ordinate values of this the rule for constructing $R_\sigma(y)$ is applied; we have for the components

$$g^*(y, \beta_k^{1/2}, \tau) = \frac{\sigma}{\sigma-1}\left[\frac{e^{-\frac{x^2}{4(\beta_k-\tau)}}}{\sqrt{4\pi(\beta_k-\tau)}} - \frac{e^{-\frac{\sigma^2 x^2}{4(\beta_k-\tau)}}}{\sqrt{4\pi(\beta_k-\tau)}}\right].$$

Then Condition II in Ch. III, § 2 is satisfied, the components being $\left(\sqrt{\beta_k-\tau}\cdot 2\sqrt{\dfrac{\log \sigma^2}{\sigma^2-1}}\right)$ unimodal, and the other conditions of applicability are also satisfied. In the end the test function is

$$b(y, \tau) = \sum_{k=1}^{N} p_k \left(\frac{\sigma}{\sigma-1}\right)\left[\frac{e^{-\frac{y^2}{4(\beta_k-\tau)}}}{\sqrt{4\pi(\beta_k-\tau)}} - \frac{e^{-\frac{\sigma^2 y^2}{4(\beta_k-\tau)}}}{\sqrt{4\pi(\beta_k-\tau)}}\right].$$

Thus the problem is reduced to another, already dealt with. A short calculation shows that the modes of the components of the latter superposition lie the nearer to 0 the nearer σ is to 1. The distance between the modes of the kth and lth component graphs is equal to $2\sqrt{\dfrac{\log \sigma^2}{\sigma^2-1}}\,(\sqrt{\beta_l-\tau} - \sqrt{\beta_k-\tau})$. This increases as $\sigma\!\downarrow\!1$. Further, the corresponding peak heights increase as $\sigma\!\downarrow\!1$. All this does not go far enough, however, to fix the value of σ in advance, and so enable us to get the best separation of the component graphs. Only experimentation with several τ and σ-values will give hints to the number of peaks in a concrete case. Therefore, in practice, the method of the simple diminishing of the formant is employed on the *measured* data of $k(x)$, and, from its result, the approximation to the graph of $b(y, \tau)$ is constructed, according to the rule for constructing the function $R_\sigma(y)$ e.g. graphically with several τ and σ-values in order to get the decomposition by comparing the separate figures.

The following example is put in a new subsection because of its extreme importance.

1.1. Decomposition of a superposition of exponential density functions

Let us take

$$(1) \qquad k(x) = \begin{cases} \sum_{k=1}^{N} p_k \dfrac{e^{-\frac{x}{\beta_k}}}{\beta_k} & (x > 0) \\[2mm] 0 & (x \leq 0) \end{cases} \qquad (0 < \beta_1 < \beta_2 < \ldots < \beta_N < \Lambda_2)$$

i.e. a superposition of exponential density functions. Its decomposition has been investigated in a number of papers; we turn to them later. Here the general idea in Ch. III, § 2, Section 1 can be applied as the exponential density function is a special Gamma density function, and we can apply the method of the diminishing of the formant, introduced in decomposing that superposition (cf. Ch. III, § 1, Subsection 1.1). That decomposition was, however, critical; thus another method of the **decomposition of a superposition of exponential density functions** is shown below.

Let us substitute x^2 for x $(x > 0)$, $p_k \sqrt{\dfrac{\pi}{\beta_k}} = q_k$ for p_k and $4B_k$ for β_k in (1) and continue it for $x < 0$ to be even. Then we get the superposition of (0) unimodal normal density functions

$$k_1(x) = \sum_{k=1}^{N} q_k \frac{e^{-\frac{x^2}{4B_k}}}{\sqrt{4\pi B_k}} \qquad \left(0 < B_1 < B_2 < \ldots < B_N < \frac{\Lambda_2}{4}\right)$$

which can be treated as described in the preceding point. The function $g(y, \omega(\beta_k), \tau)$ needed in this case can be written down easily.

In numerical practice this means that $k(x)$ should also be "measured" for relatively large x-values. This is not always possible.

Example. Let us consider a methodological example. We take

$$k(x) = \begin{cases} \dfrac{1}{\sqrt{2\pi}} e^{-2x} + \dfrac{1}{\sqrt{2\pi}} e^{-8x} & (x > 0) \\[2mm] 0 & (x \leq 0) \end{cases} \quad ,$$

the values of $k(x)$ taken from a table, being cutoff at the 4th decimal place. Here

$p_1 = \dfrac{1}{2\sqrt{2\pi}}, \quad p_2 = \dfrac{1}{8\sqrt{2\pi}}, \quad \beta_1 = \dfrac{1}{2}, \quad \beta_2 = \dfrac{1}{8}.$ In Fig. 1a the solid line shows the graph of $k(x)$. The dotted line shows $k_1(x)$, obtained with the aid of the above procedure. (Here $q_1 = 1/2$, $q_2 = 1/4$.) In Fig. 1b the dotted line shows the graph of the approximation to the function for $\tau = 1/16$ constructed by the use of the method described in Ch. V, § 2, using the simple diminishing of the formant.

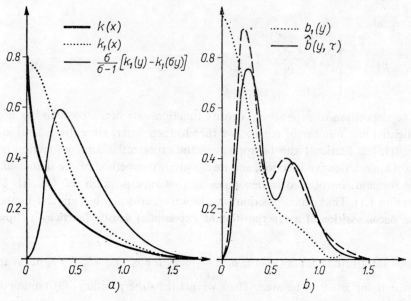

III. § 2. 1.1. Figure I

The solid line shows the test function approximation $\hat{b}(y, \tau)$, constructed from the data of the dotted line i.e. the graph of an approximation to $b(y, \tau)$ for $\tau = 1/16$, $\sigma = 3/2$, according to Ch. III, § 2, Section 1. The graph of $\hat{b}(y, \tau)$, the approximate test function, reveals fairly well that two components are present in $k_1(x)$ i.e. in $k(x)$ (let us compare it with the graph of the exact test function plotted with a dashed line in Fig. 1b). Figure 1a also shows the graph of $\dfrac{\sigma}{\sigma - 1}[k_1(y) - k_1(\sigma y)]$ plotted with a thin line, at $\sigma = 3/2$. This does not show separated peaks.

Supplements and problems to Ch. III, § 2

1. The statements concerning the $\mathscr{A}$-method included in the *Supplements and problems to* Ch. III, § 1 are valid, with appropriate alterations, for the $\mathscr{B}$-method also.

2. In the method described in Ch. III, § 2, Section 1 it is easy to give the test function directly by means of the characteristic function of the superposition to be decomposed. This expression is, however, too complicated to be e.g. the basis of a procedure described in Ch. V, § 2 to get an approximation to the test function.

3. The solution of the example in Ch. III, § 2 is a *correct* problem; the method described in Ch. III, § 2, Section 1 has, however, the great disadvantage that e.g. the numerical application of the method of the simple diminishing of the formant presents an *incorrect* problem as seen in Ch. III, § 1, Subsection 1.1. (The former one has also a disadvantage: the "measured" superposition data are to be taken from a large interval.) A hint at the possibility of introducing — at least, theoretically — a procedure for calculating the test function *leading to a correct problem* is given by the following considerations.

Let the superposition to be decomposed be

$$k(x) = \sum_{k=1}^{N} p_k \frac{1}{\beta_k^{\Omega}} f\left(\frac{x}{\beta_k^{\Omega}}\right) \qquad (0 < \beta_1 < \beta_2 < \ldots < \beta_N < \Lambda_2; \quad \Omega > 0)$$

where $f(x)$ is (0) unimodal and 1. it is an (A_1) density function or 2. it is an (A_2) density function — and, in both cases all moments of $f(x, \beta_k)$ exist; finally

$$\Omega = \begin{cases} 1/2 & \text{if } f(x) \text{ is an } (A_1) \text{ density function,} \\ 1 & \text{if } f(x) \text{ is an } (A_2) \text{ density function.} \end{cases}$$

Let the function $g(y, \omega(\beta_k), \tau)$ in Condition I in Ch. III, § 2 that corresponds to $\frac{1}{\beta_k^{\Omega}} f\left(\frac{x}{\beta_k^{\Omega}}\right)$ be defined by the relation

$$g(y, \omega(\beta_k), \tau) = \frac{1}{\tau^p} \Phi\left(\frac{\beta_k y - R}{\tau^p}\right) \qquad (0 < \tau < T_2)$$

where $R > 0$, $p > 0$ are given, $\Phi(x)$ is an integral function and a constant multiple of a (0) unimodal density function. As τ is a $(T_2, 0)$ monotone formant of the graph of this latter density function, Conditions I, II, III, in Ch. III, § 2 are fulfilled if

$$b(y) = \sum_{k=1}^{N} p_k \frac{1}{\tau^p} \Phi\left(\frac{\beta_k y - R}{\tau^p}\right)$$

is taken as test function.

The concrete form of the function $g(y, \omega(\beta_k), \tau)$ in Condition I in Ch. III, § 2 corresponding to $\dfrac{1}{\beta_k^\Omega} f\left(\dfrac{x}{\beta_k^\Omega}\right)$ can be obtained as follows. Since $\Phi(x)$ is integral and even, we have

$$\frac{1}{\tau^p} \Phi\left(\frac{\beta_k y - R}{\tau^p}\right) = \frac{1}{\tau^p} \Phi\left(\frac{R}{\tau^p} - \frac{\beta_k y}{\tau^p}\right) = \frac{1}{\tau^p} \sum_{v=0}^{\infty} \Phi^{(v)}\left(\frac{R}{\tau^p}\right)\left(-\frac{\beta_k y}{\tau^p}\right)^v \left(\frac{1}{v!}\right).$$

Further

$$\int_{-\infty}^{\infty} \frac{1}{\beta_k^\Omega} f\left(\frac{x}{\beta_k^\Omega}\right) x^n\, dx = \beta_k^{\Omega n} \int_{-\infty}^{\infty} x^n f(x)\, dx =$$

$$\begin{cases} \beta_k^{n/2}\mu_n & \text{if } f(x) \text{ is even and } n \text{ is even,} \\ 0 & \text{if } f(x) \text{ is even and } n \text{ is odd,} \\ \beta_k^n \mu_n & \text{if } f(x) \text{ vanishes in } (-\infty, 0), \end{cases}$$

where μ_n is the nth moment of $f(x)$. Thus

$$\frac{1}{\tau^p} \Phi\left(\frac{\beta_k y - R}{\tau^p}\right) = \frac{1}{\tau^p} \sum_{v=0}^{\infty} \Phi^{(v)}\left(\frac{R}{\tau^p}\right)\left(-\frac{1}{\tau^p}\right)^v \frac{y^v \beta_k^v}{v!} =$$

$$\frac{1}{\tau^p} \sum_{v=0}^{\infty} \Phi^{(v)}\left(\frac{R}{\tau^p}\right)\left(-\frac{1}{\tau^p}\right)^v \frac{y^v}{v!\,\mu_{v/\Omega}} \int_{-\infty}^{\infty} \frac{x^{v/\Omega}}{\beta_k^\Omega} f\left(\frac{x}{\beta_k^\Omega}\right) dx =$$

$$\int_{-\infty}^{\infty} \frac{1}{\beta_k^\Omega} f\left(\frac{x}{\beta_k^\Omega}\right)\left(\frac{1}{\tau^p} \sum_{v=1}^{\infty} \Phi^{(v)}\left(\frac{R}{\tau^p}\right)\left(-\frac{1}{\tau^p}\right)^v \frac{x^{v/\Omega} y^v}{v!\,\mu_{v/\Omega}}\right) dx,$$

supposing that integration and summation can be interchanged. This yields the definition of $g(y, \omega(\beta_k), \tau)$: it represents an integral transformation

$$g(y, \omega(\beta_k), \tau) = \int_{-\infty}^{\infty} K(x^{1/\Omega} y, R, \tau) \frac{1}{\beta_k^\Omega} f\left(\frac{x}{\beta_k^\Omega}\right) dx$$

where

$$K(x^{1/\Omega} y, R, \tau) = \frac{1}{\tau^p} \sum_{n=0}^{\infty} \Phi^{(v)}\left(\frac{R}{\tau^p}\right)\left(-\frac{1}{\tau^p}\right)^n \frac{x^{n/\Omega} y^n}{n!\,\mu_{n/\Omega}} \qquad (R > 1, \quad 0 < \tau < T)$$

$\left(\mu_n = \displaystyle\int_{-\infty}^{\infty} x^n f(x)\, dx\right)$. The test function suitable for the decomposition of $k(x)$ is given by

$$b(y) = \sum_{k=1}^{N} p_k\, g(y, \omega(\beta_k), \tau) = \int_{-\infty}^{\infty} K(x^{1/\Omega} y, R, \tau)\, k(x)\, dx,$$

since $k(x)$ is even if its components are, and conversely.

The kernel function is to be investigated more precisely in every concrete case.

Theoretically the problem is thus solved, as the performance of this integral transformation is a *correct* problem. *Theoretically* the greater R is and the smaller τ is, the better the separation of the components in the graph of the (approximate) test function. One of these parameters can be fixed and the other varied. The constants $\Phi^{(v)}\left(\dfrac{R}{\tau^p}\right)$ can be taken from existing tables of special functions in several cases.

When this procedure is carried out *numerically*, however, difficulties arise. The integrals of infinite limits would not create difficulties; they can be handled like those described in Ch. V, § 4, Section 1; it is the fact that the integral of the kernel $K(x^{1/\Omega} y, R, \tau)$ is, in general, divergent that brings difficulties, as it greatly increases the error included in the "measured" values of $k(x)$; for this error is also transformed by the procedure. This indicates the care needed when experimenting with this idea.

As to the theoretical interest of the method described, the case of the *decomposition of a superposition of* (0) *symmetrical normal density functions*, when

$$k(x) = \sum_{k=1}^{N} p_k \frac{e^{-\frac{x^2}{4\beta_k}}}{\sqrt{4\pi\beta_k}}$$

can be deduced mechanically from the preceding. We should take above $\Omega = 1/2$,

$$f(x) = \frac{e^{-\frac{x^2}{4}}}{\sqrt{4\pi}}$$

$p = 1/2$. Here $\mu_{2n+1} = 0$, $\mu_{2n} = \dfrac{(2n)!}{n!}$ $(n = 0, 1, ...)$. Let us take

$$\Phi(x) = \frac{e^{-\frac{x^2}{2}}}{\sqrt{2\pi}};$$ this fulfils the conditions put on $\Phi(x)$. We should then have the test function

$$b(y) = \sum_{k=1}^{N} p_k \frac{e^{-\frac{(\beta_k y - R)^2}{2\tau}}}{\sqrt{2\pi\tau}} = \int_{-\infty}^{\infty} K(x^2 y, R, \tau)\, f(x)\, dx \qquad (0 < \tau < T)$$

where

$$K(x^2 y, R, \tau) = \frac{1}{\sqrt{\tau}} \sum_{n=0}^{\infty} \Phi^{(n)}\left(\frac{R}{\sqrt{\tau}}\right)\left(-\frac{1}{\sqrt{\tau}}\right)^n \frac{x^{2n} y^n}{(2n)!}$$

(here all the conditions are fulfilled). Since

$$\Phi^{(n)}(x) = (-1)^n \frac{e^{-\frac{x^2}{2}}}{\sqrt{2\pi}} He_n(x)$$

$\left(He_n(x) \text{ is the } n\text{th Hermite polynomial generated with } e^{-\frac{x^2}{2}}\right),$

$$K(x^2 y, R, \tau) = \frac{e^{-\frac{R^2}{2\tau}}}{\sqrt{2\pi\tau}} \sum_{n=0}^{\infty} He_n\left(\frac{R}{\sqrt{\tau}}\right)\left(\frac{1}{\sqrt{\tau}}\right)^n \frac{x^{2n} y^n}{(2n)!}.$$

In the case of the *decomposition of a superposition of exponential density functions* when

$$k(x) = \begin{cases} \sum_{k=1}^{N} p_k \dfrac{e^{-\frac{x^2}{\beta_k}}}{\beta_k} & (x > 0) \\ 0 & (x \leq 0) \end{cases} \qquad (0 < \beta_1 < \beta_2 < \ldots < \beta_N < \Lambda_2)$$

the above general case is specialized as follows: we have to take $\Omega = 1$, $f(x) = \begin{cases} e^{-x} & (x > 0) \\ 0 & (x \leq 0) \end{cases}$; consequently $\mu_n = n!$; and $p = 1/2$. Let us put $\Phi(x) = \dfrac{e^{-\frac{x^2}{2}}}{\sqrt{2\pi}}$ again. Then the test function is

$$b(y) = \sum_{k=1}^{N} p_k \frac{e^{-\frac{(\beta_k y - R)^2}{2\tau}}}{\sqrt{2\pi\tau}} = \int_0^{\infty} K(xy, R, \tau) k(x)\, dx,$$

$$K(\xi, R, \tau) = \frac{1}{\sqrt{\tau}} \sum_{n=0}^{\infty} \Phi^{(n)}\left(\frac{R}{\sqrt{\tau}}\right)\left(-\frac{1}{\sqrt{\tau}}\right)^n \frac{\xi^n}{(n!)^2} =$$

$$\frac{e^{-\frac{R^2}{2\tau}}}{\sqrt{2\pi\tau}} \sum_{n=0}^{\infty} He_n\left(\frac{R}{\sqrt{\tau}}\right)\left(-\frac{1}{\sqrt{\tau}}\right)^n \frac{\xi^n}{(n!)^2},$$

all the conditions needed for the construction of $b(y)$ being fulfilled here.

Instead of $\Phi(x)$ above, other function types, e.g. the Gamma density function

$$\Psi(x) = \begin{cases} \dfrac{\beta_k^{\frac{1}{\tau}+1} x^{\frac{1}{\tau}} e^{-\frac{\beta_k x}{\tau}}}{\tau^{\frac{1}{\tau}+1} \Gamma(1/\tau+1)} & (x > 0) \\ 0 & (x \leq 0) \end{cases} \qquad \left(\frac{1}{\tau} \text{ is integer}\right)$$

can also be taken whose unique mode is at $x = \dfrac{1}{\beta_k}$ for any τ and whose graph is the "narrower" the smaller τ is. This is shown e.g. by its characteristic function $\psi(t) = \dfrac{1}{(1 - i\tau t/\beta_k)^{1/\tau+1}}$, tending to $e^{\frac{it}{\beta_k}}$ as $\tau \to 0$. The corresponding kernel function yielding the test function can be calculated easily by analogy with the procedure for obtaining $K(x^{1/\Omega}y, R, \tau)$ above; it will not, however, depend on $x^{1/\Omega}y$ only. If in case of the *decomposition of a superposition of exponential density functions* this density function $\Psi(x)$ is taken for $\Phi(x)$, the kernel of the integral transformation providing the test function will be

$$K(x, y, \tau) = \frac{(xy/\tau)^{\frac{1/\tau+1}{2}}}{(1/\tau)!\,y} J_{1/\tau+1}\left(2\sqrt{xy/\tau}\right)$$

$\left(\dfrac{1}{\tau} \text{ is integer}\right)$ where $J_\mu(x)$ is a Bessel function of the Ist order. The practical value of this is also limited, as stated in Ch. III, § 2, Section 1.

4. It is theoretically interesting that in the case of the *decomposition of a superposition of exponential density functions*

$$k(x) = \begin{cases} \displaystyle\sum_{k=1}^{N} p_k \frac{e^{-\frac{x}{\beta_k}}}{\beta_k} & (x > 0) \\ 0 & (x \leq 0) \end{cases} \qquad (0 < \beta_1 < \beta_2 < ... < \beta_N < \Lambda_2)$$

a superposition of harmonic functions also can be taken to correspond to $k(x)$, instead of one of the above test functions. This superposition of harmonic functions will then be handled by the tools of *periodogram analysis* in order to determine the unknown parameters.

In particular, one needs an integral transformation that transforms the exponential density function into a cosine function (Medgyessy 1971b, 1972b). More exactly, one determines the function $K(x, y)$ for which

$$\int_0^\infty K(x, y) \frac{e^{-\frac{x}{\beta_k}}}{\beta_k} dx = \cos \beta_k y.$$

This is easy by means of the technique of Point 3 above, since

$$\cos \beta_k y = \sum_{v=0}^\infty (-1)^v \frac{(\beta_k y)^{2v}}{(2v)!}$$

and

$$\int_0^\infty x^\mu \frac{e^{-\frac{x}{\beta_k}}}{\beta_k} dx = \mu!\,\beta_k^\mu,$$

$$\cos \beta_k y = \sum_{v=0}^\infty \frac{(-1)^v y^{2v}}{[(2v)!]^2} \int_0^\infty x^{2v} \frac{e^{-\frac{x}{\beta_k}}}{\beta_k} dx = \int_0^\infty \left(\sum_{v=0}^\infty \frac{(-1)^v (xy)^{2v}}{[(2v)!]^2}\right) \frac{e^{-\frac{x}{\beta_k}}}{\beta_k} dx,$$

the interchanging of the integration and the summation being justified. But

$$\sum_{v=0}^{\infty} \frac{(-1)^v (xy)^{2v}}{[(2v)!]^2} = \text{ber } 2\sqrt{xy}$$

where the right-hand side is the ber function of Thomson. Thus

$$\int_0^{\infty} \text{ber} \left(2\sqrt{xy}\right) \frac{e^{-\frac{x}{\beta_k}}}{\beta_k} dx = \cos \beta_k y$$

and

$$K(x, y) = K(xy) = \text{ber} \left(2\sqrt{xy}\right).$$

Finally, we have

$$b(y) = \int_0^{\infty} K(xy) k(x) dx = \int_0^{\infty} \text{ber} \left(2\sqrt{xy}\right) k(x) dx = \int_0^{\infty} \text{ber} \left(2\sqrt{xy}\right) \left(\sum_{k=1}^{N} p_k \frac{e^{-\frac{x}{\beta_k}}}{\beta_k}\right) dx,$$

that is,

$$b(y) = \sum_{k=1}^{N} p_k \cos \beta_k y.$$

From the graph of this special "test function" N and, eventually, the other parameters are to be determined. This is a classical problem of *periodogram analysis* to be solved by means of the relevant methods (Serebrennikov, Pervozvanskiĭ 1965), e.g. by forming the function

$$U(\omega) = \frac{1}{M+1} \sum_{p=0}^{M} b(\varrho h) \cos \varrho h \omega$$

where M is a sufficiently large natural number, since the graph of $U(x)$ has, in general, sharp maxima at the points $\omega = \beta_k$ and their heights are proportional to p_k; thus this is the graph which is convenient for the determination of the unknown parameters of the starting superposition. Numerically, this method is very difficult to apply for reasons similar to those described in the application of the transformation in Point 3 above that lead to $b(y)$.

An integral transformation of

$$k(x) = \begin{cases} \sum_{k=1}^{N} p_k \dfrac{e^{-\frac{x}{\beta_k}}}{\beta_k} & (x > 0) \\[2ex] 0 & (x \leqq 0) \end{cases}$$

leading to a superposition of normal density functions as test function can be obtained by applying to the above $b(y)$ an appropriate *second* integral transformation.

5. The *decomposition of a superposition of exponential density functions* has become increasingly interesting in the last few years in several experimental sciences. Relevant ideas such as those described above arose earlier. One of them is due to Brownell, Callahan (1963).

They, too, introduced a special integral transform in order to obtain a test function; more precisely, if a superposition

$$k(x) = \begin{cases} \sum_{k=1}^{N} p_k \dfrac{e^{-\frac{x}{\beta_k}}}{\beta_k} & (x > 0) \\ 0 & (x \leq 0) \end{cases} \qquad (0 < \beta_1 < \beta_2 < \ldots < \beta_N < \Lambda_2)$$

is given, they applied the integral transformation of product kernel

$$\int_0^\infty sx \sin sx \cdot k(x)\,dx = G(s).$$

It is easily proved that

$$G(s) = \sum_{k=1}^{N} p_k \frac{2s^2 \beta_k^2}{(s^2 \beta_k^2 + 1)^2}.$$

For $s > 0$ the graphs of the components of this test function have peaks at $s = \dfrac{1}{\beta_k}$, and if we have $p_k \beta_k = p_l \beta_l$ ($k \neq l$), the graphs of the components will appear separated.

Thus $G(s)$ may be appropriate to be considered as a test function. If it is computed numerically by means of the measured values $\hat{k}(x)$ of $k(x)$, several difficulties arise. For instance, in practice the graph of $k(x)$ is given in an interval $(0, x_{\max})$ only. The authors proceeded at this point as follows: for small s-values they applied an extrapolation (with fitted exponential functions) for the values of $k(x)$ ($x > x_{\max}$). This resulted in the approximation

$$G_1(s) = \int_0^{x_{\max}} sx \sin sx \cdot k(x)\,dx + \int_{x_{\max}}^{\pi/s} sx \sin sx \cdot B e^{-\frac{x}{x_{\max}}}\,dx +$$

$$+ \int_{\pi/s}^\infty sx \sin sx \cdot A e^{-\frac{Ax}{\pi}}\,dx$$

where $B = e k(x_{\max})$, $A = e^2 k(x_{\max})$. For large s-values the authors took the approximation

$$G_2(s) = \int_0^{\pi/s} sx \sin sx \cdot k(x)\,dx + \int_{\pi/s}^\infty sx \sin sx \cdot A_1 e^{-\frac{sx}{\pi}}\,dx \qquad \left(A_1 = e k\!\left(\frac{\pi}{s}\right) \right).$$

With these formulae they interpolated between data points $\hat{k}(x)$ with polynomials of the 2nd degree and then integrated numerically. They found that if some separation was to be attained the β_k-values had to be in a proportion of 4:1 to each other.

In Fig. 1 where both curves are taken from Brownell, Callahan (1963), the solid line shows the graph of an approximation of the test function $G(s)$, computed as described above, if

$$k(x) = e^{-cx} + 10\,e^{-10cx} + 100\,e^{-100cx}$$

plus a 10 per cent normally distributed random error (c is a given constant). Here the criterion of optimal separation, $p_k\beta_k = p_l\beta_l = \dfrac{1}{c}\ (c \neq l)$ is fulfilled. Three peaks pointing to the presence of three components are discernible. The dashed line shows the graph of the exact test function $G(s)$, in order to give a picture of the accuracy of the procedure in this methodological example.

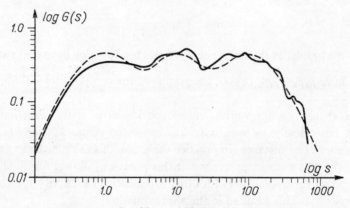

Supplements and problems to Ch. III, § 2. Figure 1

Although this procedure seems to diminish the disadvantageous effect that the integral of the kernel function $sx \sin sx$ is infinite (clearly, in practice a function $\hat{k}(x) = k(x) + \varepsilon(x)$ is integrated where $\varepsilon(x)$ is the error whose integral $\int_0^\infty sx \sin sx \cdot \varepsilon(x)\,dx$ can also be infinite when $\int_0^\infty sx \sin sx \cdot dx$ is so); the influence of the "measurement" errors on $k(x)$ makes this method rather delicate.

Other relevant methods used in biological research are cited in the paper of the above-mentioned authors.

The critical points of such methods are described in Landahl (1963).

6. In the case of the *decomposition of a superposition of exponential density functions*

$$k(x) = \begin{cases} \sum_{k=1}^{N} p_k \dfrac{e^{-\frac{x}{\beta_k}}}{\beta_k} & (x > 0) \\ \\ 0 & (x \leqq 0) \end{cases} \qquad (0 < \beta_1 < \beta_2 < ... < \beta_N < \Lambda_2)$$

the determination of an approximate test function $b(y)$, in a sense similar to the described $\mathcal{B}$-method, was described earlier by Gardner, Gardner, Laush, Meinke (1959). In their method, which became popular, the test function $b(y)$ was obtained immediately; the preliminary introduction of some integral transformation was avoided. Let us write $\dfrac{p_k}{\beta_k} = p_k^*$, $\dfrac{1}{\beta_k} = \beta_k^*$ in the superposition $k(x)$. Then

$$k(x) = \sum_{k=1}^{N} p_k^* e^{-\beta_k^* x} = \int_0^\infty e^{-\lambda x} g(\lambda) \, d\lambda \qquad (x > 0)$$

where — *formally* —

$$g(x) = \sum_{k=1}^{N} p_k^* \delta(x - \beta_k^*);$$

here $\delta(x)$ denotes the "Dirac delta function".

The solving of the integral equation

$$k(x) = \int_0^\infty e^{-\lambda x} g(\lambda) \, d\lambda$$

for $g(\lambda)$ would give, in principle, the values p_k and, consequently, also N. In order to obtain $g(\lambda)$ Gardner, Gardner, Laush, Meinke (1959) introduced the new variables y and v by putting $\lambda = e^{-y}$, $x = e^v$, and obtained the convolution integral equation

$$e^v f(e^v) = \int_{-\infty}^\infty e^{-e^{(v-y)}} e^{(v-y)} g(e^{-y}) \, dy,$$

which can be solved with the aid of the Fourier transform. Evidently

$$g(e^{-y}) = \frac{1}{2\pi} \int_{-\infty}^\infty e^{-iyt} \frac{F(t)}{K(t)} \, dt,$$

where

$$F(t) = \mathbf{F}[e^y f(e^y); t], \quad K(t) = \mathbf{F}[e^{-e^y} e^y; t].$$

$K(t)$ can be expressed by the Gamma function of complex argument. The authors tried to deal with the integral equation for $g(e^{-y})$ numerically, assuming that $g(e^{-y})$ could be considered as some "approximation" (with sharp peaks) to the transformed sum of "Dirac delta functions", with regard to the errors included in the values of the left-hand side of the integral equation, to be used in a numerical procedure. After some transformations of the integral in the integral equation and taking into account that only finite limits of integration can be used,

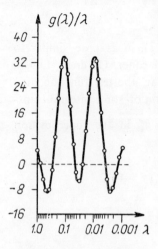

the authors obtained, in the end, the approximate result by two subsequent numerical integrations carried out by means of the trapezoidal formula. They also gave a detailed investigation of the numerical procedure and several examples proving the correctness of the method.

From the practical point of view, even if no "Dirac delta functions" were introduced, the solution of that integral equation represents, seemingly, an *incorrect* problem (cf. Ch. V, § 1). This therefore introduces, in advance, great difficulties but in spite of this, this method can be assumed as one of the most widespread.

*Supplements and problems
to Ch. III, § 2. Figure 2*

In Fig. 2, taken from Gardner, Gardner, Laush, Meinke (1959), the circles show the approximate values of $\dfrac{g(\lambda)}{\lambda}$ which are obtained more easily by the

described appropriate numerical procedure than those of $g(\lambda)$, the solution of the above integral equation in case of

$$k(x) = 1000\,e^{-0.1x} + 100\,e^{-0.01x}\ (x > 0);$$

the ordinate values of the latter were distorted randomly in order to simulate a decay curve with scatter. The two dominating peaks point to the two components in this methodological example.

In any case it can be shown that the use of the "Dirac delta functions" can be avoided and, thus, the setting can be made exact (Medgyessy 1971b, 1972b). Let us realize that

$$e^{-\beta_k^* x} = \int_{-\infty}^{\infty} e^{-yx - \tau x^2}\,\frac{e^{-\frac{(y - \beta_k^*)^2}{4\tau}}}{\sqrt{4\pi\tau}}\,dy \qquad (x > 0,\ \tau > 0)$$

whence we have that

$$\sum_{k=1}^{N} p_k^*\,e^{-\beta_k^* x} = \int_{-\infty}^{\infty} K(x, y, \tau)\left(\sum_{k=1}^{N} p_k^*\,\frac{e^{-\frac{(y - \beta_k^*)^2}{4\tau}}}{\sqrt{4\pi\tau}}\right) dy$$

where
$$K(x, y, \tau) = e^{-y-\tau x^2} \qquad (x > 0).$$

Then, solving the integral equation
$$k(x) = \int_{-\infty}^{\infty} K(x, y, \tau) g^*(y) \, dy$$

for $g^*(y)$ with a sufficiently small τ, we should get the test function

$$g^*(y) = \sum_{k=1}^{N} p_k \frac{e^{-\frac{(y-\beta_k^*)^2}{4\tau}}}{\sqrt{4\pi\tau}},$$

suitable to determine N (and, eventually, β_k^*) if τ is sufficiently small. For $\tau \downarrow 0$, the integral equation for $g^*(y)$ goes over (formally) in that for $g(\lambda)$ as $g^*(y)$ goes in the above $g(y)$. We can also write

$$e^{\tau x^2} k(x) = \int_{-\infty}^{\infty} e^{-xy} g^*(y) \, dy;$$

here we should have to carry out the numerical *inversion of a two-sided Laplace transformation*.

Although the problem has now an exact setting, the solution of the last two integral equations remains an *incorrect* problem which also influences the numerical treatment.

The method of Gardner, Gardner, Laush, Meinke (1959) has been simplified significantly by means of a numerical Laplace transform technique by Papoulis (1973).

7. The **decomposition of a superposition of Gamma density functions of the same order**, when

$$k(x) = \begin{cases} \sum_{k=1}^{N} p_k \dfrac{x^{\gamma-1} e^{-\frac{x}{\beta_k}}}{\beta_k^\gamma \Gamma(\gamma)} & (x > 0) \\ 0 & (x \le 0) \end{cases} \qquad (0 < \beta_1 < \beta_2 < \ldots < \beta_N < \Lambda_2)$$

(γ is given) can be reduced to the problem in Ch. III, §2, Subsection 1.1 by putting
$$k_1(x) = \frac{k(x)}{x^{\gamma-1}}, \quad q_k = \frac{p_k}{\beta_k^{\gamma-1} \Gamma(\gamma)}; \quad \text{then we have}$$

$$k_1(x) = \begin{cases} \sum_{k=1}^{N} q_k \dfrac{e^{-\frac{x}{\beta_k}}}{\beta_k} & (x > 0) \\ 0 & (x \le 0) \end{cases}$$

i.e. a superposition of exponential density functions whose decomposition can be performed by any method in Ch. III, § 2, Subsection 1.1. No additional numerical difficulties appear. The change of p_k into q_k may, of course, disturb the picture.

§ 3. *AD HOC* METHODS

This paragraph enumerates some *ad hoc* methods of decomposition of superpositions of density functions. They differ essentially from those described earlier in the present Chapter but they can be and are used widely in practice. Their classification according to the titles below is somewhat awkward because the ideas appear mixed.

(A) *Methods of graphical character*

1. Here the basic idea is as follows. We suppose that we possess such a long section of the graph of a superposition of density functions that both tails of the graph approach close to the abscissa axis. We suppose further that in the right- (left-) hand side of the graph, the graph of the last (first) component dominates i.e. the graphs of its neighbour components do not, for all practical purposes, influence it. Then we take several curve points that seem to belong — knowing the analytical form of the components — to a component (e.g. Gaussian) curve on the one end and, considering them as points of the graph of a *single* component, we *compute* by means of its coordinates *the parameters* of this component. Then they will be regarded as the approximations to the unknown parameters of the last component.

With these approximate parameter values the *whole* graph of the component can be constructed and it is considered as an approximation to the graph of the true lateral component. Then the latter graph is subtracted graphically from the original one and the above procedure is iterated, which, in the end, yields also the number of the components. If several components appear in the graph, the method can be applied to any of them.

In the case of a superposition of normal density functions, such a method, with some of its modifications, was published e.g. by Schellenberg (1932); Kiss, Sándorfy (1948); Wallner, Ulke (1952); Idu, Cucu (1959); Ruff (1965) (with the additional use of the slope at lateral curve points).

Example. In Fig. 1, taken from the paper of Schellenberg (1932), the solid line shows the relative intensity distribution of the ultraviolet emission spectrum of CaOAg, supposed to be a superposition of normal density functions. The component graphs obtained by the method described above were plotted with dashed lines; they were in good accordance with the physical background.

For a particular case of superpositions of distorted normal density functions, a modification of the method has been given by Magos, Medgyessy (1954).

For the critical case of a superposition of exponential density functions, special analogous — and rather trivial — methods were given as well.

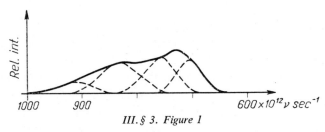

III.§ 3. Figure 1

2. In several cases the preceding method can be modified in such a way that, avoiding the reading-off of curve coordinates, the lateral parts of the graph of the superposition, considered as parts of separate component graphs, are followed "by eye" or by *fitting* curves of appropriate function curves to them, e.g. by mirroring lateral curve sections around conjectured peak height lines into the interior of the graph section to yield a whole component graph, and so on, by subtracting each "constructed" component graph graphically. Sometimes, the graphs of almost all components can be plotted in this manner, using various techniques to be hardly surveyed. In cases when the components appear fairly well, the graph is dissected simply by means of vertical lines located at the "minimum" points of the graph. Correction methods appear also here. Cases of these methods were described for a superposition of normal density functions, e.g. by Wiedemann (1947); Idu, Cucu (1959); Morison Smith, Bartlet (1961); Kaplan, Gurvič (1963). One of the simple methods was "computerized" by Poulik, Pinteric (1955).

The results of such a procedure (mirroring lateral parts and fitting "by eye" graphs of normal density functions) is shown with a dotted line, in Fig. 4b in Ch. III, § 1, Subsection 1.1.1.1, representing a section of a curve obtained during a blood serum electrophoresis, with the aid of Tiselius' apparatus (Medgyessy 1953, 1954c).

3. The lateral components may also be identified by using transparent patterns of the component curves of different parameters; their pictures will be *projected* on the ends of the given graph, the parameters of that with the best fit will be considered as those required. By a certain optical enlarging technique, the "stock" of component curves to be fitted can be increased. Then the above graphical plotting in and subtraction is applied, which is very easy here as the graph to be subtracted can be seen immediately. The iteration of the whole procedure

is also possible. In the case of a superposition of normal density functions — where patterns of Gaussian curves are needed — a relevant apparatus was invented by Labhart (1947). In this, diapositives of Gaussian curves of identical height but of different variance are used; the change of height is reached by means of changing the magnitude of the projected image.

4. From this it was but a natural step to *generate electronically the graph* of a component of a given superposition on the screen of a cathode ray tube; the corresponding parameters will be equivalent to some data of circuit elements and can be altered in certain intervals. It is also easy to produce a superposition of such curves on the screen. This picture should be *fitted* to the graph of the superposition to be decomposed, which is drawn on transparent paper and put over the screen, by varying the parameters. In the end, the parameters of the best fit will be regarded as approximations of the unknown parameters of the given superposition; here the number of the components will be obtained immediately. For the case of a superposition of normal as well as Cauchy density functions, such an apparatus was built by Noble, Hayes, Jr., Eden (1959).

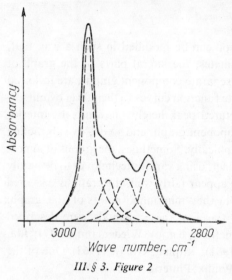

III. § 3. Figure 2

Another device for this purpose, for normal density functions was invented by Sydow, Dittmann (1963). Both articles contain illustrative examples.

In Fig. 2, taken from the article of Noble, Hayes, Jr., Eden (1959), the solid line shows the absorbancy distribution curve of iso-octane (2, 2, 4-trimethylpentane), assumed to represent a superposition of Cauchy density functions ("**Lorentz distributions**"). The result of a decomposition carried out by the described device is shown by a dashed line. Four components have been established.

We notice that the best *fitting* used for the determination of the number of components in the described procedure does not always yield their true number.

5. Special *coordinate papers* have also been widely used in graphical decomposition. Here the whole graph is plotted on the paper, constructed for the type under consideration, which carries out such a transformation of the original graph

that the graphs of the transformed components will become simple curves (straight lines, parabolas, hyperbolas etc.). The reading of data from them will be easier — by an appropriate fitting of these simple curves — than from the original graph since the different components often appear very remarkably. For instance, the superposition of normal density functions is transformed, under certain conditions, into a superposition of parabolas by a semilogarithmic paper. The superposition of normal distribution functions is transformed by a Gaussian paper into a chain of almost straight-line graphs.

Naturally, the same result is obtained by plotting data on a simple coordinate paper in linear scale following an appropriate *transformation* of the abscissa and ordinate values of the original graph of the superposition. Generally, the graph of the side components appear well; their subtraction makes the procedure iterative and easier.

A representative of relevant methods is described e.g. by Szigeti (1947). Figure 3, taken from this paper, shows the logarithms of the measurement data (circles) of the intensity distribution of the emission spectrum of zinc-beryllium-silicates, plotted as a certain quadratic function of the frequency on a linear paper. After the described manipulations three straight lines could be fitted to the data points. This supported the presence of a superposition of three normal density functions; this type of superposition was assumed on theoretical grounds.

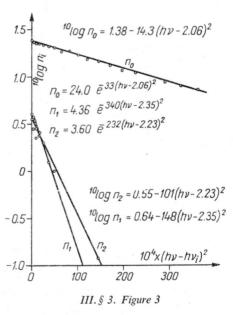

$$^{10}log\, n_0 = 1.38 - 14.3\,(h\nu - 2.06)^2$$

$$n_0 = 24.0\; \bar{e}^{\,33(h\nu - 2.06)^2}$$

$$n_1 = 4.36\; \bar{e}^{\,340(h\nu - 2.35)^2}$$

$$n_2 = 3.60\; \bar{e}^{\,232(h\nu - 2.23)^2}$$

$$^{10}log\, n_2 = 0.55 - 101(h\nu - 2.23)^2$$

$$^{10}log\, n_1 = 0.64 - 148(h\nu - 2.35)^2$$

$$10^4 \times (h\nu - h\nu_i)^2$$

III. § 3. *Figure 3*

A great number of similar methods are described, implicitly, in papers on *mathematical statistics* dealing with statistical samples originating from a population, in which the density function of the measured quantity is a superposition of normal density functions. These papers work, in general, with the histogram made from sample data, i.e. with an approximation to the graph of the relevant superposition. Thus they use methods of the type we are considering.

A good representantative of such methods that can be applied to decompose superpositions of normal density functions can be found in a paper by

Bhattacharya (1967). In Fig. 4, taken from this paper, the circles show, theoretically, values of an approximation (formed by a special histogram) of $\log f(x+h) - \log f(x)$ where $f(x)$ is a superposition of an unknown number of normal density functions and h is an appropriate constant. In this case a single Gaussian curve would yield a straight line on the paper. With this knowledge, the straight lines I—IV could be fitted with ease to the data points; the graphical subtraction of the data yielded by the right-hand side straight line revealed the presence of one more component (straight line V). Altogether five components in the superposition could be seen.

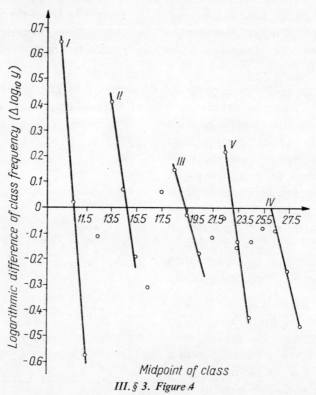

III. § 3. Figure 4

As to the technique, papers by Harding (1949) (e.g. for a superposition of lognormal density functions; cf. Ch. III, § 1, Section 3, (a); Cassie (1954); Daeves, Beckel (1958) may be considered also as belonging to our sphere of interest.

The case of the *superposition of exponential density functions* (plotting the data on a semilogarithmic paper giving a straight line for every component, etc.) is dealt with by a method which is much better than the usual ones in Defares, Sneddon (1960) (see also Peter, Peter 1960).

All the above methods can be subject also to modifications of various type. However, they become *ineffective* in general e.g. if the graphs of the components of the investigated superposition lie too near to each other. In spite of this they, too, are very popular nowadays.

(B) *Methods of algebraic character*

1. Some of these methods essentially *approximate* the superposition to be investigated by a superposition of a *given* number N ($N=2, 3, ...$) of components belonging to the same type as the components of the given superposition. The unknown parameters of the approximating function are determined by means of the method of least squares, the method of moments etc., using data taken from the graph of the given superposition. The procedure is carried out with $N=2$, $N=3$, ... etc. If for some value of N the fit *(in some sense)* is the best, and, at the same time, the "smoothest" then this value of N will be taken to be the number of the components of the superposition, and the parameters of this best fit as the required parameters.

This fitting is carried out mostly by the method of moments as it is the easiest. It may be illustrated by the case of a superposition of normal density functions. The relevant procedure for the case of *equal* variances was first given by Doetsch (1928), using the method of moments. He calculated, fixing N, the successive moments M_v ($v=0, 1, 2, ...$) of

$$f(x) = \sum_{k=1}^{N} p_k \frac{e^{-\frac{(x-\alpha_k)^2}{4\beta}}}{\sqrt{4\pi\beta}} \qquad (\alpha_i \neq \alpha_j \ (i \neq j); \quad \beta > 0)$$

which are algebraic expressions of p_k, α_k, β and made them equal to the approximations $\hat{M}_v$ of the moments (about the origin) of the given superposition, computed from the graph of the given superposition numerically, graphically or by means of special mathematical instruments (e.g. moment planimeters). Thus he had the system of algebraic equations of higher order

$$\sum_{k=1}^{N} p_k = \hat{M}_0,$$

(1)
$$\sum_{k=1}^{N} p_k \alpha_k = \hat{M}_1,$$

$$\sum_{k=1}^{N} p_k \alpha_k^2 + 2 \sum_{k=1}^{N} p_k \beta = \hat{M}_2.$$

$$\therefore$$

The value of N being fixed, as many equations of the former type have to be formed as are sufficient for the determination of p_k, α_k, β. The latter ones are considered as the unknown parameters. The whole procedure has to be repeated with another value of N, and so on. If an upper bound N_1 of N is known, $N=N_1$ is taken and some parameters obtained by the solution of the system of equations (1) are expected to turn to zero; this hints at the exact value of N.

If β is known (e.g. from the graph of the last or first component) then the system of equations (1) becomes equivalent to a system of equations belonging to the type

$$(2) \qquad \sum_{k=1}^{N} y_k x_k^v = D^v \qquad (v = 0, 1, ..., n, \quad n \geqq 2N)$$

for the unknowns y_k, x_k $(k=1, ..., N)$. This is a classical system of nonlinear equations; the technique of its solution is well known, originating with C. F. Gauss (Gauss 1866 pp. 165—196; Runge, König 1924). The solution is obtained by solving an algebraic equation of order N and two systems of N linear equations. If the D_v are "measured" values containing errors, it may often present *incorrect* problems (cf. Ch. V, § 1). The solution has been generalized, to some extent, by Medgyessy (1954a, 1961a pp. 217—219). In the case of $N=2$ or $N=3$ the system (1) or (2) can be solved fairly easily.

Doetsch (1928) analysed the H_α (doublet) intensity distribution curve, investigated by Sen (1922), by means of the above moment method on the assumption of its being a superposition of two Gaussian functions. This gave an uninterpretable result, and was thought to have been caused by the presence of a third spectrum line. Therefore, the same curve was considered in accordance with the exact decomposition method of Doetsch (1928) (i.e. with the simple diminishing of the formant) described in Ch. III, §1, Subsection 1.1.1.1. Being insufficient, the procedure was not repeated with $N=3$. However, an intensity distribution curve of H_α obtained with a finer experimental technique was considered in accordance with the same moment method, on the assumption $N = 3$; three components were obtained readily. Nevertheless the investigation was repeated with the exact decomposition method (cf. Ch. III, § 1, Subsection 1.1.1.1). Three components appeared; all this supported the above statements.

Schellenberg (1932) worked out the moment method of Doetsch (1928) for different β_k-values. But he did not work with it; he only compared the moments of a superposition, calculated on the basis of the parameter values obtained by some primitive method from the graph of the superposition, with the moments directly obtained from the graph of the superposition in order to obtain a means of checking.

Sticker (1930a), (1930b) decomposed superpositions of normal density functions by the moment method of Doetsch (1928), assuming that $\beta_1=\beta_2=...=\beta_N$.

He carried out the procedure with several N-values which seemed reasonable. The value of N at which the superposition, constructed with the parameters obtained during the decomposition procedure, showed the best and, at the same time, the "smoothest" fit to the experimental data, was considered as the actual value of N (and the corresponding parameter values as the actual values). In Fig. 5 taken from Sticker (1930a), the circles show data of the colour frequency distribution curve of 1404 stars of a certain ensemble. The curve was considered as a superposition of normal density functions. The dashed lines show the graphs of the components obtained by means of the moment method, if N is taken equal to 5. The solid line shows the graph of the superposition of these Gaussian curves. As it fits well the measurement data, the assumption $N=5$ was accepted and the parameters obtained in this way were considered as characteristic of the real situation.

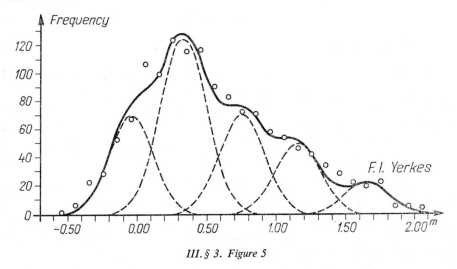

III. § 3. Figure 5

Naturally if the above method of fitting, using the method of moments, is applied, any method of *mathematical statistics* for giving moment estimators for the parameters of a mixture population density function belonging to a given type can be used, because moments can be calculated not only from a sample but also from the graph of a superposition. A great number of papers can thus be utilized in this context by using their analytical kernels. A classical example is a famous article by Pearson (1894) on the approximation of a population density function by means of a mixture of two normal density functions. His result was based on the solution of a nonic and the technique was never simplified significantly. In recent years newer techniques have been elaborated with fairly good results, at least for small values of N, providing newer hints for the analytical details of the method dealt with in the present discussion.

We point out that the *decomposition of a superposition of exponential density functions* can also be dealt with by the above algebraic method, i.e. by means of the solution of a system of equations of the type (2) (the ideas can be taken from works on mathematical statistics (Gumbel 1939; Rider 1961)). Namely let

$$f(x) = \begin{cases} \sum_{k=1}^{N} p_k \dfrac{e^{-\frac{x}{\beta_k}}}{\beta_k} & (x > 0) \\ \\ 0 & (x \leq 0). \end{cases} \qquad (0 < \beta_1 < \beta_2 < \ldots < \beta_N < \Lambda_2)$$

We have, writing down the moments, the system of algebraic equations of higher degree

$$m_\varrho = \int_0^\infty x^\varrho f(x)\, dx = \sum_{k=1}^{N} \varrho!\, p_k \beta_k^\varrho \qquad (p = 0, 1, \ldots)$$

which belongs to type (2).

Ideas of similar methods but using the least squares technique can be found in several articles (e.g. Giaccardi 1939; Morozova 1970; Morozova, Balabanova 1970; Molodenkova, Kovalev 1972 (the last three papers use an approximation with a superposition of suitable density functions, if N is, or is supposed to be, *known*); Kindler 1969, 1973 (approximation with a superposition of exponential density functions)).

2. The same idea as in Point 1 can be realized not only by computing moments but by constructing a system of equations with equidistant "measured" ordinate values of the given graph of a superposition, the solution of which yields, then, the unknown parameters. In several cases it is made easier because the components satisfy some recurrence equation.

For the latter, the classical example is the *decomposition of a superposition of exponential density functions, with a known number N of components*. Let the values of

$$f(x) = \begin{cases} \sum_{k=1}^{N} p_k \dfrac{e^{-\frac{x}{\beta_k}}}{\beta_k} & (x > 0) \\ \\ 0 & (x \leq 0) \end{cases} \qquad (0 < \beta_1 < \beta_2 < \ldots < \beta_N < \Lambda_2)$$

be taken at the equidistant points $x_i = x_1 + (i-1)h$ $(x_1, h$ $(h > 0)$ are fixed). Let us put $f(x_i) = y_i$ $(i = 1, 2, \ldots, 2N)$,

$$p_k \frac{e^{-\frac{x_k}{\beta_k}}}{\beta_k} = P_k, \quad e^{-\frac{h}{\beta_k}} = z_k \qquad (k = 1, \ldots, N).$$

Then we have the system of equations

$$
\begin{aligned}
y_1 &= P_1 & + P_2 & & + \ldots + P_N \\
y_2 &= P_1 z_1 & + P_2 z_2 & & + \ldots + P_N z_N \\
y_3 &= P_1 z_1 & + P_2 z_2^2 & & + \ldots + P_N z_N^2 \\
& \quad \ddots & \ddots & \quad \ddots & \qquad \ddots \\
y_{2N} &= P_1 z_1^{2N-1} & + P_2 z_2^{2N-1} & + \ldots & + P_N z_N^{2N-1}
\end{aligned}
$$

(3)

for the unknowns P_k, z_k. This system again belongs to type (2). It will be substituted by another containing the measured values of y_i. From its solution, the unknowns p_k, β_k can also be determined, approximately. This is the so-called *de Prony's method*.

There are very many variants of this method based partly on the knowledge of approximate values of the unknowns and subsequent iterations (Lánczos 1957 pp. 272—280). Extensions for the case of an *unknown N* have also been published (Householder 1950 pp. 28—32).

The values y_i also satisfy a set of recurrence equations. This fact can also be the starting point for setting up a system of equations yielding, at least implicitly, approximations to the unknown parameters (Varga 1966*b*).

The solution of the above system of equations is not at all simple in practice; (3) may represent an *incorrect* problem, too, which is partly the cause of particular difficulties. Aleksandrov (1970) treated this solution by means of the regularization method of A. N. Tihonov (see Ch. V, § 1, Section 4) and obtained better results than those based on earlier techniques. In the case of an unknown N he carried out the calculations for some N_1 that could have been assumed, on the basis of supplementary informations, to be greater than N. He expected to get approximately zero weights for components not existing in fact. Numerical experiments seem to support these ideas.

3. The ideas of Point 1 can also be realized by making use of the existence of a special differential equation for the components of a superposition of density functions (Thionet 1966). For instance, a superposition

$$
f(x) = p_1 \frac{e^{-\frac{(x-\alpha_1)^2}{2}}}{\sqrt{2\pi}} + p_2 \frac{e^{-\frac{(x-\alpha_2)^2}{2}}}{\sqrt{2\pi}}
$$

satisfies the differential equation

(4) $\qquad f''(x) + (2x - \alpha_1 - \alpha_2) f'(x) + [1 + (x - \alpha_1)(x - \alpha_2)] f(x) = 0.$

Knowing $f(x)$, $f'(x)$, $f''(x)$ at a sufficient number of points $x=x_\nu$ ($\nu=1, 2, ...$) we write (4) for $x=x_1, x=x_2, ...$. Thus we obtain a system of equations whence the unknowns α_1, α_2 can be determined. A new (obvious) system of equations will yield p_1 and p_2. Since $f(x_\nu)$ is known with an error only, and $f'(x_\nu)$, $f''(x_\nu)$ are calculated approximately from the "measured" $f(x_\nu)$-values, the result of the whole procedure may be loaded with a great error. The phrasing for the general case is obvious.

4. Let us consider the *decomposition of a superposition of Cauchy density functions*

(5)
$$k(x) = \sum_{k=1}^{N} p_k \frac{1}{\pi\beta_k} \frac{1}{1+(x-\alpha_k)^2/\beta_k^2}.$$
$$(\alpha_i \neq \alpha_j \ (i \neq j), \quad 0 < \beta_1 \leq \beta_2 \leq ... \leq \beta_N < \Lambda_2).$$

Here a method of decomposition can be given by means of the following theorem (Medgyessy 1961a p. 162).

Theorem 1. In order that an irreducible rational function of real coefficients $R(x)=\dfrac{P(x)}{Q(x)}$ *be representable in the form*
$$\frac{P(x)}{Q(x)} = \sum_{k=1}^{N} p_k \frac{1}{\pi\beta_k} \frac{1}{1+(x-\alpha_k)^2/\beta_k^2}$$
($p_k>0$, $\beta_k>0$, α_k *are real, identical pairs* (α_k, β_k) *do not occur*) *it is necessary and sufficient that*

1. $Q(x)$ *have simple, complex roots* z_k ($k=1, 2, ...$) *only;*

2. *the order of* $P(x)$ *is less than the order of* $Q(x)$;

3. $\operatorname{Re}\dfrac{P(z_k)}{Q'(z_k)} = 0 \qquad (k = 1, 2, ...)$;

4. $\operatorname{Im}\dfrac{P(z_k)}{Q'(z_k)} \cdot \operatorname{Im}\bar{z}_k > 0 \qquad (k = 1, 2, ...)$.

(*If it is not necessary for* p_k *to be positive,* 4. *can be dropped.*)

For the proof see Medgyessy (1961a) pp. 162—163.

Let some section of the graph of a superposition (5) be given. Assuming successively $N=2, 3, ...$, an approximation by rational functions of $k(x)$ is represented for every N, by means of ordinate values "measured" from the graph section. This can be done by certain special methods (see e.g. Solodovnikov 1952 pp. 219—234). Possessing the rational function approximation giving the smallest deviation from the starting superposition it will be broken into partial fractions belonging to type $\dfrac{T_k}{x^2+2U_k x+V_k}$ $(T_k>0)$ whence T_k, U_k, V_k may be determined. Then the parameters of the original superposition can be determined from them.

The delicate point of this method is, evidently, that a good approximation can be obtained for an N-value other than the true one. Thus this method can only be applied with great care.

Analogous considerations are valid if instead of the decomposition of a superposition of Cauchy density functions, the *decomposition of a superposition of density functions given by irreducible rational functions* is considered.

(C) *A method of analytical character*

Let us consider some physical, biological etc. systems, yielding at the moments t_j the theoretical "output" values

$$Q_j = \sum_{k=1}^{R} \beta_k^{(j)} y_k(t_j) \quad (j = 1, 2, ..., K);$$

where $\beta_k^{(j)}$ is given. Only the "measured" values b_j of Q_j are known. Let the functions $y_k(t)$ satisfy the system of differential equations,

$$\dot{y}_1(t) = f_1[y_1(t), ..., y_R(t)]$$
$$\dot{y}_2(t) = f_2[y_1(t), ..., y_R(t)]$$

$$\dot{y}_N(t) = f_N[y_1(t), ... y_R(t)] \quad (N \leq R)$$
$$\dot{y}_{N+1}(t) = 0$$

$$\dot{y}_R(t) = 0$$

under the initial conditions $y_k(0)=c_k$ $(k=1, 2, ..., R)$ and $f_v(\xi_1, ..., \xi_R)$ $(v=1, ..., N)$ is a given function. Now c_k is to be determined by means of the condition

$$\sum_{j=1}^{K} \left\{ \sum_{k=1}^{R} \beta_k^{(j)} y_k(t_j) - b_j \right\}^2 = \min.$$

This problem is, by some specialization, equivalent to the fitting of a superposition belonging to a given type (and having a given number of components) to the measured values b_j. Bellman, Kagiwada, Kalaba (1965), in considering this problem introduced a method called *"differential approximation"* for its solution (see also Bellman, Kalaba 1965). This method can also yield the solution of the problem of approximating a given ("measured") function by a certain superposition similarly as described in (B). By this step they consider the problem of decomposition as being solved.

(D) *A method due to R. Bellman*

Bellman (1960) gave the following method for the *decomposition of a superposition of exponential density functions*

$$f(x) = \begin{cases} \sum\limits_{k=1}^{N} p_k \dfrac{e^{-\frac{x}{\beta_k}}}{\beta_k} & (x > 0) \\ 0 & (x \leqq 0) \end{cases} \qquad (0 < \beta_1 < \beta_2 < \ldots < \beta_N < \Lambda_2).$$

It is as follows. Let us write, at a convenient scaling,

$$f(n) = f_n = \sum_{k=1}^{N} \frac{p_k}{\beta_k} (e^{-\beta_k})^n \qquad (n = 0, 1, \ldots),$$

i.e.

(6) $$f_n = \sum_{k=1}^{n} c_k B_k^n \qquad \left(c_k = \frac{p_k}{\beta_k}, \quad B_k = e^{-\beta_k} \right).$$

Let us now consider the sequence of Casorati determinants

$$C_r(i) = \begin{vmatrix} f_i & f_{i+1} & \cdots & f_{i+r} \\ f_{i+1} & f_{i+2} & \cdots & f_{i+r+1} \\ \vdots & \vdots & & \vdots \\ f_{i+r} & f_{i+r+1} & \cdots & f_{i+2r} \end{vmatrix} \qquad (i = 0, 1, \ldots; \quad r = 1, 2 \ldots).$$

Starting from f_n satisfying the difference equation

$$f_{i+N} = a_N f_i + a_{N-1} f_{i+1} + \ldots + a_1 f_{i+N-1} \qquad (i = 0, 1 \ldots; \quad v = 1, \ldots, N)$$

where a_v is an appropriate constant, one deduces that

$$C_r(i) \quad \begin{cases} \neq 0 & \text{for} \quad r < N \\ = 0 & \text{for} \quad r \geqq N \end{cases}$$

i.e. the index of the *first* determinant having the value zero in the sequence $C_0(i), C_1(i), \ldots$ gives the value of N. All this is approximately true if "measured" values of f_i are taken. The subsequent steps of decomposition are trivial.

Medgyessy (1966b) showed that any superposition of density functions $k(x) = \sum\limits_{k=1}^{N} p_k f(x, \beta_k)$ which is reducible for an appropriate sequence $x = a_n$ to the form

$$k(a_n) = \sum_{k=1}^{N} q_k \Phi(n) A_k^n,$$

where $\Phi(n)$ depends on n only, and q_k, A_k $(q_k > 0, A_k > 0)$ are functions of β_k, can be treated by this method, if $\dfrac{k(a_n)}{\Phi(n)}$ is written instead of f_n in (6). For example, in the *decomposition of a superposition of* (0) *symmetrical normal density functions*

$$k(x) = \sum_{k=1}^{N} p_k \frac{e^{-\frac{x^2}{4\beta_k}}}{\sqrt{4\pi\beta_k}}$$

let us put $a_n = \sqrt{n}$, $A_k = e^{-\frac{1}{4\beta_k}}$, $\Phi(n) \equiv 1$, $q_k = \dfrac{p_k}{\sqrt{4\pi\beta_k}}$. Then

$$k(\sqrt{n}) = \sum_{k=1}^{N} q_k A_k^n$$

belongs to the type of (6). This can be treated by the present method, as it can readily be seen.

In practice only "measured" values $\hat{f}_n$ exist for f_n, rendering this elegant method critical because the errors ε_n included in the "measured" values $\hat{f}_n$ of f_n ($\hat{f}_n = f_n + \varepsilon_n$) may cause great oscillations in the values

$$\hat{C}_r(i) = \begin{vmatrix} f_i + \varepsilon_i & f_{i+1} + \varepsilon_{i+1} & \cdots & f_{i+r} + \varepsilon_{i+r} \\ f_{i+1} + \varepsilon_{i+1} & f_{i+2} + \varepsilon_{i+2} & \cdots & f_{i+r+1} + \varepsilon_{i+r+1} \\ \vdots & \vdots & & \vdots \\ f_{i+r} + \varepsilon_{i+r} & f_{i+r+1} + \varepsilon_{i+r+1} & \cdots & f_{i+2r} + \varepsilon_{i+2r} \end{vmatrix}$$

$$(i = 0, 1, \ldots; \quad r = 1, 2, \ldots)$$

corresponding to $C_r(i)$. If e.g. $|f_n| \leq M$, $|\varepsilon_n| \leq \varepsilon$ for any n, Hadamard's inequality gives that

$$|\hat{C}_r(i) - C_r(i)| \leq (M\sqrt{r+1})^{r+1} \left[\left(1 + \frac{\varepsilon}{M}\right)^{r+1} - 1 \right]$$

increasing rapidly with r, and it hints at the possible deviation of $\hat{C}_r(i)$ from zero even if $C_r(i) = 0$ for some r. Thus the criterion may fail.

The present method was developed, with a reduction of the disturbing computational errors, by Parsons (1968, 1970), which has led to its increasing importance.

12*

(E) *Two special methods of the decomposition of a superposition of Gamma density functions of different order*

1. Let us consider the decomposition of a superposition of Gamma density functions of different order

$$k(x) = \begin{cases} \sum_{k=1}^{N} p_k \dfrac{x^{k-1}e^{-x}}{(k-1)!} & (x > 0) \\ 0 & (x \le 0). \end{cases}$$

N can be determined by considering the members of the sequence

$$\left\{ \int_0^\infty k(x) L_n(x)\, dx \right\} \qquad (n = 0, 1, \ldots)$$

where $L_n(x)$ is the n th Laguerre polynomial, as

$$\int_0^\infty k(x) L_n(x)\, dx \quad \begin{cases} = 0 & \text{if} \quad n \ge N \\ \ne 0 & \text{if} \quad n < N. \end{cases}$$

This follows from the orthogonal properties of the Laguerre polynomials. The index of the Laguerre polynomial for which the integral equals 0 *for the first time* yields N. p_k will be determined from an appropriate system of linear equations. In practice the vanishing of the integral would appear only approximately, making the method delicate.

2. An idea for the decomposition of a superposition of Gamma density functions of different order

$$k(x) = \begin{cases} \sum_{k=1}^{N} p_k \dfrac{B^k x^{k-1} e^{-Bx}}{(k-1)!} & (x > 0) \\ 0 & (x \le 0) \end{cases} \qquad (B > 0)$$

is as follows. Let us form the Laplace transform, denoted by $\mathbf{L}[k(x); s]$ of $k(x)$:

$$\mathbf{L}[k(x); s] = \sum_{k=1}^{N} p_k \frac{1}{(1 - s/B)^k}.$$

This becomes infinite for $s = B$. Then, B known, $k(x)e^{Bx}$ will be expanded as a polynomial; this will have the form $\sum_{v=0}^{N-1} a_v x^v$. Since $a_v = \dfrac{p_{v+1} B^{v+1}}{v!}$, p_v is also determined (Medgyessy 1954c).

Practically, the method is delicate and it will not be investigated further here.

(F) *A special method of the decomposition of a superposition of exponential density functions*

Let us consider the decomposition of a superposition of exponential density functions

$$k(x) = \begin{cases} \displaystyle\sum_{k=1}^{N} p_k \frac{e^{-\frac{x}{\beta_k}}}{\beta_k} & (x > 0) \\ 0 & (x \leqq 0) \end{cases} \qquad (0 < \beta_1 < \beta_2 < \ldots < \beta_N < \Lambda_2).$$

The Laplace transform of $k(x)$ is

$$\int_0^\infty e^{-sx} k(x)\,dx = G(s) = \sum_{k=1}^{N} \frac{p_k}{s + \beta_k} \qquad (\mathrm{Re}\, s \geqq \beta_1).$$

This is a meromorphic function of s defined in the complex s-plane, having simple poles at $0 = \beta_k$ ($k = 1, \ldots, N$). Then the Padé approximation of $G(s)$ constructed by means of "measured" values of $k(x)$ will reveal the places of these poles (e.g. the number of the roots of the denominator of the Padé approximation gives N), whence N, p_k, β_k can be obtained approximately (Blomer 1974).

(G) *A method related to the method of formant changing*

An interesting extension of the method of formant changing not included in the framework of Chapter III, is due to Medgyessy (1955a, 1961a pp. 107—121) and is the following. Let us consider a superposition

$$k(x) = \sum_{k=1}^{N} p_k\, f(x - \alpha_k, \beta_k) \qquad (\alpha_i \neq \alpha_j \quad (i \neq j); \quad \Lambda_2 < \beta_1 \leqq \beta_2 \leqq \ldots \leqq \beta_N < \Lambda_2)$$

in which β_k is a $(\Lambda_2, 0)$ monotone formant of $\mathscr{G}f(x - \alpha_k, \beta_k)$. Let us suppose that a test function

$$b(y, \lambda) = \sum_{k=1}^{N} p_k f(x - \alpha_k, \beta_k - \lambda) \qquad (0 < \lambda < \beta_1)$$

(cf. Ch. III, § 1, Subsection 1.1.1) *cannot* be constructed for $k(x)$; but let us suppose that a test function belonging to the type

$$b_1(y, \lambda) = \sum_{k=1}^{N} p_k \frac{1}{2}[f(x - \alpha_k, \beta_k - \lambda) + f(x - \alpha_k, \beta_k + \lambda)]$$

can be constructed, in which λ is a $(0, \beta_1)$ *monotone formant in the wider sense* of the graphs of the components of $b_1(y, \lambda)$ (cf. Ch. II, § 4). Then $b_1(y, \lambda)$ is suitable, in fact, for use as a test function.

To specify, let such superpositions

$$k^*(x) = \sum_{k=1}^{N} p_k s_A\, (x - \alpha_k, B, \beta_k)$$

be considered, in which the components are stable density functions belonging to the type $s_A(x, B, c)$ with characteristic function $\varphi(t) = e^{-c|t|^A\{1+i\beta\,\text{sgn}\,t\cdot\Omega(A)\}}$ $\left(c>0, 0<A \leq 2, |B| \leq 1, \Omega(A) = \tan\dfrac{\pi A}{2}\,(A \neq 1), \Omega(1) = 0\right)$ (cf. Ch. II, §8, Sections 3 and 4) with one of the parameter pairs (A, B):

A	1/2	1/2	1	3/2	3/2
B	1	-1	0	1	-1
A_2	-2	2	-1	-2	2
m	1	1	2	3	3

From Section 3 in Ch. II, § 8 it follows that in these cases, and *only* in these cases, A_2 and m being given in the above table,

$$\frac{\partial^2 s_A}{\partial c^2} - A_2 \frac{\partial^m s_A}{\partial x^m} = 0.$$

Evidently, the same partial differential equation holds for $0 \leq \lambda < \beta_1$, for the function

$$b^*(y, \lambda) = \sum_{k=1}^{N} p_k \frac{1}{2} [s_A(x-\alpha_k, \beta_k-\lambda) + s_A(x-\alpha_k, \beta_k+\lambda)] \qquad (0 \leq \lambda < \beta_1).$$

If λ is a $(0, \beta_1)$ monotone formant, in the wider sense, of the graphs of the components of $b^*(y, \lambda)$, $b^*(y, \lambda)$ will be used as a test function for the decomposition of $k^*(x)$. Now we have to determine $b^*(y, \lambda)$ with the aid of $k^*(x)$.

In our special case, Theorem 6 of Ch. II, § 8 can be applied with the restriction that λ can, except for the case $A=1$, $B=0$, vary from 0 up to $\dfrac{\beta_1}{\sqrt{2}}$ only. Then we have

$$b^*(y, \lambda) = \sum_{n=0}^{\infty} \frac{\lambda^{2n}}{(2n)!} A_2^n \frac{d^{mn}}{dx^{mn}} k^*(x).$$

In practice this formula is of little use because numerical differentiation is an *incorrect* problem (cf. Ch. V, § 1), and the enormous errors resulting make an estimation of the errors valueless. The extension e.g. of the regularization method (cf. Ch. V, § 1, Section 4) used in dealing with incorrect problems, to the above expansions would probably bring a change in this situation; at present this is, however, an unsolved **problem**.

Furthermore, methodological examples prove that the present method can be applied in certain cases. Such a case is that of the **decomposition of a superposition** of Cauchy density functions; here $A=1$, $B=0$ above, and writing $k_1^*(x)$ for the $k^*(x)$ figuring here

$$k_1^*(x) = \sum_{k=1}^{N} p_k \frac{1}{\pi \beta_k} \frac{1}{1+(x-\alpha_k)^2/\beta_k^2}$$

$$(\alpha_i \neq \alpha_j \ (i \neq j); \quad 0 < \beta_1 \leq \beta_2 \leq \ldots \leq \beta_N < \Lambda_2).$$

Let us consider the corresponding test function

$$b^*(y, \lambda) =$$

$$\sum_{k=1}^{N} p_k \frac{1}{2} \left[\frac{1}{\pi(\beta_k-\lambda)} \frac{1}{1+(x-\alpha_k)^2/(\beta_k-\lambda)^2} + \frac{1}{\pi(\beta_k+\lambda)} \frac{1}{1+(x-\alpha_k)^2/(\beta_k+\lambda)^2} \right].$$

We recall that in Ch. II, § 4 we saw that λ is a $(0, \beta_1)$ monotone formant, in the wider sense, of $\mathscr{G}b^*(y, \lambda)$. (Intuitively, if λ increases, then the graph of

$$\frac{1}{\pi(\beta_k-\lambda)} \frac{1}{1+(x-\alpha_k)^2/(\beta_k-\lambda)^2}$$ becomes "narrower" and higher: that of

$$\frac{1}{\pi(\beta_k+\lambda)} \frac{1}{1+(x-\alpha_k)^2/(\beta_k+\lambda)^2}$$ becomes "flatter"; in the end the whole picture is

a higher and, from a certain height on, "narrower" graph.) Since $m=2$, $\Lambda_2=-1$, here

$$b^*(y, \lambda) = \sum_{n=0}^{\infty} \frac{\lambda^{2n}}{(2n)!} (-1)^n \frac{d^{2n}}{dx^{2n}} k^*(x).$$

Let the approximation to $b^*(y, \lambda)$ be shown by a methodological

Example. In Fig. 6 taken from Medgyessy (1955a, 1961a p. 120), the solid line shows the graph of the superposition of Cauchy density functions

$$k_1^*(x) = \frac{1}{1+(x-0.5)^2} + \frac{1}{1+(x+0.5)^2}$$

i.e. $p_1=p_2=\pi$, $\beta_1=\beta_2=1$, The dotted line shows an approximation to the relevant test function $b^*(y, \lambda)$ for $\lambda=3/4$, constructed with equidistant "measured" values of $k_1^*(x)$ (the cutting off of certain decimal figures representing the "errors") and summing up to the 5th term in the expansion of $b^*(y, \lambda)$, the derivatives being computed from 11 mesh points of distance 0.5 by means of central difference formulas. The thin line shows the exact result

$$b^*(y, 3/4) = \frac{1}{0.25} \frac{1}{1+(y-0.5)^2/0.25^2} + \frac{1}{0.25} \frac{1}{1+(y+0.5)^2/0.25^2} +$$

$$\frac{1}{1.75} \frac{1}{1+(y-0.5)^2/1.75^2} + \frac{1}{1.75} \frac{1}{1+(y+0.5)^2/1.75^2}.$$

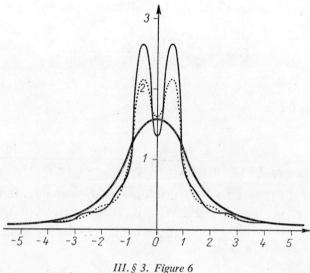

III. § 3. *Figure 6*

The approximation clearly shows the two peaks of the exact test function graph, revealing in this way the two components in the superposition $k_1^*(x)$, although $k_1^*(x)$ itself is unimodal. Cf. Example 2, (a) in Ch. III, § 1, Subsection 1.1.1, where the same superposition was decomposed by another method.

Evidently if the "measured" data of $k_1^*(x)$ are full of errors, such a satisfying result cannot be expected.

The described method also can be applied, in principle, to the **decomposition of a superposition of density functions of type V of Pearson.** Here $A=1/2$, $B=1$, $m=1$, $A_2=2$ and the general formula is

$$b^*(y, \lambda) = \sum_{n=0}^{\infty} \frac{\lambda^{2n}}{(2n)!} 2^n \frac{d^n}{dx^n} k^*(y) \qquad \left(0 < \lambda < \frac{\beta_1}{\sqrt{2}}\right)$$

if λ is a monotone formant in the wider sense of the graph of $b^*(y, \lambda)$ (Medgyessy 1961a pp. 110—121).

Since the above statements on the change of the places of the peaks, made in Example 2, (b) in Ch. III, § 1, Subsection 1.1.1, are valid also here, the numerical differentiation, as well as the other steps are delicate; thus the use of this procedure is dubious. Therefore, further details will not be given here.

Supplements and problems to Ch. III, § 3

1. A preliminary narrowing of the components by the $\mathcal{A}$-method (e.g. by means of the results in Ch. III, § 2, Subsection 1.1) would improve the graphical methods for the decomposition of superpositions.

2. Details of the application of special *coordinate papers* or plotting-in procedures can be found e.g. in the following works on mathematical statistics: Buchanan-Wollaston, Hodgeson (1929); Daeves (1933); Husung (1938); Knoll (1942); Daeves, Beckel (1948); Ramsthaler (1949); Daeves (1951); Hald (1952) pp. 155—158; Oka (1954); Graf, Henning (1960) pp. 66—70; Weichselberger (1961); Tanaka (1962).

3. An interesting *graphical procedure* consisting of manipulating the graph of the superposition, suitable for determining some parameters if certain other ones are given, was shown by Papoulis (1955). Its extension to a general method for decomposing superpositions of density functions is still an unsolved **problem**.

4. In applying an algebraic method, it is possible, of course, to take the "measured" ordinate values of the given superposition at sufficiently many given abscissae $x = x_k$ and write them into the superposition expression, thus obtaining a *system of non-linear equations* for the unknown parameters; some value of N should, naturally, be assumed. The solution of this may, however, be very complicated. A simple example of this idea will be considered later.

5. The system of non-linear equations described by Doetsch (1928) and solved by means of earlier results, is one of the *system of non-linear equations* that can be solved explicitly. It would be very useful to search for and collect such systems of equations; at present this is, however, an unsolved **problem**.

6. To the preceding it should be added that in the considerations in (B) 1. one might take some more general integrals $\int_{-\infty}^{\infty} \omega_i(x) k(x)\, dx$ instead of moments $\int_{-\infty}^{\infty} x^\varrho k(x)\, dx$ ($\varrho = 0, 1, 2, \ldots$) choosing the $\omega_\varrho(x)$ so that they will give a system of equations which can be solved more simply than that constructed with the moments. Finding such $\omega_\varrho(x)$ is in general an unsolved **problem**.

7. The moment method introduced by Doetsch (1928) as well as its modifications are mentioned in Sticker (1930*a*), (1930*b*); Schellenberg (1932). Here we mention in connection with the problem of approximation by means of normal density functions, that Stone (1927), Tricomi (1935), (1936) showed that, under certain conditions, an *arbitrary* function can be *approximated* on $(-\infty, \infty)$ by

linear combinations of Gaussian functions belonging to the type $e^{-\lambda_i(x-\mu_i)^2}$. This fact seems to diminish the value of the relevant method. It is interesting that an apparatus of stochastic character having in principle another objective, was built to carry out the computations needed in solving the system of equations (1) in Ch. III, § 3, (B) (Bernstein 1932). F. Bernstein used this method in discussing the same H_α doublet problem as considered by Doetsch (1928).

8. In connection with the use of the moment estimating methods of *mathematical statistics* and the investigations of Pearson (1894), we notice that, in contrast to some statements by Medgyessy (1961a), a thorough examination shows that the decomposition of a superposition of normal density functions *cannot* be assumed to have been K. Pearson's objective. His chief aim was the mentioned *approximation*. K. Pearson expected to obtain a characterization of the asymmetry of the considered empirical density function by approximating to it in such a way. The presence of a mixture population density function in the background did not motivate much the investigation. A similar aim is seen in Meszéna, Scherf (1960).

For simplifications and modifications of the considerations described by Pearson (1894) see Charlier (1905); Pearson, Lee (1909); Pearson (1915a); Crum (1923); Ščigolev (1924); Charlier, Wicksell (1925); Brown (1933); Burrau (1934); Pollard (1934); Strömgren (1934); Gottschalk (1948). With regard to the method based on the moments, newer techniques were published by Rao (1948), (1952) pp. 300—304; Preston (1953); relating to the preceding ones Cohen, Jr. (1967).

9. The system of equations appearing in de Prony's method occurs, among others, in the calculation of the unknown parameters of a superposition of normal density functions of the same variance, based on the moments and described above. As to de Prony's method and its improvements see de Prony (1795); Willers (1923) pp. 74—84; Runge, König (1924) pp. 231—235; Whittaker, Robinson (1949) pp. 369—371; Willers (1950) pp. 243—247; Householder (1950) pp. 28—32; Hildebrand (1956) pp. 378—379; Lánczos (1957) pp. 272—280; Buckingham (1957) pp. 329—333; Bellman, Kagiwada, Kalaba (1956) pp. 907—910; Bellman, Kalaba (1965).

10. All the methods in (B) 1.—3. have the disadvantage that the number N of the components has to be assumed. To find a definite *test* for deciding whether one particular value of N is better than any other is still an unsolved **problem.**

It is worth mentioning that Pearson (1894), using the method of moments in approximations, considered that approximation to be the best which gave an $(n+1)$th moment being the nearest to the $(n+1)$th sample moment, provided n moments were used in the approximation procedure. However, this principle was criticized several times.

The famous example of Lánczos (1957 pp. 272—280) is recalled here. He showed that the superposition of *three* exponential density functions

$$k(x) = 0.0951\,e^{-x} + 0.8607\,e^{-3x} + 1.5576\,e^{-5x}$$

could be approximated excellently by the similar superposition

$$f(x) = 2.202\,e^{-4.45x} + 0.305\,e^{-1.581x}$$

with only two components, if the initial data had a "measurement" error in the third decimal figure; it was impossible to give a test by means of which one could find a better approximating superposition of three components, even with certain improvements in the method. Thus, in practice, one must be very cautious in using the described methods and in interpreting their results if the experimental background gave no hint for accepting them.

11. The basic idea of the method given by Bellman (1960) described in Ch. III, § 3 (D) had already been considered by Kühnen (1909). The error of the method when applied in practice can be estimated only roughly. A *probabilistic* investigation of the error seems, however, to give better estimates, since determinants with random variables as elements have been studied extensively, e.g. in stochastic programming, in recent years.

12. The **decomposition of a superposition of χ density functions of different order** (in particular, Maxwell velocity "distributions") having the form

$$k(x) = \begin{cases} \displaystyle\sum_{k=1}^{N} p_k x^k e^{-x^2} & (x > 0) \\ 0 & (x \leqq 0) \end{cases}$$

can be solved in a way similar to that in Ch. III, § 3, (E) 1. by taking Hermite polynomials instead of Laguerre polynomials.

Generalizations with regard to other superpositions of density functions in which weight functions of orthogonal systems occur are obvious.

13. The approximating difference formulae for the derivatives needed in the Example in Ch. III, § 3 (G) and based on 11 mesh point values can be found in the works of Medgyessy (1955*a*, 1961*a* pp. 216—217). In principle an approximation to the test function $b^*(y, \lambda)$ which occurs there could also be obtained by solving numerically the trivial partial differential equation holding for $b^*(y, \lambda)$; it seems, however, that there is an *incorrect* problem in the background. Thus it might be possible to give a finite difference approximation to the equation in order to get a solution; we will not, however, be sure of obtaining a usable result because of the accumulation of the errors. Therefore, this idea will not be considered further.

14. If in the superposition to be decomposed, investigated in Ch. III, § 3 (G), $\mathbf{F}[f(x, \beta_k); t] = \psi(t)^{\beta_k}$ where $\psi(t)$ is known, then $k(x)$ and $b_1(y, \lambda)$ are connected

by a simple relation. Namely

$$\mathbf{F}\left[\frac{1}{2}\{f(x,\beta_k-\lambda)+f(x,\beta_k+\lambda)\};t\right]=$$

$$\frac{1}{2}[\psi(t)^{\beta_k-\lambda}+\psi(t)^{\beta_k+\lambda}]=\psi(t)^{\beta_k}\frac{1}{2}[\psi(t)^{-\lambda}+\psi(t)^{\lambda}].$$

Consequently

$$\mathbf{F}[k(x);t]=\frac{1}{2}[\psi(t)^{-\lambda}+\psi(t)^{\lambda}]\,\mathbf{F}[b_1(y,\lambda);t]\qquad(0\leq\lambda<\beta_1).$$

However, no integral equation can be written down for $b_1(x,\lambda)$ because, if $\psi(t)\neq e^{it}$,

$$\mathbf{F}^{-1}\left[\frac{1}{2}[\psi(t)^{-\lambda}+\psi(t)^{\lambda}];x\right]$$

does not exist. The presented special case belongs hereto. It is easily seen that $b_1(x,\lambda)$ can be represented, sometimes, by means of a function series formed with the derivatives of $k(x)$. Moreover, we are again faced with an incorrect problem; as a result the situation becomes more complicated, since $b_1(y,\lambda)$ is not an ideal test function at all. Thus in practice this method should be used cautiously; furthermore, the earlier considerations valid for the superpositions $k^*(x)$ (in Ch. III, § 3 (G)) seem to give more concrete results than the present statements. For a detailed treatment of the theme the reader is referred to Medgyessy (1961a) pp. 105—110.

An approximation to the special test function $b^*(y,\lambda)$ introduced in Ch. III, § 3 (G) for the special stable density function superpositions is likely to be obtained by a similar "noise filtering" technique as shown in Ch. V, § 2. It is an unsolved **problem**, however, whether this procedure is *correct* or not; so are the relevant analytical calculations.

The method detailed in Ch. III, § 3 (G) for the decomposition of the superposition

$$k^*(x)=\sum_{k=1}^{N}p_k\,s_A(x-\alpha_k,B,\beta_k)$$

can be utilized, by Point 7 in *Supplements and problems to* Ch. II, § 8, also in decomposing a superposition

$$k_1^*(x)=\sum_{k=1}^{N}p_k\int_{-\infty}^{\infty}s_A(x-y-\alpha_k,B,\beta_k)\,A(y)\,dy$$

where $A(y)$ is a density function not containing α_k,β_k.

15. In carrying out numerically many of the decomposition methods dealt with up to now, we have got the *approximation* to the test function $b(y)$ as an expression of the form $b^*(y)=\sum_{j=-m}^{m}a_j k(x+jh)$, where $k(x)$ is the superposition

to be decomposed and h is a constant. Then the question arises *whether such an expression can yield — also exactly — a test function*. If it can, we should obtain by this the least inexact — and, simultaneously, correct — decomposition procedure, because it would be free of neglections which cannot be circumvented in numerical methods (cf. the end of the *Example* in Ch. III, § 2).

A hint for such investigations may be given by the **decomposition of a superposition of Laplace density functions** in the following way.

Let

$$k(x) = \sum_{k=1}^{N} p_k \frac{e^{-\frac{|x - \alpha_k|}{\beta_k}}}{2\beta_k}$$

$$(\alpha_i \neq \alpha_j \ (i \neq j); \ 0 < \beta_1 \leq \beta_2 \leq \ldots \leq \beta_N < \Lambda_2)$$

be a superposition of Laplace density functions. The function

$$b(y) = \sum_{j=-1}^{1} a_j k(y + j\lambda)$$

$$\left((a_0 > 0, \quad a_1 = a_{-1} < 0, \quad \sum_{j=-1}^{1} a_j = 1 \right)$$

can be taken, according to Ch. III, § 1, for a test function corresponding to $k(x)$ Namely it is easily seen that if a_1 is arbitrary and $0 < \lambda < \xi \beta_1$, where $\xi = \mathrm{Ar\,ch}\left(1 + \frac{1}{2|a_1|}\right)$ and $a_0 = 1 - 2a_1$, then $b(y)$ is a superposition of (α_k) unimodal density functions $(k = 1, \ldots, N)$ and, if $\lambda \uparrow \xi \beta_1$, the variance of the kth component of $b(y)$ is less than the variance of the kth component of $k(x)$ (although it *does not* attain 0); consequently — *characterizing the "narrowness" of the components with their variances* — the components of $k(x)$ become "narrower" in the construction of $b(y)$, that is, they present themselves, possibly, more separated in the graph of $b(y)$. Essentially it is again the $\mathscr{A}$-method that appears here. It can be calculated easily that in the case of $\lambda = \xi \beta_1$, the variance of the first component of $b(y)$ equals the variance of the first component of $k(x)$ times $1 - |a_1| \, \mathrm{Ar\,ch}^2\left(1 + \frac{1}{2|a_1|}\right)$; further that a strong separation is obtained with great $|a_1|$ (and, consequently, small λ) values.

In practice — as β_1 is unknown — we construct, having fixed a_1, with increasing λ-values the graphs of the corresponding $b(y)$ and compare them. During the procedure we *subtract*, also here, certain quantities from the variances of the components of $k(x)$ similarly as it has been done, generally, in former methods (Medgyessy 1972b).

The general investigation of the described method is, at present, an unsolved **problem**. We notice that its basic idea can be applied also to the *decomposition of a superposition of* ch *density functions* as well as to the *decomposition of a superposition of Cauchy density functions* (Medgyessy 1973, 1974b).

IV. DECOMPOSITION OF SUPERPOSITIONS, II. DECOMPOSITION OF SUPERPOSITIONS OF DISCRETE DISTRIBUTIONS

§ 1. FIRST TYPE OF SUPERPOSITION. THE $\mathscr{C}$-METHOD

This chapter deals with Problem 2 of Ch. I, § 1, i.e. with the different methods of (numerical) decomposition of a superposition

$$\{k_n\} = \left\{ \sum_{k=1}^{N} p_k f_n(\gamma_k, \delta_k) \right\} \qquad (n = 0, 1, \dots)$$

in that practically fundamental case when **the superposition $\{k_n\}$ to be decomposed belongs to the type**

$$\{k_n\} = \left\{ \sum_{k=1}^{N} p_k f_{n-\gamma_k}(\delta_k) \right\} \qquad (n = 0, 1, \dots ; \; \Lambda_1 < \delta_1 \leqq \delta_2 \leqq \dots \leqq \delta_N < \Lambda_2)$$

where $\gamma_k = 0, 1, 2, \dots$, and $\{f_n(\delta_k)\}$ is an (A_2) (or (A_3)) distribution (cf. Ch. II, § 5) defined for $n = 0, 1, \dots, M$ (M can also be $+\infty$), $\{f_{n-\gamma_k}(\delta_k)\}$ is the same distribution with origin γ_k instead of 0; $\{f_n(\delta_k)\}$ is unimodal for all values of δ_k and **the abscissae of the peaks of the graphs of $\{f_{n-\gamma_k}(\delta_k)\}$ are all different natural numbers** ($k = 1, \dots, N$) (see Fig. 1).

The present paragraph deals with a method of decomposition of $\{k_n\}$ called the $\mathscr{C}$-*method*. Its basic idea is due to Medgyessy (1954a, 1954c, 1961a Ch. II).

Essentially, the method is as follows.

Let the so-called **test distribution**

$$\{b_m\} = \left\{ \sum_{k=1}^{N} p_k g_m(\gamma_k, \delta_k) \right\} \qquad (m = 0, 1, \dots)$$

correspond to the superposition $\{k_n\}$ represented by its graph. For this test distribution let the following *conditions* hold:

I. $\{g_m(\gamma_k, \delta_k)\}$ ($m = 0, 1, \dots$) is a unimodal transform of increased narrowness of $\{f_{n-\gamma_k}(\delta_k)\}$ ($k = 1, \dots, N$) (cf. Ch. II, § 7) and the implied well-defined analytical relation between $\{g_m(\gamma_k, \delta_k)\}$ and $\{f_{n-\gamma_k}(\delta_k)\}$ is independent of γ_k, δ_k.

II. The distance between the abscissae of the peaks of $\{g_m(\gamma_k, \delta_k)\}$ and $\{g_m(\gamma_l, \delta_l)\}$ ($k \neq l$) is at most less by 1 than the distance between the abscissae of the peaks of $\{f_{n-\gamma_k}(\delta_k)\}$ and $\{f_{n-\gamma_l}(\delta_l)\}$ (see Fig. 1).

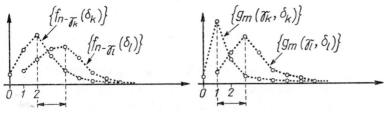

IV.§ 1. Figure 1

III. $\{b_m\}$ can be constructed by the help of $\{k_n\}$ only without the knowledge of $N, \gamma_k, p_k, \delta_k$ and utilizing, eventually, the analytical relation between $\{g_m(\gamma_n, \delta_n)\}$ and $\{f_{n-\gamma_k}(\delta_k)\}$.

Evidently $\{b_m\}$ is the superposition of the component distributions $\{g_m(\gamma_k, \delta_k)\}$.

In consequence of the character of $\{g_m(\gamma_k, \delta_k)\}$ we have that if we were able to plot the graph of $\{b_m\}$ then, *probably,* the graphs of the different components of it would present themselves *more separated* in this than in the graph of $\{k_n\}$ (see Fig. 2). If the separation is sufficiently great the graphs of the components will appear almost without disturbing one another. Then from the graph of $\{b_m\}$ the number N of the components — and, eventually, also the approximate value of certain parameters — can be determined.

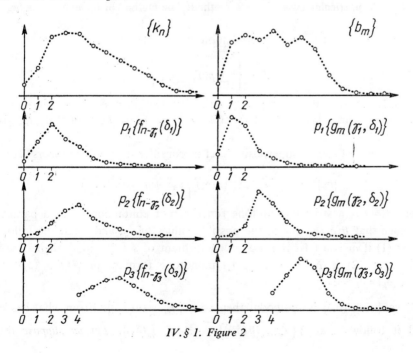

IV.§ 1. Figure 2

Let us consider the following steps:

(A) Determination of the described distribution $\{g_m(\gamma_k, \delta_k)\}$ with the help of $\{f_{n-\gamma_k}(\delta_k)\}$.

(B) Determination, with the help of (A), of the test distribution $\{b_m\}$, with the aid of $\{k_n\}$, serving as the basis for the decomposition method.

(C) On the basis of (B) and of the "*measured*" values of $\{k_n\}$, the elaboration of a numerical method for the construction of an *approximation* to the test distribution $\{b_m\}$.

For the graph $\mathscr{G}_2$ of this approximation to $\{b_m\}$ the statements concerning the evaluation of the graph of the "exact" test function are approximately true.

Definition 1. The construction of the graph $\mathscr{G}_2$ of the test distribution *approximation*, followed by the determination of the number and of the abscissae etc. of the peaks of $\mathscr{G}_2$ is called the $\mathscr{C}$-**method** yielding the (numerical) decomposition of the superposition $\{k_n\}$.

In the following, test distributions $\{b_m\}$ will be determined for certain types of superpositions $\{k_n\}$ utilizing, of course, the results of Ch. III, § 5.

1. First particular case of the $\mathscr{C}$-method: The method of formant changing

A very important type of superposition

$$\{k_n\} = \left\{ \sum_{k=1}^{N} p_k f_{n-\gamma_k}(\delta_k) \right\}$$

$(n = 0, 1, \ldots; \; \gamma_k = 0, 1, \ldots; \; \gamma_i \neq \gamma_j \; (i \neq j); \; \Lambda_1 < \delta_1 \leqq \delta_2 \leqq \ldots \leqq \delta_N < \Lambda_2)$ is that one in which δ_k is a (Λ_1, Λ_2) (or (Λ_2, Λ_1)) *monotone formant* of $\mathscr{G}\{f_{n-\gamma_k}(\delta_k)\}$ $(k = 1, \ldots, N)$.

In the case of such a superposition let us consider the set of numbers

$$\{b_m\} = \left\{ \sum_{k=1}^{N} p_k f_{m-\gamma_k}(\theta(\delta_k)) \right\} \qquad (m = 0, 1, \ldots)$$

where $\theta(\delta_k)$ is a strictly monotone function not containing δ_k as a parameter and such that $\theta(\delta_k)$ is one of the possible values of δ_k, and for it the following holds: (1) if δ_k is a (Λ_1, Λ_2) monotone formant of $\mathscr{G}\{f_{n-\gamma_k}(\delta_k)\}$, then $\Lambda_1 \leqq \delta_k < \theta(\delta_k) < \Lambda_2$; (2) if δ_k is a (Λ_2, Λ_1) monotone formant of $\mathscr{G}\{f_{n-\gamma_k}(\delta_k)\}$ then $\Lambda_1 < \theta(\delta_k) < \delta_k \leqq \Lambda_2$.

Since $\{f_{n-\gamma_k}(\delta_k)\}$ is unimodal then so is $\{f_{m-\gamma_k}(\theta(\delta_k))\}$. From what has been said it follows that $\mathscr{G}\{f_{m-\gamma_k}(\theta(\delta_k))\} \overset{w}{\prec} \mathscr{G}\{f_{n-\gamma_k}(\delta_k)\}$. *Let us suppose* that a well-defined analytical relation, independent of γ_k, δ_k holds between $\{f_{m-\gamma_k}(\theta(\delta_k))\}$ and $\{f_{n-\gamma_k}(\delta_k)\}$. Then Condition I in Ch. IV, § 1 is fulfilled for $\{f_m-\gamma_k(\theta(\delta_k))\}$ i.e. $\{f_m-\gamma_k(\theta(\delta_k))\}$ can be considered as the unimodal transform of increased narrowness of $\{f_{n-\gamma_k}(\delta_k)\}$ of the *same* type. If the abscissa of the peak of $\mathscr{G}\{f_n(\delta_k)\}$ is *independent* of δ_k, then Condition II in Ch. IV, § 1 is fulfilled for $\{f_{m-\gamma_k}(\theta(\delta_k))\}$; in the converse case *let us suppose* that this condition is fulfilled. Finally *let us suppose* that $\{b_m\}$ can be constructed with the help of $\{k_n\}$ without knowledge of N, p_k, γ_k, δ_k utilizing, eventually, the relation between $\{f_{m-\gamma_k}(\theta(\delta_k))\}$ and $\{f_{n-\gamma_k}(\delta_k)\}$. Thus Condition III in Ch. IV, § 1 is also fulfilled. Thus $\{b_m\}$ can be taken as the test distribution corresponding to $\{k_n\}$; and carrying out the steps (A), (B), and (C) in Ch. IV, § 1, the $\mathscr{C}$-method can be applied.

The nearer $\theta(\beta_k)$ is to Λ_2 (or Λ_1) the more peaked the components of $\{b_m\}$ are, and the better the decomposition can be expected to be.

In the case of the present type of superposition $\{k_n\}$, the $\mathscr{C}$-method will also be called — for obvious reasons — the **method of formant changing**. The idea first appeared in Medgyessy (1954*a*), (1954*c*), (1961*a* pp. 121—129).

In the following, concrete cases of the considered type will be treated.

1.1. The simple diminishing of the formant

The most frequent type of the superposition

$$\{k_n\} = \left\{ \sum_{k=1}^{N} p_k\, f_{n-\gamma_k}(\delta_k) \right\}$$

$(n = 0, 1, \ldots;\ \gamma_k = 0, 1, \ldots;\ \gamma_i \neq \gamma_j\, (i \neq j);\ \Lambda_1 < \delta_1 \leqq \delta_2 \leqq \ldots \leqq \delta_N < \Lambda_2)$

dealt with in Ch. IV, § 1 is that in which $0 < \delta_1 \leqq \ldots \leqq \delta_N < \Lambda_2$ where δ_k $(k = 1, \ldots, N)$ is a $(\Lambda_2, 0)$ monotone formant of $\mathscr{G}\{f_n(\delta_k)\}$ and $\theta(\delta_k) = \delta_k - \lambda$ where λ is a parameter and $0 \leqq \lambda < \delta_1$; this latter condition is *crucial*. This $\theta(\delta_k)$ satisfies the restriction put on $\theta(\delta_k)$ in Ch. IV, § 1, Section 1. Here $\mathscr{G}\{f_{m-\gamma_k}(\delta_k - \lambda)\} \gtrless \mathscr{G}\{f_{n-\gamma_k}(\delta_k)\}$. *Let us suppose* that a well-defined analytical relation, independent of α_k, β_k, holds between $\{f_{m-\gamma_k}(\delta_k - \lambda)\}$ and $\{f_{n-\gamma_k}(\delta_k)\}$. Then Condition I in Ch. IV, § 1 is fulfilled for $\{f_{m-\gamma_k}(\delta_k - \lambda)\}$ i.e. $\{f_{m-\gamma_k}(\delta_k - \lambda)\}$ can be considered as the unimodal transform of increased narrowness of $\{f_{n-\gamma_k}\gamma_k(\delta_k)\}$, of the *same* type. If the abscissa of the peak of $\mathscr{G}\{f_n(\delta_k)\}$ is *independent* of δ_k, then Condition II in Ch. IV, § 1 will be fulfilled, too, for $\{f_{m-\gamma_k}(\delta_k - \lambda)\}$; in the converse case *let us suppose* that this condition is fulfilled. Finally *let us suppose* that $\{b_m\}$ can be constructed with the help of $\{k_n\}$ and without knowledge of N, p_k, γ_k, δ_k, utilizing eventually the relation between $\{f_{m-\gamma_k}(\delta_k-\lambda)\}$ and $\{f_{n-\gamma_k}(\delta_k)\}$. Then Condition III in Ch. III, § 1 is also fulfilled. Writing now $\{b_m(\lambda)\}$ instead of $\{b_m\}$,

$$\{b_m(\lambda)\} = \left\{ \sum_{k=1}^{N} p_k f_{m-\gamma_k}(\delta_k - \lambda) \right\} \qquad (m = 0, 1, \ldots; \quad 0 < \lambda < \delta_1)$$

can be taken as the test distribution corresponding to the superposition

$$\{k_n\} = \left\{ \sum_{k=1}^{N} p_k f_{n-\gamma_k}(\delta_k) \right\}$$

and, carrying out the steps (A), (B), and (C) in Ch. IV, § 1, the $\mathscr{C}$-method can be applied.

In the present case the $\mathscr{C}$-method is called the **method of the (simple) diminishing of the formant**. (This, as a special case of the formant changing, appears earlier than the latter: Medgyessy (1954a), (1954c), (1961a pp. 121—127.)

The nearer λ lies to δ_1 the narrower the graphs of the components of $\{b_m\}$ are (moreover, the component belonging to δ_1 will tend to 1 at $n=\gamma_1$ and to 0 otherwise); consequently, the better a decomposition can be expected to be. In the case of a numerical decomposition δ_1 is not known; λ must, however, be less than this. Therefore it is advisable to carry out the numerical procedure for a whole sequence of λ-values, increasing monotonely from 0 onwards and then to compare the results. While the used λ-value is less than δ_1, the different components will appear more and more separated in the graphs of the results while the result obtained with a λ-value lying the nearest to δ_1 will show the best separation of the components of the test distribution. The character of numerical methods implies that if a λ-value is taken equal to or greater than δ_1, the increasing separation will not necessarily fail. Experience shows that for $\lambda=\delta_1$ the component with δ_1, though remaining unimodal at the first mode, *does not* tend to a degenerated distribution, and eventual negative values of $\{b_m(\lambda)\}$ at points where the $\{k_n\}$ is at about 0 do not influence either the significance of its peaks at other places or the utility of the whole procedure. At any rate a greater oscillation into negative ordinate values of $\{b_m(\lambda)\}$ points to the rejection of the used λ-values. It also has to be taken into account that little peaks may appear, as a consequence of the numeri-

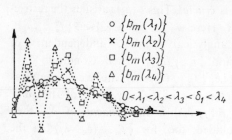

$$\circ \ \{b_m(\lambda_1)\}$$
$$\times \ \{b_m(\lambda_2)\}$$
$$\square \ \{b_m(\lambda_3)\}$$
$$\triangle \ \{b_m(\lambda_4)\}$$
$$0<\lambda_1<\lambda_2<\lambda_3<\delta_1<\lambda_4$$

IV.§ 1. 1.1. Figure 1

cal methods during the decomposition procedure. Hence we can see at the same time the avoiding of the disturbing effect of the condition that λ has to be less than δ_1 (cf. Fig. 1).

The construction of $\{b_m(\lambda)\}$ by means of $\{k_n\}$ is to be determined in every concrete case.

Example. **Decomposition of a superposition of binomial distributions.** Let

$$\{f_n(\delta_k^*)\} = \left\{\binom{M}{n}\delta_k^{*n}(1-\delta_k^*)^{M-n}\right\} \qquad (n = 0, 1, ..., M;\ 1/2 \leq \delta_k^* < 1)$$

i.e. a binomial distribution *with* $1/2 \leq \delta_k < 1$, and let us consider the superposition of binomial distributions

$$\{k_n^*\} = \left\{\sum_{k=1}^{N} p_k \binom{M}{m}\delta_k^{*n}(1-\delta_k^*)^{M-n}\right\} \qquad (n = 0, 1, ..., M)$$

where the components all start from $n=0$ and M ($M \geq 2$) is a not necessarily given integer *and* $M \geq 2N-1 \geq 2$ i.e. the superposition is identifiable (cf. Ch. II, § 3). Let us write δ_k^* in the form $\delta_k^* = e^{-\delta_k}$ ($0 < \delta_k \leq \log 2;\ k = 1, ..., N$) and in the following let us investigate the superposition

$$\{k_n\} = \left\{\sum_{k=1}^{N} p_k \binom{M}{n}e^{-\delta_k n}(1-e^{-\delta_k})^{M-n}\right\}$$

$$(n = 0, 1, ..., M;\ 0 < \delta_1 < \delta_2 < ... < \delta_N \leq \log 2).$$

We shall see that after the introduction of the new parameters δ_k this superposition can also be included in the group dealt with in the present discussion.

Let us suppose that the abscissae of the peaks of the graphs of the components are *different* integers. δ_k is a $(\log 2, 0)$ monotone formant of the kth component of $\{k_n\}$ (cf. Ch. II, § 5). Then

$$\{b_m(\mu)\} = \left\{\sum_{k=1}^{N} p_k \binom{M}{m}e^{-(\delta_k-\mu)m}(1-e^{-\delta_k-\mu})^{M-m}\right\}$$

$$(m = 0, 1, ..., M;\ 0 < \mu < \delta_1)$$

can be taken as a test distribution, as here the Conditions I, II and III in Ch. IV, § 1 are fulfilled. Namely the components of $\{k_n\}$ as well as the components of the test distribution are unimodal, and as $\delta_k-\mu$ is one possible value of the monotone formant δ_k,

$$\mathscr{G}\left\{\binom{M}{m}e^{-(\delta_k-\mu)m}(1-e^{-(\delta_k-\mu)})^{M-m}\right\} \overset{w}{\lesssim} \mathscr{G}\left\{\binom{M}{n}e^{-\delta_k n}(1-e^{-\delta_k})^{M-n}\right\}.$$

13*

The relation between the distance of the abscissae of the peaks of the graph of $\left\{\binom{M}{n}e^{-\delta_k n}(1-e^{-\delta_k})^{M-n}\right\}$ and $\left\{\binom{M}{n}e^{-\delta_l n}(1-e^{-\delta_l})^{M-n}\right\}$, i.e. $\left|[(M+1)e^{-\delta_k}]-[(M+1)e^{-\delta_l}]\right|$ ([ ] denotes the greatest contained integer), and the distance between the abscissae of the peaks of the graphs of $\left\{\binom{M}{m}e^{-(\delta_k-\mu)m}(1-e^{-(\delta_k-\mu)})^{M-m}\right\}$ and $\left\{\binom{M}{m}e^{-(\delta_l-\mu)m}(1-e^{-(\delta_l-\mu)})^{M-m}\right\}$ respectively, i.e. $\left|[(M+1)e^{-(\delta_k-\mu)}]-[(M+1)e^{-(\delta_l-\mu)}]\right|$, can hardly be established; it can be proved, however, that the first can be at most greater by 1 than the second.

As to the relation between the components of $\{k_n\}$ and the test distribution $\{b_m(\mu)\}$ we have

$$\mathbf{G}\left[\left\{\binom{M}{n}e^{-\delta_k n}(1-e^{-\delta_k})^{M-n}\right\}; z\right] = [e^{-\delta_k}z + (1-e^{-\delta_k})]^M = g_k(z),$$

$$\mathbf{G}\left[\left\{\binom{M}{m}e^{-(\delta_k-\mu)m}(1-e^{-(\delta_k-\mu)})^{M-m}\right\}; z\right] = [e^{-(\delta_k-\mu)}z + (1-e^{-(\delta_k-\mu)})]^M = h_k(z).$$

Then

$$h_k(z) = g_k(e^\mu z + 1 - e^\mu) \qquad (0 < \mu < \delta_k \leqq \log 2),$$

the new argument $e^\mu z + 1 - e^\mu$ introduced here being *independent of k*. Hence the relation between $\{b_m(\mu)\}$ and $\{k_n\}$ is

$$\mathbf{G}[b_m(\mu); z] = \mathbf{G}[\{k_n\}; 1 - e^\mu + e^\mu z]$$

i.e.

$$\sum_{m=0}^{\infty} b_m(\mu)z^m = \sum_{n=0}^{\infty} k_n[1 - e^\mu + e^\mu z]^n;$$

this relation is independent of N, p_k, δ_k. By comparing coefficients, we get

$$\{b_m(\mu)\} = \left\{e^{\mu\mu m} \sum_{\varrho=m}^{M} \binom{\varrho}{m}(1-e^\mu)^{\varrho-m}k_\varrho\right\} \qquad (m = 0, 1, ..., M; 0 \leqq \mu < \delta_1 < \log 2)$$

or,

$$b_m(\lambda) = \sum_{\varrho=m}^{M} e^{\mu m}\binom{\varrho}{m}(1-e^\mu)^{\varrho-m}k_\varrho \qquad (m = 0, 1, ...; \quad 0 < \mu < \delta_1).$$

These formulae were also obtained formerly in an elementary way (see Medgyessy 1954a, 1954c; as well as in another context: Medgyessy 1961a pp. 129—133), with the difference that there we wrote binomial distributions in the usual form, i.e. with δ_k^* instead of $e^{-\delta_k}$ ($1/2 \leqq \delta_1^* < \delta_2^* < ... < \delta_N^* < 1$) and λ instead of e^μ. The above expression for $b_m(\mu)$ was replaced there by $b_m^*(\lambda)$ with

$$b_m^*(\lambda) = \lambda^m \sum_{\varrho=m}^{M} \binom{\varrho}{m}(1-\lambda)^{\varrho-m}k_\varrho \qquad (1/2 \leqq \delta_N \leqq \lambda < 1).$$

In the case of a numerical decomposition, if only the "measured" value $\hat{k}_\varrho$ of k_ϱ is known and $|k_\varrho - \hat{k}_\varrho| \leq \varepsilon$ and only the $\hat{k}_\varrho$-values can be used, then the approximate test distribution members $\hat{b}_m(\mu)$ formed with the $\hat{k}_\varrho$ according to the expression of $b_m(\lambda)$ will be used. For the error $E = |b_m(\mu) - \hat{b}_m(\mu)|$ committed here we shall have

$$E \leq \varepsilon\, e^{\mu m} \sum_{\varrho=m}^{M} \binom{\varrho}{m} (e^\mu - 1)^{\varrho-m},$$

increasing with the increase of μ. The whole problem is *correct:* $E \to 0$ if $\varepsilon \to 0$. From the viewpoint of the application of the $\mathscr{C}$-method it is important that $b_m(\mu)$ can be used directly in numerical cases without approximation of analytical expressions.

The restriction $1/2 \leq \delta_1^* < \delta_2^* < ... < \delta_N^* < 1$ or, $0 < \delta_1 < \delta_2 < ... < \delta_N \leq \log 2$ is not essential in cases when, during the described decomposition procedure, from some $\mu = \mu_0$ (or $\lambda = \lambda_0$) on, all the values of the parameters $\delta_k - \mu$ (or $\delta_k^* \lambda$, as shown by a short consideration) lie between 0 and $\log 2$ (or $1/2$ and 1). In this case all parameters $\delta_k - \mu$ satisfy the conditions of applicability of the present method ($k = 1, ..., N$). Even in other cases some decomposition effect may be expected. This is very useful in dealing with practical decomposition problems.

Two *numerical examples* are given.

Example 1. Let us consider first a *methodological* example. In Fig. 2, the circles show the graph of the superposition of binomial distributions

$$\{k_n\} = \left\{ \binom{10}{n} 0.4^n \cdot 0.6^{10-n} + 2\binom{10}{n} 0.5^n \cdot 0.5^{10-n} \right\} \quad (n = 0, 1, ..., 10),$$

IV.§ 1. 1.1. *Figure 2*

i.e. $N=2$, $p_1=1$, $p_2=2$, $\delta_1=0.4$, $\delta_2=0.5$, $\gamma_1=\gamma_2=0$. The crosses show the graphs of the components. The test distribution $\{b_m^*(\lambda)\}$ was computed, for the decomposition of $\{k_n\}$, by means of the above formula with $\lambda=10/6$ (its graph is shown by the crosses in Fig. 2b). Here the graph of $\{k_n\}$ was plotted again (circles). The graph of $\{b_m^*(\lambda)\}$ reveals the two components, although the graph of $\{k_n\}$ is unimodal. Moreover some parameter values can also be read off from it (Medgyessy 1954a, 1961a pp. 132—133).

Example 2. In Fig. 3a the circles show the graph of Example 3 in Ch. I, § 1, that of the material distribution in a counter-current distribution experiment. It was supposed to be a superposition of binomial distributions. Two peaks are to be seen. Taking the measured ordinate values as $\{\hat{k}_n\}$ and computing the approximation to the test distribution $\{b_m^*(\lambda)\}$ with $\lambda=1.073$ and $\hat{k}_n$, using again the above formula we get the graph shown by crosses. More than three components appear; one of them is almost degenerate. Omitting this and repeating the decomposition with $\lambda=1.026$ on the remaining graph the graph shown by squares in Fig. 3b is obtained. For comparison the preceding phase was also plotted (crosses). At least three components appear in the superposition, and parameter values can also be read off. Here the practice justified the obtained data (Medgyessy 1954a, 1967b). Actually the condition $1/2\leq\delta_1^*<\delta_2^*<...<\delta_N^*<1$ was not satisfied in either example; but no major difficulty arose so its non-fulfilment might be neglected.

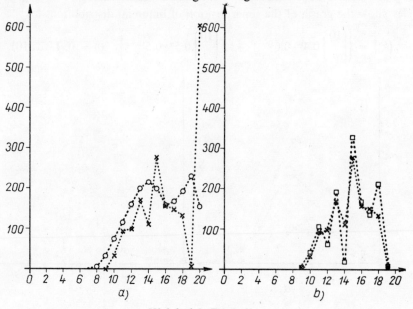

IV.§ 1. 1.1. Figure 3

1.1.1. A particular case

It is worthwhile considering separately as a special case of Ch. IV, § 1, Subsection 1.1 the superposition

$$\{k_n\} = \left\{ \sum_{k=1}^{N} p_k f_{n-\gamma_k}(\delta_k) \right\}$$

$(n = 0, 1, \ldots; \gamma_k = 0, 1, \ldots; \gamma_i \neq \gamma_j \ (i \neq j); \ 0 < \delta_1 \leq \delta_2 \leq \ldots \leq \delta_N < \Lambda_2)$

of the above type when the generating function of $\{f_n(\delta_k)\}$ has the form

(1) $$G[\{f_n(\delta_k)\}; z] = g(z, \delta_k) = h(z)^{\delta_k}$$

where δ_k is a $(\Lambda_2, 0)$ monotone formant of the graph of $\{f_n(\delta_k)\}$, and $h(z)$ is a generating function. (It appears first in Medgyessy 1954a, 1954c; see also Medgyessy 1961a pp. 72—76.)

For the test distribution

$$\{b_m(\lambda)\} = \left\{ \sum_{k=1}^{N} p_k f_{m-\gamma_k}(\delta_k - \lambda) \right\} \qquad (m = 0, 1, \ldots; \ 0 < \lambda < \delta_1)$$

introduced already above, there holds now from the conditions enumerated in Ch. IV, § 1, Subsection 1.1 that a well-defined analytical relation independent of γ_k, δ_k exists between $\{f_{n-\gamma_k}(\delta_k - \lambda)\}$ and $\{f_{n-\gamma_k}(\delta_k)\}$ because

$$G[\{f_{m-\gamma_k}(\delta_k - \lambda)\}; z] = z^{\gamma_k} h(z)^{\delta_k - \lambda} = \frac{z^{\gamma_k} h(z)^{\delta_k}}{h(z)^{\lambda}} = \frac{G[\{f_{n-\gamma_k}(\delta_k)\}; z]}{h(z)^{\lambda}}.$$

Condition I in Ch. IV, § 1 is fulfilled. Condition II in Ch. IV, § 1 is fulfilled, further Condition III in Ch. IV, § 1 is also fulfilled as in consequence of the last equality

(2) $$G[\{b_m(\lambda)\}; z] = \frac{1}{h(z)^{\lambda}} G[\{k_n\}; z].$$

Hence $b_m(\lambda)$ can be calculated by comparing the coefficients in the power series of the generating functions, and the relation between $\{b_m(\lambda)\}$ and $\{k_n\}$ is, by $G[\{f_n(\lambda); z] = h(z)^{\lambda}$,

(3) $$\{k_n\} = \left\{ \sum_{\varrho=0}^{n} f_{n-\varrho}(\lambda) b_\varrho(\lambda) \right\} \qquad (n = 0, 1, \ldots; \ 0 < \lambda < \delta_1);$$

that is from the implied system of linear equations the members of the test distribution, $b_m(\lambda)$, can be calculated recursively (Medgyessy 1954a, 1954c, 1961a pp. 72—76).

In the case of a numerical decomposition of $\{k_n\}$ the solution of (3) may be also an *incorrect* problem. Though it can be handled with certain methods in Ch. V, § 1, it will not be examined any further.

After all this, the $\mathscr{C}$-method can be applied.

We have the very important special case when

$$\frac{1}{h(z)^\lambda} = \sum_{v=0}^{\infty} h_v(\lambda)\, z^v.$$

Then, as it is easily seen,

$$\{b_m(\lambda)\} = \left\{ \sum_{\varrho=0}^{m} h_{m-\varrho}(\lambda)\, k_\varrho \right\}.$$

Now $\{b_m(\lambda)\}$, the test distribution corresponding to $\{k_n\}$, can be obtained *directly* in practice as

$$b_m(\lambda) = \sum_{\varrho=0}^{m} h_{m-\varrho}(\lambda)\, k_\varrho \qquad (m = 0, 1, \ldots;\quad 0 < \lambda < \delta_1).$$

Then the problem is, evidently, *correct*.

Let the error of the "measured" value $\hat{k}_n$ of k_n be $|k_n - \hat{k}_n| = \varepsilon_n$ and let $\varepsilon_n \leq \varepsilon$. Using the values $\hat{k}_n$, by the preceding relation, the values $\hat{b}_m(x)$ deviating from the exact values $b_m(\lambda)$ are obtained. The corresponding error E is, then,

$$E = |b_m(\lambda) - \hat{b}_m(\lambda)| \leq \sum_{\varrho=0}^{m} |k_\varrho\, h_{m-\varrho}(\lambda) - \hat{k}_\varrho\, h_{m-\varrho}(\lambda)|.$$

Thus

$$E \leq \varepsilon \sum_{\varrho=0}^{m} |h_\varrho(\lambda)|.$$

Example. **Decomposition of a superposition of Poisson distributions.** Let

$$\{f_n(\delta_k)\} = \left\{ e^{-\delta_k} \frac{\delta_k^n}{n!} \right\} \qquad (n = 0, 1, \ldots;\quad 0 < \delta_k < \Lambda_2)$$

i.e. a Poisson distribution. Let us consider the superposition of Poisson distributions

$$\{k_n\} = \left\{ \sum_{k=1}^{N} p_k e^{-\delta_k} \frac{\delta_k^{n-\gamma_k}}{(n-\gamma_k)!} \right\} \qquad (n = 0, 1 \ldots;\quad 0 < \delta_1 < \delta_2 < \ldots < \delta_N < \Lambda_2)$$

where the γ_k $(k = 1, \ldots, N)$ are natural numbers (if $n - \gamma_k < 0$, the corresponding member is defined to be identically 0). Let us suppose that the abscissae of the peaks are *different* natural numbers. This superposition belongs to the above type because δ_k is a $(\Lambda_2, 0)$ monotone formant of $\mathscr{G}\left\{ e^{-\delta_k} \frac{\delta_k^n}{n!} \right\}$ (cf. Ch. II,

§ 5); on the other hand, $\mathbf{G}\left[\left\{e^{-\delta_k}\dfrac{\delta_k^n}{n!}\right\}; z\right] = e^{\delta_k(z-1)} = (e^{z-1})^{\delta_k}$, moreover $\dfrac{1}{h(z)^\lambda} =$

$e^{-\lambda(z-1)} = e^\lambda \displaystyle\sum_{\varrho=0}^{\infty} \dfrac{(-\lambda)^\varrho}{\varrho!} z^\varrho$. Thus the test distribution yielding the basis of the decomposition is given by

$$\{b_m(\lambda)\} = \left\{ \sum_{\varrho=0}^{m} e^\lambda \dfrac{(-\lambda)^{m-\varrho}}{(m-\varrho)!} k_\varrho \right\}$$

i.e.

$$b_m(\lambda) = e^\lambda \sum_{\varrho=0}^{m} \dfrac{(-\lambda)^{m-\varrho}}{(m-\varrho)!} k_\varrho \qquad (m = 0, 1, \ldots; \quad 0 < \lambda < \delta_1).$$

This was obtained formerly in an elementary way (see Medgyessy 1954a, 1954c, and also in another context: Medgyessy 1961a pp. 77—79).

As to the conditions of the applicability of the method of the diminishing of the formant they are fulfilled here; e.g. as to Condition II in Ch. IV, § 1, the distance between the peaks of the graphs of $\{f_{n-\gamma_k}(\delta_k)\}$ and $\{f_{n-\gamma_l}(\delta_l)\}$ is $\left|([\delta_l]-\gamma_l)-([\delta_k]-\gamma_k)\right|$ (here [ ] denotes the greatest contained integer) and the distance between the peaks of the graphs of $\{f_{n-\gamma_k}(\delta_k-\lambda)\}$ and $\{f_{n-\gamma_l}(\delta_l-\lambda)\}$ is $\left|([\delta_l-\lambda]-\gamma_l)-([\delta_k-\lambda]-\gamma_k)\right|$; the latter is at most less by 1 than the former.

In the case of a numerical decomposition, if the absolute value of the error of k_ϱ is $\leq \varepsilon$, an estimate of the absolute value E_1 of the error in $b_m(\lambda)$ can be given by the aid of the above inequality for E:

$$E_1 \leq \varepsilon e^\lambda \sum_{\varrho=0}^{m} \dfrac{\lambda^{m-\varrho}}{(m-\varrho)!}.$$

This increases with the increase of λ (Medgyessy 1954a, 1954c, 1961a pp. 77—79, 1971b, 1972b).

No numerical example will be given here.

2. Second particular case of the $\mathscr{C}$-method

Let us consider the newer type of the above superposition

$$\{k_n\} = \left\{ \sum_{k=1}^{N} p_k f_{n-\gamma_k}(\delta_k) \right\}$$

$(n = 0, 1, \ldots; \; \gamma_k = 0, 1, \ldots; \; \gamma_i \neq \gamma_j \; (i \neq j); \; \Lambda_1 < \delta_1 \leq \delta_2 \leq \ldots \leq \delta_N < \Lambda_2)$

in which $f_n(\delta_k)$ is characterized by the fact that if

(1) $\mathbf{G}[\{f_n(\delta_k)\}; z] = g(z, \delta_k) \qquad (k = 1, \ldots, N),$

then 1. $g(z, 0) = 1$; 2. $g(z, \delta_k)$ is the generating function of an infinitely divisible distribution; and 3. $\dfrac{g(z, \delta_k)}{g(z, \lambda)}$ $(0 < \lambda < \delta_1)$, in which $g(z, \lambda)$ is, clearly, well-defined, is the generating function of some infinitely divisible, unimodal distribution $\{f_m^*(\delta_k, \lambda)\}$, where λ is a 0, δ_k *monotone formant of* $\mathscr{G}\{f_m(\delta_k, \lambda)\}$. This type was dealt with first by Medgyessy (1961a pp. 58—63).

Let us consider the set of numbers

$$\{b_m(\lambda)\} = \left\{ \sum_{k=1}^{N} p_k \, f_{m-\gamma_k}^*(\delta_k, \lambda) \right\} \qquad (m = 0, 1, \ldots).$$

For this, Condition I of Ch. IV, § 1 is fulfilled by the present assumption as $\{f_{m-\gamma_k}^*(\delta_k, \lambda)\}$ is also unimodal and λ is a $(0, \delta_k)$ monotone formant of $\mathscr{G}\{f_{m-\gamma_k}^*(\delta_k, \lambda)\}$; thus $\mathscr{G}\{f_{m-\gamma_k}^*(\delta_k, \lambda)\} \precsim \mathscr{G}\{f_{n-\gamma_k}(\delta_k)\}$, i.e. $\{f_{m-\gamma_k}^*(\delta_k, \lambda)\}$ can be considered as the unimodal transform of increased narrowness of $\{f_{n-\gamma_k}(\delta_k)\}$, and there holds the relation, independent of $\gamma_k, \delta_k,$

(2) $\qquad \mathbf{G}[\{f_{m-\gamma_k}^*(\delta_k, \lambda)\}; z] = z^{\gamma_k} \dfrac{g(z, \delta_k)}{g(z, \lambda)} = \dfrac{\mathbf{G}[\{f_{n-\gamma_k}(\delta_k)\}; z]}{g(z, \lambda)}$

between $\{f_{m-\gamma_k}^*(\delta_k, \lambda)\}$ and $\{f_{n-\gamma_k}(\delta_k)\}$. If the abscissa of the peak of $\mathscr{G}\{f_{m-\gamma_k}^*(\delta_k, \lambda)\}$ is independent of δ_k, then Condition II in Ch. IV, § 1 is also fulfilled. In the converse case the fulfilment of Condition II in Ch. IV, § 1 is *to be supposed* additionally. Condition III in Ch. IV, § 1 is, however, fulfilled, as by (2)

$$\mathbf{G}[\{b_m(\lambda)\}; z] = \frac{1}{g(z, \lambda)} \mathbf{G}[\{k_n\}; z];$$

hence $\{b_m(\lambda)\}$ can be calculated by comparing the coefficients of the power series of the generating functions. By

$$\mathbf{G}[b_m(\lambda); z] = \sum_{m=0}^{\infty} b_m(\lambda) \, z^m,$$

$$\mathbf{G}[\{f_n(\lambda)\}; z] = g(z, \lambda) = \sum_{n=0}^{\infty} f_n(\lambda) \, z^n$$

the relation between $\{b_m(\lambda)\}$ and $\{k_n\}$ is

(3) $\qquad \{k_n\} = \left\{ \sum_{\varrho=0}^{m} f_{n-\varrho}(\lambda) \, b_\varrho(\lambda) \right\} \qquad (n = 0, 1, \ldots; \quad 0 < \lambda < \delta_1),$

that is from the implied system of linear equations $b_\varrho(\lambda)$, the members of the test distribution can be calculated recursively (Medgyessy 1961a pp. 58—63).

The solution of this system of linear equations can also be an *incorrect* problem in a numerical case, to be handed by the method in Ch. V, § 1.

After all this the $\mathscr{C}$-method can be applied.

Often we observe the important fact that

$$\frac{1}{g(z, \lambda)} = \sum_{v=0}^{\infty} a_v(\lambda) z^v.$$

Then, as it is easily seen,

(4) $\qquad \{b_m(\lambda)\} = \left\{ \sum_{\varrho=0}^{m} a_{m-\varrho}(\lambda) k_\varrho \right\} \qquad (m = 0, 1, ..., \quad 0 < \lambda < \delta_1).$

Now $\{b_m(\lambda)\}$, the test distribution corresponding to $\{k_n\}$, can be obtained *directly* in practice as

$$b_m(\lambda) = \sum_{\varrho=0}^{m} a_{m-\varrho}(\lambda) k_\varrho \qquad (m = 0, 1, ..., \quad 0 < \lambda < \delta_1).$$

Then the problem is, evidently, *correct*.

Let the error of the "measured" value $\hat{k}_n$ of k_n be $|k_n - \hat{k}_n| = \varepsilon_n$ and let $|\varepsilon_n| \leqq \varepsilon$. On the basis of the values $\hat{k}_n$ we get, by the preceding relation, certain values $\hat{b}_m(\lambda)$ instead of $b_m(\lambda)$. The error E, committed here, is

$$E = |b_m(\lambda) - \hat{b}_m(\lambda)| = \left| \sum_{\varrho=0}^{m} a_{m-\varrho}(\lambda) k_\varrho - a_{m-\varrho}(\lambda) \hat{k}_\varrho \right|;$$

finally

$$E \leqq \varepsilon \sum_{\varrho=0}^{m} |a_\varrho(\lambda)|.$$

The superpositions dealt with in Ch. IV, § 1, Subsection 1.1.1 are, in fact, particular cases of the present type: they could, however, be included in a simpler category.

Example. **Decomposition of a superposition of negative binomial distributions.** Let

$$\{f_n(\delta_k)\} = \left\{ \binom{R-1+n}{R-1} (1-\delta_k)^R \delta_k^n \right\} \qquad (n = 0, 1, ...; \quad 0 < \delta_k < 1; \quad R = 1, 2, ...)$$

i.e. a negative binomial distribution of order R (R is given). Let us consider the superposition of negative binomial distributions

$$\{k_n\} = \left\{ \sum_{k=1}^{N} p_k \binom{R-1+n}{R-1} (1-\delta_k)^R \delta_k^{n-\gamma_k} \right\}$$

$$(n = 0, 1, ...; \quad 0 < \delta_1 < \delta_2 < ... < \delta_N < 1)$$

where γ_k ($k=1, ..., N$) is a natural number and if $n-\gamma_k< 0$, the corresponding member is defined as 0. *Let us suppose* that the abscissae of the peaks of the components are **different** natural numbers; by this, the *superposition of geometric distributions* ($R=1$) *is excluded* (cf. Ch. IV, § 1, Section 3). This superposition belongs to the above type because its components are unimodal, and the distribution is infinitely divisible (Gnedenko, Kolmogorov 1954 pp. 73—75) and, writing

$$\mathbf{G}\left[\left\{\binom{R-1+n}{R-1}(1-\delta_k)^R\delta_k^n\right\}; z\right] = \left(\frac{1-\delta_k}{1-\delta_kz}\right)^R = g(z, \lambda),$$

$g(z, 0) = 1$; further

$$\frac{g(z, \delta_k)}{g(z, \lambda)} = \left(\frac{1-\delta_k}{1-\lambda}\right)^R\left(\frac{1-\lambda z}{1-\delta_kz}\right)^R \qquad (0 \leqq \lambda \leqq \delta_k < 1),$$

as proved in Ch. II, § 2, is the generating function of the distribution

$$\{f_m^*(\delta_k, \lambda)\} = \left\{\left(\frac{1-\delta_k}{1-\lambda}\right)^R\left(\frac{\lambda}{\delta_k}\right)^R\left[1+\delta_k^m\sum_{v=1}^{R}\binom{R}{v}\left(\frac{\delta_k}{\lambda}-1\right)^v\binom{v-1+m}{v-1}\right]\right\}$$

which is infinitely divisible (cf. Ch. II, § 8, Section 2) and in addition unimoda (cf. Ch. II, § 5). λ is a $(0, \delta_k)$ monotone formant of $\mathscr{G}\{f_m^*(\delta_k, \lambda)\}$ (cf. Ch. II, § 5).

Let us consider here the set of numbers

$$\{b_m(\lambda)\} = \left\{\sum_{k=1}^{N}p_k f_{m-\gamma_k}^*(\delta_k, \lambda)\right\} \qquad (m = 0, 1, ...).$$

As to Condition II of Ch. IV, § 1 we *cannot follow* here the change of the distances between the abscissae of the peaks because of analytical difficulties. In practice this is, however, not essential either; we generally investigate only whether the graph of the test distribution shows separated peaks or not. Therefore *we accept* the above method also here. Having

$$\frac{1}{g(z, \lambda)} = \left(\frac{1-\lambda z}{1-\lambda}\right)^R = \frac{1}{(1-\lambda)^R}\sum_{\varrho=0}^{R}\binom{R}{\varrho}(-\lambda)^\varrho z^\varrho,$$

the distribution accepted for test distribution is given by

$$\{b_m(\lambda)\} = \left\{\sum_{\varrho=0}^{m}\frac{1}{(1-\lambda)^\varrho}\binom{R}{\varrho}(-\lambda)^{m-\varrho}k_\varrho\right\},$$

i.e.

$$b_m(\lambda) = \frac{1}{(1-\lambda)^R}\sum_{\varrho=0}^{m}\binom{R}{\varrho}(-\lambda)^{m-\varrho}k_\varrho \qquad (m = 0, 1, ...; 0 \leqq \lambda \leqq \delta_1 < 1)$$

found formerly in another way (Medgyessy 1961a pp. 66—71). If the procedure is successful we can say, nevertheless, that the relation between $\{b_m(\lambda)\}$ and $\{k_n\}$ is of the required type, even if we cannot speak here exactly of the application of the $\mathscr{C}$-method.

It *should be understood* here that the peaks of the components $\{f^*_{m-\gamma_k}(\delta_k, \lambda)\}$ are all *different* natural numbers. This is not really essential. The graph of the test distribution can be used only if it presents separate peaks; if this is not so, the individual behaviour of the graphs of the components is uninteresting.

In the case of a numerical decomposition, if the absolute value of the error of k_ϱ is $\leq \varepsilon$, then an estimate of the absolute value E_2 of the error of $b_m(\lambda)$ is given, by the aid of the above inequality for E:

$$E_2 \leq \frac{\varepsilon}{(1-\lambda)^R} \sum_{\varrho=0}^{m} \binom{R}{m-\varrho} \lambda^{m-\varrho}.$$

This increases with the increase of λ.

No numerical example will be given here.

3. Decomposition of superpositions of the first type by means of the transformation of the superposition

A number of well-known types of superposition do not belong to the superpositions of (A_2) (or (A_3)) distributions dealt with above. Thus sometimes a superposition of such distributions

$$\{k_n\} = \left\{ \sum_{k=1}^{N} p_k\, f_{n-\gamma_k}(\delta_k) \right\}$$

$$(n = 0, 1, \ldots; \ \gamma_i \neq \gamma_j \ (i \neq j); \ \Lambda_1 < \delta_1 \leq \delta_2 \leq \ldots \leq \delta_N < \Lambda_2)$$

cannot be decomposed by the described methods. However, by a convenient linear transformation it can be transformed into a superposition of discrete distributions $\{k^*_m\}$ $(m=0, 1, \ldots)$ that can already be decomposed by one of the methods considered hitherto — or, alternatively, the introduction of a new method of decomposition becomes possible. An example is now given of such a transformation (not influencing the numerical treatment of the decomposition). The ideas given here can of course be applied elsewhere.

Example. Let us consider the problem of the **decomposition of a superposition of (0) unimodal geometrical distributions**. Let

$$\{k_n\} = \left\{ \sum_{k=1}^{N} p_k(1-\delta_k)\delta_k^n \right\} \qquad (n = 0, 1, \ldots; \ 1/2 \leq \delta_1 < \delta_2 < \ldots < \delta_N < 1).$$

For any value of δ_k $(k=1, ..., N)$, the components are (0) *unimodal* geometrical distributions. Theoretically the second particular case of the $\mathscr{C}$-method (cf. Ch. IV, § 1, Section 2), more precisely the method of the decomposition of a superposition of negative binomial distributions, would be applicable here; however, all the components of the corresponding test distribution would remain (0) unimodal in the test distribution and the above method of decomposition of negative binomial distributions cannot be applied.

Let us suppose that a linear transformation $\sum\limits_{n=0} K_n^{(m)}(1-\delta_k)\delta_k^n$ represented by a matrix $(K_n^{(m)})$ $(n=0, 1, ...; m=0, 1, ...)$ and applied to the elements of $(1-\delta_k)\delta_k^n$ exists for which

$$(1) \qquad \sum_{n=0} K_n^{(m)}(1-\delta_k)\delta_k^n = \binom{M}{m}\delta_k^n(1-\delta_k)^{M-n} \qquad (m = 0, 1, ... M),$$

where the upper limit of the summation, as well as M and the explicit form of $K_n^{(m)}$, are to be investigated or fixed later. ($\{K_n^{(m)}\}$ $(m=0, 1, ...)$ need not be a distribution.) Let us apply this transformation to $\{k_n\}$. We would have

$$\left\{\sum_{n=0} K_n^{(m)}k_n\right\} = \left\{\sum_{k=1}^{N} p_k\binom{M}{m}\delta_k^m(1-\delta_k)^{M-m}\right\}$$

i.e. a *superposition of binomial distributions* of parameter δ_k. If M is sufficiently large then the method of the decomposition of binomial distributions (cf. Ch. IV, § 1, Subsection 1.1) will be applicable here; i.e. by the above transformation we get a new superposition of *non- (0) unimodal* distributions to be treated by a former method. Eventually *if M is large enough, the graph of* $\left\{\sum\limits_{n=0} K_n^{(m)}k_n\right\}$ *shows separated peaks* and it can thus be used as a test distribution in the decomposition. This is analogous with the method described in Ch. III, § 2. Refinements of neither will, however, be considered here.

Thus only $K_n^{(m)}$ has to be determined. Several integer values should be given to M, in general, expecting that one or other value of M will yield the most effective test distribution (a superposition of binomial distributions). We have from (1)

$$\sum_{n=0} \delta_k^n K_n^{(m)} = \binom{M}{m}\delta_k^m(1-\delta_k)^{M-m-1} = \sum_{\varrho=0}^{M-m-1}\binom{M}{m}\binom{M-m-1}{\varrho}(-1)^\varrho\delta_k^{\varrho+m} =$$

$$\sum_{n=m}^{M-1}\binom{M}{m}\binom{M-m-1}{n-m}(-1)^{n-m}\delta_k^n \qquad (m = 0, 1, ..., M-1)$$

i.e. the summation goes up to $M-1$ and

$$K_n^{(m)} = \begin{cases} \binom{M}{m}\binom{M-m-1}{n-m}(-1)^{n-m} & (n = m, m+1, ..., M-1; m = 0, 1, ..., -1) \\ 0 & (n < m). \end{cases}$$

For $m=M$ no suitable transformation exists, however. *Allowing* this deficiency, and fixing now also the upper limit of the summation we have

$$\{k_m^*\} = \left\{ \sum_{n=0}^{M-1} K_n^{(m)} k_n \right\} = \left\{ \sum_{n=m}^{M-1} \binom{M}{m}\binom{M-m-1}{n-m}(-1)^{n-m} k_n \right\} \quad (m = 0, 1, ..., M-1).$$

We notice that $\{K_n^{(m)}\}$ $(m=0, ..., M-1)$ does not consist of members of a discrete distribution.

As has been mentioned a *finite* number of k_n values was needed, and the transformation can be performed *directly* in practice. The exclusion of the case where $m=M$ may cause difficulties; however, in general it is not essential. The error estimation is trivial (Medgyessy 1971b, 1972b).

It would be possible to extend the present idea to other types of superpositions (e.g. to a *superposition of negative binomial distributions*) and the formulae of the decomposition of a superposition of binomial distributions could be applied to the above transformation in order to get a test distribution (a test superposition of binomial distributions) directly from $\{k_n\}$. However, as a way of working with test distributions is always less effective than the work with test functions, this means of extending decomposition methods will not be treated any further and the next paragraph will deal with another general method.

Supplements and problems to Ch. IV, § 1

1. Obviously a decomposition procedure can also be applied to decide whether a discrete distribution is a *superposition of discrete distributions* of given type or not.

2. Behind the concept "narrower", occurring in the definition of the narrowness-increasing transformation, the "compression" of the relevant graphs of distributions hides less than the "compression" of the graphs of the density functions in the case of density functions: the graphs of distributions do not change "steadily" with the change of the monotone formant. The peaks of the graphs of distributions cannot move steadily either. Hence it follows that in the graphs of test distributions we may never expect such a degree of separation of the components

in the course of the decomposition procedure as in the graphs of test functions. This disadvantage cannot be eliminated; it does not, however, greatly influence the applicability of the procedures in general.

3. If in a decomposition procedure the number N of the components has already been determined, the determination of the other parameters is equivalent to the *solution of a certain system of nonlinear algebraic equations,* or to a problem of fitting; both will be mentioned in Ch. IV, § 3.

4. If during the decomposition procedure the graph of some component changes into a very sharp peak, then it can be subtracted graphically from the initial graph. After the elimination of a component in this way we can *iterate* the decomposition procedure on what remains. We will not, however, generally consider such iterations in the present book.

5. The $\mathscr{C}$-method, and also the other methods to be treated later, can be applied evidently also in the case when N is *known.* Then they serve as a special procedure for the determination of (some of) the unknown parameters of the initial superposition. Publications on decomposition methods (Medgyessy 1954a, 1961a pp. 129—133) pointed out the use of the method in this sense.

6. With the simple diminishing of the formant, the separation of the components of the test distribution will not be "steady" but "oscillating", for the shapes of the component graphs do not change "steadily" either. However, this is unavoidable. Further, we cannot eliminate the fact that sometimes we do not know in advance whether the abscissae of the peaks of the superposition to be decomposed are different natural numbers, i.e. whether we will get a decomposition at all. Much will be seen only after having performed the decomposition procedure.

7. It can easily be shown that the described method for the *decomposition of a superposition of binomial distributions* is also applicable if the superposition has the form

$$\{k_n\} = \left\{ \sum_{k=1}^{N} p_k \binom{M_k}{n} e^{-\delta_k n}(1-e^{-\delta_k})^{M-n} \right\}$$

$$(n = 0, ..., 1; \; (M_i, \delta_i) \neq (M_j, \delta_j) \; (i \neq j); \; 0 < \delta_1 < \delta_2 < ... < \delta_N < \log 2)$$

(i.e. the parameters M_k $(k=1, ..., N)$ are *different* and only $\max_{1 \leq k \leq N} M_k$ is given). In this case the formula for the test distribution is the same as in the case of $M_1 = ... = M_N = M$, only the summation is to be taken up to $\max_{1 \leq k \leq N} M_k$; i.e.

$$\{b_m(\lambda)\} = \left\{ e^{\mu m} \sum_{\varrho=m}^{\max_{1 \leq k \leq N} M_k} \binom{\varrho}{m} (1-e^{\mu})^{\varrho-m} k_\varrho \right\} \qquad (0 \leq \mu < \delta_1)$$

(Medgyessy 1954a, 1954c, 1961a pp. 129—133). This, however, rarely appears in practice.

8. The substitution $e^{\mu} z + 1 - e^{\mu}$ occurring in the *decomposition of a superposition of binomial distributions* can be considered as the generating function of a "distribution" of members e^{μ}, $1 - e^{\mu}$; one of the members of this "**distribution**" is **negative**, and its "**variance**" calculated in the same way as in the case of a true distribution, is **negative**. Thus the mentioned substitution can be regarded as the substitution of the generating function of this "distribution" into a (true) generating function. We recall that in the case of true generating functions this procedure again gives a generating function. Thus here we meet the introduction of a distribution of negative "variance" for technical purposes; this procedure *diminishes* a true variance and *renders* the graph *narrower* (cf. *Supplements and problems to* Ch. III, § 1, 29). However, we will not deal further with the exploitation of this phenomenon in the decomposition of superpositions here; it is still an unsolved **problem** (cf. Medgyessy 1966c).

9. Evidently the method in Ch. IV, § 1, Subsection 1.1.1 can be generalized trivially in cases where

$$\mathbf{G}[\{f_n(\delta_k)\}; z] = h_1(z) h(z)^{\delta_k};$$

here $h_1(z)$ is some generating function not depending on δ_k and the conditions are modified conveniently.

10. At the *decomposition of a superposition of Poisson distributions*, $\dfrac{1}{h(z)^{\lambda}} = e^{-\lambda(z-1)}$ can also be considered as the generating function of a "Poisson distribution with negative parameter"; certain members of this "**distribution**" are **negative** and possess — defining the variance in the same way as in the case of true distributions — a **negative** "**variance**". Then the test distribution can be defined as the convolution of the superposition to be decomposed and of the above distribution of negative "variance" and this convolution will *diminish* a true variance and make the graph of the distribution "narrower". However, we will not deal here with the above described possibility of introducing a distribution of negative "variance" for technical reasons and with the exploitation of this idea in decomposing superpositions as they present as yet unsolved **problems** (cf. Medgyessy 1966c).

11. The condition that the *quotient* of two *generating functions* $\dfrac{g(z, \delta_k)}{g(z, \lambda)}$ $(\Lambda_1 < \delta_1 \leqq \delta_2 \leqq \ldots \leqq \delta_N < \Lambda_2; 0 \leqq \lambda < \delta_1)$ $(g(z, \delta_k)$ $(k = 1, \ldots, N)$ is the generating function of an infinitely divisible distribution) is again a generating function of

an infinitely divisible distribution, can be stated equivalently in the form of a condition imposed on the distribution figuring in the canonical representation of $g(z, \delta_k)$ (cf. Ch. I, § 8, Section 2).

12. The *second particular case of the $\mathscr{C}$-method* can, evidently, be applied if only the *generating function* of the superposition to be decomposed is known.

13. In applying the described particular case of the $\mathscr{C}$-method, the separation of the components will not be steady; it will have "oscillations" because of the discrete character of the distribution. This cannot, however, be eliminated. As with the lack of information related to the abscissae of the peaks as well as to the change of the distances between the peaks of the components in the test distribution, this can, in general, vitiate the correctness of the decomposition. In practice only the graph of the test distribution shows whether a separation of the components takes place or not.

14. In the *decomposition of a superposition of negative binomial distributions* we met the function

$$\frac{1}{g(z, \lambda)} = \left(\frac{1 - \lambda z}{1 - \lambda}\right)^R = \left(\frac{1}{1 - \lambda} - \frac{1}{1 - \lambda} z\right)^R,$$

the reciprocal of a generating function. This can also be considered as the generating function of a "**distribution**" possessing also **negative** members and a **negative** "**variance**" (cf. Point 8 above). Then the corresponding test distribution can be interpreted as the convolution of the superposition to be decomposed with the above distribution of negative "variance"; if we apply it, a true variance will be *diminished* and the relevant graph made *narrower*. Here we work again with a distribution of negative" variance", for technical reasons. However, the possibility of introducing such a distribution and the exploitation of this technique in order to be used in decomposition problems will not be dealt with further; they are still unsolved **problems** (cf. Medgyessy 1966c).

15. In the case of a decomposition carried out by means of the transformation of the superposition, theoretically it is possible to unify this transformation and the operation to be applied after the action of that transformation, which produces the decomposition. However, this leads to very complicated formulae; thus such calculations will not be carried out.

16. The *transformation* of a discrete distribution into another without knowledge of the parameters is, in many cases, equivalent to the *mixing* of distributions, belonging to the type

$$\{k_m\} = \left\{\sum_{n=0}^{\infty} K_n^{(m)} w_n\right\} \qquad (m = 0, 1, \dots)$$

where $\{K_n^{(m)}\}$ $(m=0, 1, ...)$ is a distribution containing the parameter n and $\{w_n\}$ $(n=0, 1, ...)$ is a discrete distribution. Although in our problem $\{w_n\}$ rather than $\{K_n^{(m)}\}$ was the primary distribution, the situation is formally the same. In general $\{K_n^{(m)}\}$ need not be supposed to be a discrete distribution.

However, the determination of such a transformation, in general the **determination of the unknown matrix of a linear transformation** (where this matrix must not contain unknown parameters), and even the proof of its existence, is an unsolved **problem** (cf. Medgyessy 1961a p. 156)

17. In several cases methods simpler than those described may be found to obtain an appropriate test distribution. For instance, let us consider the *decomposition of a superposition of* (0) *unimodal Poisson distributions*, where the superposition is

$$\{k_n\} = \left\{ \sum_{k=1}^{N} p_k e^{-\delta_k} \frac{\delta_k^n}{n!} \right\} \qquad (n = 0, 1, ...; \quad 0 < \delta_1 < \delta_2 < ... < \delta_N < 1);$$

by $0 < \delta_k < 1$ $(k = 1, ..., N)$ all components are (0) *unimodal* Poisson distributions. Then let the transformation that is to be applied on the components of $\{k_n\}$ be defined by the rule that the mth member of the set obtained by means of the transformation is found by multiplying the mth member of the distribution by R^m where R is given i.e. it is $e^{-\delta_k} \dfrac{\delta_k^m R^m}{m!}$ $(R > 1)$. Then applying this transformation to $\{k_n\}$ the result is $\left\{ \sum_{k=1}^{N} q_k e^{-\delta_k R} \dfrac{(\delta_k R)^m}{m!} \right\}$ (having introduced the new weights $q_k = p_k e^{(R-1)\delta_k}$). Thus we have a superposition of Poisson distributions in which the parameters can be made, by a convenient choice of R, sufficiently large; this is a convenient superposition to which the method of the decomposition of a superposition of Poisson distributions (cf. Ch. IV, § 1, Subsection 1.1.1) can be applied in order to obtain the number of components. In disadvantageous cases this technique may, of course, also fail (Medgyessy 1971b, 1972b).

18. It is possible to reduce the problem of the **decomposition of a superposition of exponential density functions**, considered in Ch. III, § 2, Subsection 1.1, to the decomposition of a superposition of binomial distributions, by the same idea as that described above. Namely, let us consider a superposition

$$k(x) = \begin{cases} \displaystyle\sum_{k=1}^{N} p_k \dfrac{e^{-\frac{x}{\beta_k}}}{\beta_k} & (x > 0) \\ 0 & (x \leqq 0) \end{cases} \qquad (0 < \beta_1 < \beta_2 < ... < \beta_N < \Lambda_2)$$

and take the values $k(n) = k_n^*$ $(n = 0, 1, ...)$. With $\dfrac{p_k}{\beta_k} = p_k^*$, $e^{-\frac{1}{\beta_k}} = \beta_k^*$ $(k = 1, ..., N)$ we then have

$$k_n^* = \sum_{k=1}^{N} p_k^* \beta_k^{*n} \qquad \left(0 < \beta_1^* < \beta_2^* < ... < \beta_N^* < e^{-\frac{1}{A_2}} (\leqq 1)\right).$$

The set of these is equivalent to a superposition of (0) unimodal geometrical distributions and can be handled by the method described above; however, it is easier to realize that the transformation

$$\sum_{n=0}^{M} \beta_k^{*n} K_n^{(m)} = \binom{M}{m} \beta_k^{*m} (1 - \beta_k^*)^{M-m}$$

can be established by taking

$$K_n^{(m)} = \begin{cases} \binom{M}{m}\binom{M-m}{n-m} (-1)^{n-m} & (n = m, m+1, ..., M; \quad m = 0, 1, ..., M) \\ 0 & (n < m) \end{cases}$$

and thus

$$\sum_{n=m}^{M} k_n^* K_n^{(m)} = \sum_{k=1}^{N} p_k^* \binom{M}{m} \beta_k^{*m} (1 - \beta_k^*)^{M-m}.$$

The decomposition of this superposition of binomial distributions, if successful, gives the unknown N (Medgyessy 1971b, 1972b).

We remark that our superposition $k(x)$ can easily be brought — by similar means — to the form of a superposition of (0) unimodal Poisson distributions to be treated as in Point 17.

Example. We give a *methodological* example. Figure 1a shows the superposition

$$k(x) = \begin{cases} e^{-0.2x} + e^{-1.5x} & (x > 0) \\ 0 & (x \leqq 0) \end{cases}$$

at the points $x = 0, 1, ..., 5$. Figure 1b shows corresponding test distributions constructed according to the preceding method with $M = 5$, if 4 (or 3 or 2) decimals of the values $k(n)$ are accurate (curves plotted with crosses, squares and triangles respectively). The graphs shown by crosses and squares of the test distributions reveal two (binomial) distributions showing that the curve in Fig. 1a consists of two components. The values of $\beta_k^* = e^{-\frac{1}{\beta_k}}$ can also be estimated. The graph shown by triangles already has no use: the errors, increasing rapidly with M, have destroyed the test distribution. A separate decomposition of a superposition of binomial distributions is therefore needed in our example.

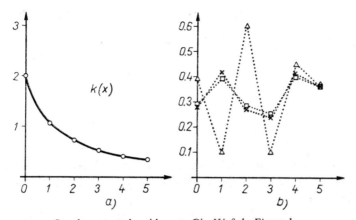

Supplements and problems to Ch. IV, § 1. Figure 1

Nevertheless, this method may be advised if a small M-value can be applied and the "measured" ordinate values are accurate enough (Medgyessy 1971*b*, 1972*b*).

19. The $\mathscr{C}$-*method* seems to be appropriate for a **generalization** of the following kind. Let us consider again a superposition

$$\{k_n\} = \left\{ \sum_{k=1}^{N} p_k\, f_{n-\gamma_k}(\delta_k) \right\}$$

$$(n = 0, 1, \ldots; \ \gamma_i \neq \gamma_j \ (i \neq j); \ \Lambda_1 < \delta_1 \leqq \delta_2 \leqq \ldots \leqq \delta_N < \Lambda_2)$$

$(\gamma_k = 0, 1, \ldots (k = 1, \ldots, N))$ of unimodal distributions $\{f_n(\delta_k)\}$ and suppose that

$$\{f_n(\delta_k)\} = \{K_n(\lambda)\} * \{q_n(\delta_k)\} \qquad (n = 0, 1, \ldots)$$

where $\{K_n(\lambda)\}$, which depends on the parameter λ $(\Lambda_1 \leqq \lambda \leqq \Lambda_2)$, does not depend on δ_k, where $\{q_n(\delta_k)\}$ is a unimodal distribution and the variances of $\{f_n(\delta_k)\}$, $\{K_n(\lambda)\}$, $\{q_n(\delta_k)\}$ i.e. $\mathbf{D}^2\{f_n(\delta_k)\}$, $\mathbf{D}^2\{K_n(\lambda)\}$, $\mathbf{D}^2\{q_n(\delta_k)\}$ all exist and characterize, in some sense, what, in future investigations, will be defined as the "narrowness" of the relevant distribution graphs. In other words $\{f_n(\delta_k)\}$ is *factorizable* (Linnik 1960, Lukács 1964), one of the factors being *unimodal* again; we notice that the possibility of a decomposition into *unimodal* factors is, at present, an unsolved **problem**. If $\mathbf{D}^2\{K_n(\lambda)\}$ increases as $\lambda \uparrow \Lambda_2$ then

$$\{b_m\} = \left\{ \sum_{k=1}^{N} p_k q_{n-\gamma_k}(\delta_k) \right\}$$

can be used as a test function because — in consequence of $\mathbf{D}^2\{f_n(\delta_k)\}=$ $\mathbf{D}^2\{K_n(\lambda)\} + \mathbf{D}^2\{q_n(\delta_k)\}$ — the graphs of its components become "narrower" when λ increases. Obviously

$$\{k_n\} = \{K_n(\lambda)\} * \{b_m\},$$

i.e.

$$k_n = \sum_{\varrho=0}^{n} K_{n-\varrho}(\lambda)\, b_\varrho \qquad (n = 0, 1, \ldots)$$

from which b_ϱ can be determined, also numerically, as a linear expression of the k_n, in several cases *without an exact knowledge of the analytical form of* $\{b_m\}$: the knowledge of the generating function should be sufficient.

Several types of superposition are of this type; the systematic investigation of the basic idea is, however, still an unsolved **problem**.

20. An interesting *outsider* of the superpositions of discrete distributions dealt with up to now in this chapter (Medgyessy 1961*a* pp. 72—77), together with remarks on its decomposition, will be considered below (see also Medgyessy 1971*b*, 1972*b*).

Let us consider the **decomposition of a superposition of binomial distributions of different order** with the same parameter, but with different values of the parameter. That is, let

$$\{k_n\} = \left\{ \sum_{k=1}^{N} p_k \binom{M_k}{n} p^n q^{M_k-n} \right\}$$

$$(n = 0, 1, \ldots;\ M_k = 1, 2, \ldots;\ 0 < M_1 < \ldots < M_N < \Lambda_2),$$

where p is given, $p>0$, $q>0$, $p+q=1$ and we may assume that $p<q$; M_k is an integer. Ch. II, § 3 showed that this superposition is *identifiable*. The decomposition of this superposition cannot be dealt with according to our methods as we cannot introduce a monotone formant. However, a *test distribution* $\{b_m(\lambda)\}$ can be found by taking

$$\{b_m(\lambda)\} = \left\{ \sum_{k=1}^{N} p_k \binom{M_k-\lambda}{m} p^m q^{M_k-\lambda-m} \right\} \qquad (m = 0, 1, \ldots, M_N - \lambda)$$

where λ *is an integer* such that $0 < \lambda < M_1$.

Namely, if $\lambda=1, 2, \ldots$ the (unimodal) graphs of the corresponding components in $\{b_m(\lambda)\}$ become "narrower", their peaks higher and they tend to the y-axis; thus a separation of the component graphs may be expected. A linear transformation acting on $\{k_n\}$ and yielding $\{b_m(\lambda)\}$ can be found easily by realizing that

$$\mathbf{G}\left[\left\{\binom{M_k-\lambda}{m}p^m q^{M_k-\lambda-m}\right\}; z\right] = (pz+q)^{M_k-\lambda} =$$

$$(pz+q)^{M_k}(pz+q)^{-\lambda} = (pz+q)^{-\lambda}\mathbf{G}\left[\left\{\binom{M_k}{m}p^m q^{M_k-m}\right\}; z\right]$$

i.e.

$$\mathbf{G}[\{b_m(\lambda)\}; z] = \frac{1}{(pz+q)^\lambda}\mathbf{G}[\{k_n\}; z].$$

Formally this is the case considered in Ch. IV, § 1, Subsection 1.1.1. By comparing the coefficients of the power series on both sides we finally obtain

$$\{b_m(\lambda)\} = \left\{\sum_{\varrho=0}^{m}\frac{1}{q^\lambda}\binom{\lambda-1+m-\varrho}{\lambda-1}\left(-\frac{p}{q}\right)^{m-\varrho}k_\varrho\right\}$$

(some terms can, eventually, become zero). Hence the required linear transformation can be read off.

In practice trials should be made with several monotonely increasing integer λ-values, constructing the graphs of the corresponding test distributions $\{b_m(\lambda)\}$ and comparing them, as in carrying out the $\mathscr{C}$-method.

It can be seen that the parameter M_k behaves here like a *monotone formant*; thus the method is an *analogue of the method of formant changing*. This indicates the possibility of defining, in certain cases, a **discrete-valued monotone formant**, together, of course, with some new definition of the concept "narrower" (cf. Ch. II, § 4 and § 5).

The described method can be extended easily to the case of the **decomposition of a superposition of negative binomial distributions of different order.**

However the systematic treatment of this field is, at present, an unsolved **problem**.

§ 2. SECOND TYPE OF SUPERPOSITION. THE $\mathscr{D}$-METHOD

In the preceding points we were mainly concerned with the decomposition of superpositions in which the abscissae of the peaks of the (unimodal) components were, from the very first, *different*. The $\mathscr{C}$-method as a method of decomposition, was based completely on this fact.

Now let us consider the **decomposition of superpositions belonging to the type**

$$\{k_n\} = \left\{ \sum_{k=1}^{N} p_k f_n(\delta_k) \right\} \qquad (n = 0, 1, \ldots; \ \Lambda_1 < \delta_1 < \delta_2 < \ldots < \delta_N < \Lambda_2)$$

in which the components are members of a family of unimodal (A_2) (or (A_3)) distributions, depending only on a *single* parameter and where the **abscissae of the peaks of the graphs of $\{f_n(\delta_k)\}$ are the same for any k $(k = 1, \ldots, N)$** (see Fig. 1).

A simple *example* of this is provided by the superposition of geometric distributions in Ch. IV, § 1, Section 3. Another example is a superposition of Poisson distributions

$$\{k_n\} = \left\{ \sum_{k=1}^{N} p_k \frac{e^{-\delta_k} \delta_k^n}{n!} \right\} \qquad (n = 0, 1, \ldots)$$

where $0 < \delta_1 < \delta_2 < \ldots < \delta_N < 1$. Here all components are (0) unimodal (cf. *Supplements and problems to* Ch. IV, § 1, 17).

In principle, the present superposition $\{k_n\}$ cannot be handled by the $\mathscr{C}$-method owing to the type of the superposition to be decomposed: there is no distance between the peaks of the components. (In the example they all occur at $x = 0$ as the parameters are between 0 and 1.) *Formally*, the $\mathscr{C}$-method can be performed often on such a superposition.

At this point a new method of decomposition for such superpositions called the $\mathscr{D}$-method is given. Its idea appears first in Medgyessy (1961*a* pp. 152—158, 1961*b*). This method *does not work, however, with a certain convenient test distribution* because of the disadvantages of the latter in decomposition problems. On the contrary, it aims at allowing a *superposition of* (possibly, symmetrical) *unimodal density functions* (i.e. a *test function*) having a regularizable graph-narrowness corresponding to the superposition of discrete distributions to be decomposed, in which the maxima of the components are different and are in some relation to the unknown parameters of the initial superposition. From this viewpoint it is related to the $\mathscr{B}$-method described in Ch. III, § 1, Section 3, and the basic idea is similar.

Let the function

$$b^*(x) = \sum_{k=1}^{N} p_k w \left(x, \eta(\delta_k), \tau \right)$$

also called here a **test function**, correspond to the superposition $\{k_n\}$ represented by its graph. For this test function let the following *conditions* hold:

I. $\eta(x)$ is a given strictly monotone function, not containing δ_k and $w(x, \eta(\delta_k), \tau)$ $(0 < \tau < T)$ is an $(\eta(\delta_k))$ *unimodal density function* (or is proportional to such a one) in which the mode $\eta(\delta_k) \neq 0$ and may depend on τ, too, and whose graph becomes, in some sense, "narrower" and its peak height increases as $\tau \downarrow 0$ (or $\tau \uparrow T$); finally, which is in a well-defined analytical relation, independent of γ_k, δ_k, with $\{f_n(\delta_k)\}$ $(k = 1, \dots, N)$.

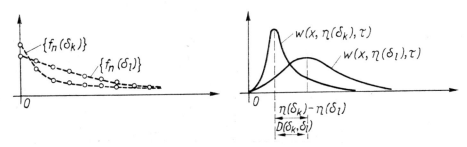

IV.§ 2. Figure 1

II. For *any* τ the distance between abscissae of the peaks of $w(x, \eta(\delta_k), \tau)$ and $w(x, \eta(\delta_k), \tau)$ i.e. $|\eta(\delta_k) - \eta(\delta_l)|$ is not less than some quantity $D(\delta_k, \delta_l)$ depending *only* on δ_k and δ_l (and being *independent* of τ) (see Fig. 1).

III. $b^*(x)$ can be constructed by the help of $\{k_n\}$ only, without the knowledge of N, p_k, δ_k and utilizing, eventually, the analytical relation between $w(x, \omega(\delta_k), \tau)$ and $\{f_n(\delta_k)\}$.

Evidently $b^*(x)$ is a superposition of density functions.

The peak height increases and, in addition, "narrowing" in I occurs if e.g. τ is a $(T, 0)$ (or $(0, T)$) monotone formant of $\mathscr{G}w(x, \eta(\delta_k), \tau)$. If τ is *sufficiently* near to 0 (or T), the kth component of this function will possess a graph with a sharp peak at the point $\eta(\delta_k)$. Further, there will be a certain minimum distance between the abscissae of the peaks of the components of $b^*(x)$ independent of τ (cf. Condition II). Thus the graph of $b^*(x)$ will, *probably,* show more and more *separated* peaks as $\tau \downarrow 0$ (or $\tau \uparrow T$). With a strong separation they appear almost individually, the abscissae of the peaks being near to the $\eta(\delta_k)$; thus the number N of the components — and, eventually, the value of $\eta(\delta_k)$ (i.e. δ_k) — can be deduced from them (see Fig. 2).

Let us consider the following steps.

(A) Determination of the described functions $w(x, \eta(\delta_k), \tau)$ with the help of $\{f_n(\delta_k)\}$.

(B) Determination, with the help of (A), of the test function $b^*(x)$ with the aid of $\{k_n\}$, serving as the basis for the decomposition method.

(C) On the basis of (B) and of the "*measured*" values of $\{k_n\}$ the elaboration of a numerical method for the construction of an *approximation* to the test function $b^*(x)$.

For the graph $\mathscr{G}_1$ of this approximation to $b^*(x)$, the statements concerning the evaluation of the graph of the "exact" test function remain approximately valid.

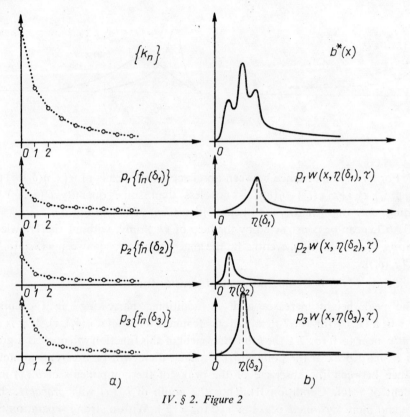

IV. § 2. Figure 2

Definition 1. The construction of the graph $\mathscr{G}_1$ of the test function *approximation*, followed by the determination of the number and of the abscissae etc. of the peaks of $\mathscr{G}_1$ is called the $\mathscr{D}$-**method**, yielding the (numerical) decomposition of the superposition $\{k_n\}$ (Medgyessy 1961a pp. 152—158, 1961b; see also Medgyessy 1971b, 1972b).

In the following a test function $b^*(x)$ will be given for a particular type of superposition $\{k_n\}$ only. From the practical viewpoint, however, it comprises several important superpositions.

1. A particular case of the $\mathscr{D}$-method

Let us consider a superposition of discrete distributions of the type

$$\{k_n\} = \left\{ \sum_{k=1}^{N} p_k f_n(\delta_k) \right\} = \left\{ \sum_{k=1}^{N} p_k \psi(n) A(\delta_k) \eta(\delta_k)^n \right\}$$

$$(n = 0, 1, \ldots; \; \Lambda_1 < \delta_1 < \delta_2 < \ldots < \delta_N < \Lambda_2)$$

where $\{f_n(\delta_k)\} = \{\psi(n) A(\delta_k) \eta(\delta_k)^n\}$ are discrete distributions, $\psi(n)$ is given and depends on n only, and $A(\delta_k)$ is some given function of δ_k; $\eta(\delta_k)$ is as above $(k = 1, \ldots, N)$.

Let us consider the following test function $b_1^*(x)$:

$$b_1^*(x) = \sum_{k=1}^{N} p_k A(\delta_k) \frac{e^{-\frac{(x - R\,\eta(\delta_k))^2}{4\tau}}}{\sqrt{4\pi\tau}},$$

$(R > 1$ is given); that is, $w(x, \eta(\delta_k), \tau)$ in Ch. IV, § 2 is

$$w(x, \eta(\delta_k), \tau) = A(\delta_k) \frac{e^{-\frac{(x - R\,\eta(\delta_k))^2}{4\tau}}}{\sqrt{4\pi\tau}}.$$

Conditions I, II, and III of Ch. IV, § 2 are, evidently, fulfilled here, taking into account — among other things — that by Taylor's formula

$$A(\delta_k) \frac{e^{-\frac{(x - R\eta(\delta_k))^2}{4\tau}}}{\sqrt{4\pi\tau}} = \sum_{n=0}^{\infty} \frac{(-R)^n \Phi^{(n)}(x) f_n(\delta_k)}{n! \, \psi(n)},$$

where $\Phi(x) = \dfrac{e^{-\frac{x^2}{4\tau}}}{\sqrt{4\pi e}}$ $(\tau > 0$ and is arbitrarily small) and, from this

$$b_1^*(x) = \sum_{n=0}^{\infty} \frac{(-R)^n \Phi^{(n)}(x)}{n! \, \psi(n)} k_n.$$

Thus $b_1^*(x)$ can be taken as the test function corresponding to the superposition $\{k_n\}$. It shows narrow peaks at the points $R\eta(\delta_k)$ if τ is sufficiently small, lying far from each other if k is large enough; thus it is convenient for the purposes of the decomposition. The alteration of the weights $\left(\text{into } p_k A(\delta_k)\right)$ has no significance (Medgyessy 1971b, 1972b).

After steps (A), (B) and (C) in Ch. IV, § 2 the $\mathscr{D}$-method can be applied.

Example. **Decomposition of a superposition of Poisson distributions.** Let us consider the superposition of Poisson distributions

$$\{k_n\} = \left\{ \sum_{k=1}^{N} p_k \frac{e^{-\delta_k} \delta_k^n}{n!} \right\} \qquad (n = 0, 1, \ldots; \; 0 < \delta_1 < \delta_2 < \ldots < \delta_N < 1).$$

As $0 < \delta_k < 1$, the components are all (0) unimodal. Let us apply the above $\mathscr{D}$-method with $A(\delta_k) = e^{-\delta_k}$, $\eta(\delta_k) = \delta_k$, $\psi(n) = \dfrac{1}{n!}$. Then for the corresponding test function $b^*(x)$ we have

$$b^*(x) = \sum_{n=0}^{\infty} \frac{(-R)^n}{n!} \Phi^{(n)}(x), \qquad k_n = \sum_{k=1}^{N} p_k e^{-\delta_k} \frac{e^{-\frac{(x-R\delta_k)^2}{4\tau}}}{\sqrt{4\pi\tau}}.$$

In practice the summation can go up to a certain index only. Thus the formula *cannot* be applied directly as this increases the error.

Supplements and problems to Ch. IV, § 2

1. The statements included in the *Supplements and problems to* Ch. IV, § 1 in connection with the $\mathscr{C}$-method are valid, with appropriate alterations, for the $\mathscr{D}$-method also where applicable.

2. Obviously the $\mathscr{D}$-method can often be applied in cases when the components of a superposition are not (0) unimodal, but other methods fail.

3. If a numerical procedure is to be worked out for the method described in Ch. 4, § 1, Section 1, the summation up to $+\infty$ involves intricate problems. A convenient *polynomial* (of a not too high degree) instead of $\Phi(x)$ would work better, providing a finite sum. The graph of such a polynomial should be similar to that of the Gaussian function in some arbitrarily long interval $(-L, L)$; and the different components of the corresponding test function should also approximate the test function components in some sufficiently long interval (L_1, L_2) (Medgyessy 1971*b*, 1972*b*). To find such a polynomial is, however, an unsolved **problem**; it might well be an interesting problem for *constructive analysis*.

4. The function $w(x, \eta(\delta_k), \tau)$ can often be obtained by making use of some well-known *mixture* relations. For example, the mixture of Gamma density functions of different orders, formed with respect to a geometric distribution, is an exponential density function, as

$$\sum_{n=0}^{\infty} \frac{x^n e^{-x}}{n!} (1-\delta_k)\delta_k^n = (1-\delta_k)e^{-(1-\delta_k)x} \qquad (0 < \delta_k < 1).$$

This relation may also be considered as a *transformation* of the geometric distribution $\{(1-\delta_k)\delta_k^n\}$ $(n=0, 1, \dots)$ into an exponential *density function* (Medgyessy 1961*a* pp. 152—158, 1961*b*; see also Medgyessy 1971*a*, 1972*c*).

Let us now consider the *decomposition of a superposition of geometric distributions*

$$\{k_n\} = \left\{ \sum_{k=1}^{N} p_k (1-\delta_k) \delta_k^n \right\} \qquad (n = 0, 1, \ldots; \; 0 < \delta_1 < \delta_2 < \ldots < \delta_N < 1).$$

We have, applying the preceding transformation to $\{k_n\}$, the function

$$b_1(x) = \sum_{n=0}^{\infty} \left(\sum_{k=1}^{N} p_k (1-\delta_k) \delta_k^n \right) \frac{x^n e^{-x}}{n!} = \sum_{k=1}^{N} p_k (1-\delta_k) e^{-(1-\delta_k)x}$$

(cf. Ch. IV, § 1, Section 3). The latter superposition is not yet suitable to serve as a test function; however, this superposition of such functions can already be decomposed by means of the method in Ch. III, § 2, Subsection 1.1. Theoretically its further transformation by means of the transformation introduced in Ch. III, § 2, Subsection 1.1 gives an appropriate test function. The two transformations can also be written as a single transformation, yielding in this way the required transformation of the above superposition of geometric distributions into an appropriate *test function*. In practice, all this is not so simple.

Generally, in transforming a discrete distribution into a density function the concrete **determination of the transformation** or merely the proof of the existence of such a transformation is an unsolved **problem** (cf. Medgyessy 1961a pp. 152—158, 1961b).

§ 3. *AD HOC* METHODS

In this paragraph we shall enumerate *ad hoc* methods for decomposing superpositions of discrete distributions, if they can be classified at all according to the viewpoints below. They differ essentially from those described in the present chapter. They can be and are, in fact, used in practice.

(A) *Methods of algebraic character*

1. Some of these methods (Medgyessy 1954a, Thionet 1966) *approximate*, essentially, the considered superposition by a superposition of a *given* number N ($N=2, 3, \ldots$) of components belonging to the same type as those of the given superposition. The unknown parameters of the approximating function are determined from equidistant "measured" ordinate values of the graph of the given superposition. The procedure will be carried out with $N=2$, $N=3$, etc. If the fit is the best, in a given sense, for $N=N_0$, then N_0 will be taken as the unknown number of components of the given superposition and the parameters obtained at $N=N_0$ as the required parameters.

The fitting will be illustrated by the case of the *decomposition of a superposition of Poisson distributions* (Medgyessy 1954a, 1961a, pp. 160—161). Here N being fixed, the successive "measured" ordinate values $\hat{k}_n$ of the graph of

$$\{k_n\} = \left\{ \sum_{k=1}^{N} p_k \frac{e^{-\beta_k}\beta_k^n}{n!} \right\} \qquad (n = 0, 1, ...; \ 0 < \beta_1 < \beta_2 < ... < \beta_N < \Lambda_2)$$

are taken and we are able to write, approximately, taking $p_k e^{-\beta_k} = G_k$, $n! \, k_n = H_n$ $(k = 1, ..., N)$

$$\sum_{k=1}^{N} G_k \beta_k^n = H_n \qquad (n = 0, 1, ...),$$

which is a system of equations of higher order for G_k, β_k and, implicitly, for p_k, β_k, also. This system belongs to the type described in Ch. III, § 3 (B) 2 and can be solved in the way described there. From the result p_k, β_k can, of course, also be calculated (Medgyessy 1954a, 1961a pp. 160—161).

The case of the *decomposition of a superposition of binomial distributions* can be treated similarly (Medgyessy 1954a, 1961a pp. 160—161). Introducing D_n,

x_k, B_k by $\dfrac{k_n}{\dbinom{M}{n}} = D_n$, $\beta_k = \dfrac{x_k}{1+x_k}$, $B_k = \dfrac{p_k}{(1+p_k)^M}$ in the superposition

$$\{k_n\} = \left\{ \sum_{k=1}^{N} p_k \binom{M}{n} \beta_k^n (1-\beta_k)^{M-n} \right\} (n = 0, 1, ..., M; \ 0 < \beta_1 < \beta_2 < ... < \beta_N < 1)$$

(where instead of k_n the measured values of this, $\hat{k}_n$, will be used) we get the system of equations

$$\sum_{k=1}^{N} B_k x_k^n = D_n \qquad (n = 0, 1, ..., M),$$

belonging to the same type and requiring a similar treatment as that above, used in the superposition of Poisson distributions.

The situation is similar also in the case of the *decomposition of a superposition of negative binomial distributions*.

Thionet (1966) extended the present idea to superpositions with components satisfying certain recurrence equations, and obtained by this method a system of equations for certain functions of the unknowns. He also treated this system of equations exhaustively in the case of a "measurement" error in the data obtained from the graph of the investigated superposition.

2. The approximation to the given superposition, described in 1, can be obtained also by the method of moments. Then any method of *mathematical statistics* for giving moment estimators for the parameters of a population distribution, provided it is a mixture of discrete distributions, can be used. This is evident, because moments can be calculated not only from a sample but also from the measured values of a discrete distribution. All this can be found in its germ already in a little known article of Pearson (1915b). In that paper he supposed that the distribution of a discrete characteristic of a population was the mixture of two Poisson distributions or two binomial distributions; further, he determined the parameters of this approximation by means of calculating the roots of algebraic equations of higher order in a way similar to that described in the paper of Pearson (1894).

In addition the analytical kernel of a great number of papers yielding relations between the moments and the parameters of a mixture of discrete distributions can be used (cf. Schilling 1947; Rider 1962; Blischke 1962, 1964, 1965; McPhee 1963; Rosenstiehl, Ghouila-Houri 1960).

Naturally, a suitable least square method can also be constructed.

All the methods described have the disadvantage that the number N of the components has to be assumed. Besides, we have no test for deciding the best value of N. Thus the methods described can be applied only with great care and their results are to be interpreted cautiously if no additional information concerning N is available.

(B) *A method of R. Bellman*

The method given by Bellman (1960) and described in Ch. III, § 3 (D) can be applied in the case of various superpositions of discrete distributions. Medgyessy (1966b) remarked that in the case of the decomposition of any superposition $\{w_n\}$ of discrete distributions belonging to the type

$$\{w_n\} = \left\{ \sum_{k=1}^{N} q_k \Phi(n) A_k^n \right\} \qquad (n = 0, 1, \ldots)$$

where $q_k > 0$, $A_k > 0$ $(k = 1, \ldots, N)$ and $\Phi(n)$ depends on n only, the unknown N can be determined by applying the idea of the method of Bellman (1960), that is, by considering the sequence of values

$$C_r(i) = \begin{vmatrix} \dfrac{w_i}{\Phi(i)} & \dfrac{w_{i+1}}{\Phi(i+1)} & \cdots & \dfrac{w_{i+r}}{\Phi(i+r)} \\[2ex] \dfrac{w_{i+1}}{\Phi(i+1)} & \dfrac{w_{i+2}}{\Phi(i+2)} & \cdots & \dfrac{w_{i+r+1}}{\Phi(i+r+1)} \\[1ex] \vdots & \vdots & & \vdots \\[1ex] \dfrac{w_{i+r}}{\Phi(i+r)} & \dfrac{w_{i+r+1}}{\Phi(i+r+1)} & \cdots & \dfrac{w_{i+2r}}{\Phi(i+2r)} \end{vmatrix} \qquad \begin{pmatrix} i = 0, 1, 2, \dots; \\ r = 1, 2, 3, \dots \end{pmatrix}$$

at some fixed i. Then the first index $r=r_0$ for which $C_{r_0}(i)=0$ (while $C_r(i)\neq 0$ if $r<r_0$) gives the unknown N.

For instance in the case of the *decomposition of a superposition of Poisson distributions*, i.e. if

$$\{w_n\} = \left\{ \sum_{k=1}^{N} p_k \frac{e^{-\beta_k}\beta_k^n}{n!} \right\} \qquad (n = 0, 1, \dots; \quad 0 < \beta_1 < \beta_2 < \dots < \beta_N < \Lambda_2),$$

then putting $q_k = p_k e^{-\beta_k}$, $\Phi(n) = \dfrac{1}{n!}$, $A_k = \beta_k$ $(k=1, \dots, N)$ we get the above type.

In practice, we take the "measured" values $\hat{w}_n$ of w_n instead of the theoretical ones and apply the procedure; however, the method then becomes awkward because errors are involved in the elements of the determinants $\hat{C}_r(i)$ formed with $\hat{w}_n$ and this may result in considerable errors related to the exact values $C_r(i)$ of the determinants. This makes it difficult to decide when the determinant becomes, theoretically, zero (Medgyessy 1966b, 1971b, 1972b).

Supplements and problems to Ch. IV, § 3

1. Let us consider, without any reference to decomposition problems, the system of nonlinear equations or, in Ch. IV, § 3 (A) 1, generally, the similar special *systems of equation of higher degree*

$$\sum_{k=1}^{N} x_k y_k^v = a_v \quad (v = 0, 1, \dots; \quad x_k > 0, y_k > 0, a_v > 0; \quad y_i \neq y_j \ (i \neq j))$$

$(x_k, y_k \ (k=1, \dots, N)$ are unknown). Then a superposition of e.g. Poisson distributions can be made to correspond to this system, by inverting the sequence of ideas in the context of Ch. IV, § 3 (A) 1. More precisely, let us put e.g.

$x_k = p_k e^{-y_k}$, $v! A_v = a_v$; then we have, instead of the above, the system of equations

$$\sum_{k=1}^{N} p_k \frac{e^{-y_k} y_k^v}{v!} = A_v \qquad (v = 0, 1, \ldots)$$

for the unknowns p_k, y_k. This is *formally* a superposition of Poisson distributions. Its decomposition, e.g. by the method in Ch. IV, § 1, Subsection 1.1.1, would give approximate values of p_k, y_k i.e. of x_k, y_k. Thus certain *decomposition methods may provide a tool for the solution of a special system of equations,* i.e. of an *algebraic* problem.

2. In (B) the error of R. Bellman's method, when applied in practice, can be estimated only roughly. A *probabilistic* investigation of the error seems, however, to be more appropriate; since determinants with random variables as elements have been studied extensively e.g. in stochastic programming in recent years.

V. APPENDIX. NUMERICAL METHODS THAT CAN BE APPLIED IN CARRYING OUT DECOMPOSITION METHODS

In this chapter a survey is given of the methods that can be utilized in the numerical solution of the problems occurring in the present book. Mainly those methods which have been published previously have been given, since to have constructed newer ones would have been beyond the scope of the present book; only certain of the results are the present author's.

As certain of the problems lead, in practice, to the numerical solution of a Fredholm integral equation of the Ist kind and of the convolution type, this should be investigated in details. However, this problem is a particular case of a more general one: the solution of operator equations of the Ist kind. Thus it is worthwhile considering this latter problem in the next point in order to see all questions clearly. As this is also a *survey,* this next point may be of some use for those who are interested in the numerical solution of integral equations of the Ist kind *in general.* It should be mentioned that certain aspects of the next point need a knowledge of the elements of functional analysis.

A more general treatment may be found in the following: Lavrent'ev (1966); Lavrent'ev, Vasil'ev (1966); Kirillova, Piontkovskiĭ (1968); Medgyessy (1971a) (with a considerable bibliography); Nedelkov (1972); Rust, Burrus (1972); Morozov (1973) (a survey); Miller (1974); Tihonov, Arsenin (1974).

§ 1. SOLUTION OF OPERATOR EQUATIONS OF THE IST KIND

1. Correct and incorrect problems. Early methods of treating incorrect problems

Several problems in mathematics, e.g. the solution of Fredholm or Volterra integral equations of the Ist kind, or of systems of linear equations, or of partial differential equations are equivalent to the solution of an operator equation of the Ist kind. This will now be considered in more detail and in an exact setting.

Let X, Y be complete linear metric spaces, each with an appropriate metric (with "distance" denoted by $\varrho(.,.)$) introduced in it, $\mathbf{A}$ a continuous, in general not linear operator, with given domain of definition and range. If $y \in Y$, the so-called *operator equation of the Ist kind*

(1) $$\mathbf{A}x = y \qquad (y \in Y)$$

can be given formally. We shall investigate the problem of solving (1), i.e. for a given y_0 ($y_0 \in Y$) we have to find an x_0 ($x_0 \in X$) for which $\mathbf{A}x_0 = y_0$.

It is well known that this problem is called **correct** (or well-posed) in the sense of Hadamard (Hadamard 1902, 1932) if, for y_0 belonging to some given set 1. there is a solution; 2. this solution is unique and 3. the solution depends "continuously" on y_0.

At the same time there are operator equations of the Ist kind, e.g. Fredholm integral equations of the Ist kind in which the problem of solution is not correct in the sense of Hadamard, in short: **incorrect** (or ill-posed, improperly posed, ill-conditioned), because for them e.g. 1 and 2 are satisfied, but 3 is not (one cannot change y_0 to such a small extent that the solution x_0 will change less than a value given in advance).

A classical example for this is the Fredholm integral equation of the Ist kind $\int_a^k K(v, u)\, x(u)\, du = y(v)$: with an "infinitesimal" change of $y(v)$ the solution $x(u)$ may change, eventually, beyond any limit. In this case the numerical solution of such an equation will be of limited use because the points of the result are situated in an oscillating and confused manner, without any possibility of surveying them.

Nevertheless the solution of incorrect problems can also be defined and different methods of solution can be constructed for them. These will be described in the following with special regard to the integral equations of the Ist kind and the aspects of numerical analysis. *However, each way is characterized by the feature that, to the given problem (e.g. to the solution of an integral equation of the Ist kind) another problem is constructed which is already correct (e.g. the solution of an integral equation of the IInd kind appears) and, in addition, the solution of this correct problem converges to the solution of the given problem with the changing of certain parameters.*

A great number of problems which are incorrect in the sense of Hadamard can be dealt with if the concept of correctness is defined *in another way*, in accordance with actual cases. This was pointed out by A. N. Tihonov (Tihonov 1944, John 1955a, 1955b; Lavrent'ev 1962; and from the viewpoint of the evaluation of measurements in physics: Maĭorov 1965).

The problem of solving (1) is called *correct in the sense of Tihonov* if the following conditions are satisfied (Tihonov 1944):

1) It is known in advance that for elements y_0 belonging to a certain set $\mathfrak{N}$ ($\mathfrak{N} \subset Y$) a solution of (1) exists and the ensemble of these solutions belong to a *given* set $\mathfrak{M}(\mathfrak{M} \subset X)$, ("set of correctness") closed in the sense of the metric introduced in X;

2) There is a unique solution to a given y_0;

3) If at the change of y_0 the corresponding new solution belongs to $\mathfrak{M}$, then the solution depends "continuously" on y_0 i.e. if $Ax_1 = y_1$, $Ax_2 = y_2$, $y_1 \in \mathfrak{N}$, $y_2 \in \mathfrak{N}$, $x_1 \in \mathfrak{M}$, $x_2 \in \mathfrak{M}$ and $\varrho(y_1, y_2) \leqq \delta$ ($\varrho > 0$ is given arbitrarily), then $\varrho(x_1, x_2) \leqq \varepsilon(\delta)$ where $\varepsilon(\delta) \downarrow 0$ if $\delta \downarrow 0$.

A. N. Tihonov proved (Tihonov 1944; John 1955a, 1955b; Lavrent'ev 1962) that if $\mathfrak{M}$ is compact and in the operator equation $Ax = y$ (see (1)) A establishes a one-to-one correspondence and is continuous, and if, further, the above conditions 1 and 2 of correctness are fulfilled, then 3 will also be fulfilled; i.e. the problem of solving (1) will be correct in the sense of Tihonov.

M. M. Lavrent'ev showed (see e.g. Lavrent'ev 1962) that if X, Y are Hilbert spaces, if A is completely continuous, if the above conditions 1 and 2 of correctness are satisfied, if $\mathfrak{M}$ is compact and if only some "measured" approximation y_δ of y ($\varrho(y, y_\delta) = \|y - y_\delta\| \leqq \delta$ ($\| \; \|$ denotes norm)) is known, then one can *construct* such an element $x_{\alpha\delta}$ depending, besides δ, on some $\alpha = \alpha(\delta)$ for which $\varrho(x, x_{\alpha\delta}) = \|x - x_{\alpha\delta}\| \to 0$ as $\delta \to 0$ and which is the solution of a Fredholm integral equation of the IInd kind $\alpha x_{\alpha\delta} + Bx_{\alpha\delta} = y_\delta$ where the operator B can be constructed with the help of A in a simple way. $x_{\alpha\delta}$ is then *considered* to be an approximate solution of (1).

From the practical point of view the following facts are very important:

1. One of the conditions for the solvability of (1), having a character correct in the sense of Tihonov, is that $\mathfrak{M}$ be compact. In a given case, the most difficult step is to decide this. A useful criterion is as follows: if conditions 1 and 2 of correctness in the sense of Tihonov are fulfilled and if $\mathfrak{M}$ is a uniformly bounded subset T of the set of functions being continuous and bounded in $(-\infty, \infty)$, for which it holds that for a given $\varepsilon > 0$ there exists a covering of $(-\infty, \infty)$ with a finite number of open sets in each of which the oscillation of any function from T is less than ε, then $\mathfrak{M}$ is compact.

2. As to the numerical solution of problems in this field, it is very important that if for the solution of (1), conditions 1 and 2 of the correctness in the sense of Tihonov are fulfilled but, instead of y, only y_δ can be "measured" ($\varrho(y, y_\delta) \leqq \delta$)

and we *know* an approximate solution $\hat{x}_{\alpha\delta}$ which has been constructed from y_δ with the tools of numerical analysis (e.g. from the analytical solution working with the exact y or by the procedure of M. M. Lavrent'ev, that we have mentioned (α being some appropriate parameter) and — supposing, eventually, some relation between α and δ — if it can be proved that $\varrho(x, x_{\alpha\delta}) \to 0$ as $\delta \to 0$ then, with regard to what has been said, this $\hat{x}_{\alpha\delta}$ can be considered as an approximate solution of the equation. This possibility for solving (1) is most advantageous because one *need not decide* whether $\mathfrak{M}$ is compact or not.

Examples illustrating this statement can be found e.g. in John (1955).

In the procedures described hitherto one of the critical points is the determination of the compactness of $\mathfrak{M}$, the difficulties of which make newer methods of solving (1) necessary.

2. The method of D. L. Phillips

For the case when the problem is incorrect in the sense of Hadamard, a new method of solving Fredholm integral equations of the Ist kind was first published by Phillips (1962) and subsequently improved (Twomey 1963; for a further development see e.g. Hunt 1970). Its basis is as follows.

Let us consider the equation

$$(1) \qquad \int_a^b K(x, y) f(y) \, dy = g(x) \qquad (a \leq x \leq b)$$

where $f(x)$, $g(x)$ are bounded and continuous, and $K(x, y)$ is such that solving the equation is an incorrect problem in the sense of Hadamard. Then the numerical treatment of the equation is virtually impossible because the "measurement" errors of $g(x)$ may yield a strongly oscillating solution. However, in this case the answer is to let (1) be rewritten in the form

$$(2) \qquad \int_a^b K(x, y) f(y) \, dy = \hat{g}(x) + \varepsilon(x)$$

where $\varepsilon(x)$ represents the "error" to be added to $\hat{g}(x)$ obtained by "measuring" the exact function $g(x)$. As to $\varepsilon(x)$, only bounds or similar rough characteristics of it are known (actually, $\varepsilon(x)$ is a realization of some stochastic process, of a "noise"). Let us suppose that the equation $\int_a^b K(x, y) f(y) dy = \hat{g}(x)$ (derived from (2)) can be solved uniquely if $\hat{g}(x)$ is given. With a given $g(x)$ and different given forms of $\varepsilon(x)$ which depend on the "measurement" effected on $g(x)$ different solutions are obtained in every "measurement". We *agree* in that from the set of all possible solutions we select *one* by means of some appropriate restriction (excluding the insufficiently "smooth" solutions etc.) and that we will consider *this* as a *solution* of (1) or (2). D. L. Phillips and S. Twomey take as such a restriction e.g. the "smoothness" condition that the solution defined in this manner minimizes the integral of the square of its second derivative. *With this restriction* and after the usual discretization, the approximation of the above-defined solution of (2) leads to the following minimization problem. Let f_i and $\hat{g}_j$ $(i=0, \ldots, n; j=0, \ldots, n)$ be the values of $f(y)$ and $\hat{g}(x)$ respectively, taken at the points y_i and x_j. The vector $\mathbf{f}_s = (f_0^{(s)}, \ldots, f_n^{(s)})$, minimizing the expression $\sum_{i=0}^{n} (f_{i+1} - 2f_i + f_{i-1})^2$, has to be determined under the conditions $\sum_{i=0}^{n} w_i K_{ji} f_i = \hat{g}_j + \varepsilon_j$, $\sum_{j=0}^{n} \varepsilon_j^2 = \varepsilon^2$ where w_i $(i=0, \ldots, n)$ are the coefficients occurring in the quadrature formula used in the discretization of the integral, $K_{ji} = K(x_i, y_j)$, $\varepsilon_j = \varepsilon(x_j)$ and ε is given. The solution of this problem is $\mathbf{f}_s = (A + \gamma B)^{-1} \hat{\mathbf{g}}$ and the error is given by $\varepsilon = -\gamma B \mathbf{f}_s$ where $A = (w_i K_{ji})$, $A^{-1} = (\alpha_{ij})$, $\boldsymbol{\varepsilon} = (\varepsilon_0, \ldots, \varepsilon_n)$, $\hat{\mathbf{g}} = (\hat{g}_0, \ldots, \hat{g}_n)$ $B = (b_{lk})$ $(k=0, 1, \ldots, n; l=0, 1, \ldots, n)$ $(A, B$ are matrices) and $b_{lk} = \alpha_{k-2,l} - 4\alpha_{k-1,l} + 6\alpha_{k,l} - 4\alpha_{k+1,l} + \alpha_{k+2,l}$, $\alpha_{n+2,l} = 0$, $\alpha_{-2,l} = 0$ and $\gamma > 0$ is a constant ("regularization parameter").

The constant γ regulates, essentially, the smoothness; its value cannot be given in advance. In practice the method is to be applied as follows. From the preceding equations first $\mathbf{f}_s$ is determined for different γ-values (forming e.g. a monotone increasing sequence); then $\boldsymbol{\varepsilon}$ is calculated. *Let us suppose* that the maximal "measurement" error of $g(x)$ is known; then it is reasonable to accept that value γ_0 as the most suitable at which the greatest component of $\boldsymbol{\varepsilon}$ is approximately equal to the maximal error of $g(x)$. It is advisable, however, to consider the graphs of the solutions $\mathbf{f}_s$ belonging to the different γ-values *simultaneously* and to take into account that practice always considers the seemingly smoothest as the most appropriate; the γ_s-value belonging to this latter is then to be compared with γ_0. Undefined steps of similar character are, unfortunately, unavoidable in our methods; the cause of this is, partly, the discretization.

By a more appropriate choice of the restrictive conditions, the method can be varied in a number of ways. From the point of view of numerical treatment the only essential point is the matrix inversion involved.

The special but important case of (1), when the integral is a convolution integral can be studied in some examples presented in Phillips (1962) and Twomey (1963).

3. The method of V. K. Ivanov

In order to avoid the difficulties of the methods described in Section 1, V. K. Ivanov defined the solution of (1) in Ch. V, § 1, Section 1 as follows (Ivanov 1962a, 1962b, 1963; see also Douglas 1960).

He adopted the correctness definition "correct in the sense of Hadamard"; but he changed the *definition of "solution"*. More precisely: let X, Y be complete linear metric spaces as above, $\mathbf{A}$ a continuous linear operator for which $\mathbf{A}^{-1}$ is not continuous. Then solving the equation

$$(1) \qquad\qquad \mathbf{A}x = y$$

($x \in X$, $y \in Y$) will present a problem incorrect in the sense of Hadamard. Let a closed, compact set $\mathfrak{M} \subset X$ be given. V. K. Ivanov calls the element x_0 ($x_0 \in \mathfrak{M}$) which minimizes $\varrho(\mathbf{A}x, y_0)$ ($y_0 \in Y$) when $x \in \mathfrak{M}$, the **quasi-solution** of (1) belonging to $y_0 \in Y$. He considers this quasi-solution as the solution of (1). It can be proved that for an arbitrary y_0 ($y_0 \in Y$) this quasi-solution exists; if $y_0 \in \mathbf{A}\mathfrak{M}$, the quasi-solution is identical with the exact solution, which does exist in this case; if the equation $\mathbf{A}x=0$ has only a null solution, if $\mathbf{A}\mathfrak{M}$ is a convex set and the spheres in the space Y are strictly convex (this holds in the case of the C or L_2 space), then (1) possesses a *unique* quasi-solution belonging to $y_0 \in Y$, which depends "continuously" on y_0, i.e. in this case the problem is correct in the sense of Hadamard, provided that *we accept the quasi-solution as the solution.*

If X, Y are Hilbert spaces and $\mathbf{A}$ is completely continuous, then the quasi-solution can be given, under certain conditions, by the solution of a Fredholm integral equation of the IInd kind in an analytical form (Ivanov 1962).

In practice the following is important: if instead of y_0 on the right-hand side of (1), only the "measured" y_δ ($\varrho(y_0, y_\delta) \leqq \delta$) is known and $\varrho(\mathbf{A}x, y_\delta)$ reaches its minimum at $x=x_\delta$ and $\varrho(\mathbf{A}x_\delta, y_\delta) \leqq \delta$, then V. K. Ivanov makes the obvious convention to consider x_δ as the approximate solution of (1) (Ivanov 1962). It can be proved that if $\delta \to 0$, then x_δ converges to one of the quasi-solutions of (1), belonging to y_0.

From the viewpoint of numerical analysis we consider, as an approximation to the quasi-solution, the element $\hat{x}_\delta$ which can be obtained from the "measured" y_δ by minimizing $\varrho(\mathbf{A}x, y_\delta)$ under the described condition (in a suitable discretization). This minimization problem is, in general, equivalent to some problem of non-linear programming and is to be solved by the methods of that discipline.

Difficulties in the application of this method are caused generally e.g. by the required properties of the set $\mathfrak{M}$.

The technique of this method can be studied in certain worked-out examples (Fridrik 1967; Liskovec 1968); similar investigations were studied by Schmaedeke (1968, 1969).

4. The regularization method of A. N. Tihonov

The so-called *regularization method* of A. N. Tihonov (Tihonov 1963a, 1963b, 1963c, 1965a, 1967) tried to eliminate the difficulties of the methods described hitherto in the following way.

Let U, Z be metric spaces as above, $\mathbf{A}$ a continuous operator and let us consider the operator equation

(1) $$\mathbf{A}z = u$$

($u \in U$, $z \in Z$). Let us suppose that the problem of solving (1) is incorrect in the sense of Hadamard; further that for elements u belonging to a certain set, the solution of (1) exists and is unique. Starting from the practical situation, let us suppose that instead of some $\bar{u} \in U$ to which a solution $\bar{z}$ exists, only some "measured" $\tilde{u}_\delta$ is known, for which $\tilde{u}_\delta \in U$ and $\varrho(\bar{u}, \tilde{u}_\delta) \leqq \delta$ ($\delta > 0$ is given).

Now, if for the given $\tilde{u}_\delta$ ($\tilde{u}_\delta \in U$) and $\delta > 0$ an element $\tilde{z}_\delta$ ($\tilde{z}_\delta \in Z$) can be constructed in some way for which $\varrho(\bar{z}, \tilde{z}_\delta) \leqq \varepsilon(\delta)$ if $\varrho(\bar{u}, \tilde{u}_\delta) \leqq \delta$, where $\varepsilon(\delta) \downarrow 0$ as $\delta \downarrow 0$ (i.e. the solution of (1) depends on its right-hand side "continuously"), then we make the *convention* to consider this element $\tilde{z}_\delta$ as the solution of (1).

In order to construct $\tilde{z}_\delta$ A. N. Tihonov introduced the so-called **regularization method** (Tihonov 1963a, 1963b, 1963c, 1965a). For this the following concepts are needed:

An operator $\mathbf{R}_\alpha$ ($\alpha > 0$) depending on the parameter α is called the *regularizer* of the operator equation (1) if

1. $\mathbf{R}_\alpha$ is defined, for an arbitrary α ($\alpha > 0$) on the whole space U and its range is Z,

2. $\mathbf{R}_\alpha$ is continuous with respect to u on U,

3. for an arbitrary $z \in Z$, $\varrho(\mathbf{R}_\alpha \mathbf{A}z, z) \to 0$ as $\alpha \to 0$ (the convergence is not uniform).

The ensemble of the elements $\mathbf{R}_\alpha u = z^{(\alpha)}$ depending on α is called the *regularized family* of approximate solutions. By our conditions $z^{(\alpha)}$ tends to the exact solution if $\alpha \to 0$.

Before considering the construction of $\tilde{z}_\delta$ we mention in advance that in the case of a given operator equation we, first of all, have to determine $\mathbf{R}_\alpha$ or $z^{(\alpha)}$.

A. N. Tihonov showed a particular way of determining $z^{(\alpha)}$. With respect to its applications it will be described here for a *special* (and not the most general) case.

Namely let (1) be the *integral equation*

(2) $$\int\limits_a^b K(x, s)z(s)\,ds = u(x)$$

where $Z \equiv C^{(n+1)}[a, b]$ i.e. the space of functions of one variable that are smooth in the nth order in the interval $[a, b]$, $U = L_2[c, d]$ i.e. the space of quadratically integrable functions in the interval $[c, d]$. Tihonov (1963b) proved that if $K(x, s)$ is continuous with respect to both x and y and, in (2), to $u(x)=0$ the unique solution $z(x)=0$ belongs then $z^{(\alpha)}$ is the $n+1$ times continuously differentiable function which minimizes the so-called **smoothing functional**

$$M_n^{(\alpha)}[z, u] = \int\limits_c^d \left[\int\limits_a^b K(x, s)\, z(s)\,ds - u(x) \right]^2 dx + \alpha \int\limits_a^b \left\{ \sum_{i=0}^{n+1} K_i(s)\, z^{(i)}(s)^2 \right\} ds,$$

where $K_i(s) > 0$ $(i=0, 1, \ldots, n+1)$ is some continuous non-negative function $(K_i(s)=1$ can also be taken$)$ and α is the above parameter called **regularization parameter**. The above facts are equivalent to asserting that $z^{(\alpha)}$ is the solution of the Euler equation

$$L_n^{(\alpha)}[z] = \alpha \sum_{i=0}^{n+1} (-1)^{i+1} \frac{d^i}{ds^i} \left[K_i(s) \frac{d^i z}{ds^i} \right] - \left\{ \int\limits_a^b \overline{K}(s, v)z(v)\,dv - b(s) \right\} = 0$$

(the existence of the derivatives being assumed) where

$$\overline{K}(s, v) = \int\limits_c^d K(\xi, s)\, K(\xi, v)\,d\xi, \quad b(s) = \int\limits_c^d K(\xi, s)\, u(\xi)\,d\xi$$

and the boundary conditions are

$$\pi^l(s) = \sum_{i=l+1}^{n+1} (-1)^{i-l-1}[K_i(s)z(s)^i]^{(i-l-1)} = 0 \quad (l = 1, \ldots, n+1)$$

if $s=a$ or $s=b$ (by means of the Green's function of the boundary conditions the Euler equation can be transformed into a Fredholm integral equation of the IInd kind; the solution of this exists and consequently this also yields $z^{(\alpha)}$).

If $K(x, s)$ can be written in the form

$$K(x, s) = \int\limits_C^D \hat{K}(\xi, x)\hat{K}(\xi, s)\,d\xi$$

(C, D are given) and $a=c$, $b=d$ then the Euler equation and the boundary conditions will take the form

$$L_n^{(\alpha)}(z) = \alpha \sum_{i=0}^{n+1} (-1)^{i+1} \frac{d^i}{ds^i} [K_i(s)z^{(i)}(s)] - \left\{ \int\limits_a^b K(s, v)z(v)\,dv - u(s) \right\} = 0$$

and $\pi^l(a)=\pi^l(b)=0$ respectively, which is simpler than the former system. This is the case if, for example, $K(x, s)=S(x-s)$ where $S(x)$ is a stable density function (cf. Ch. III, § 1, Theorem 8) e.g. a normal density function or a Cauchy density function etc.

In the case of a general operator equation the statements are similar; details will be omitted here.

Now we can turn to the construction of $\tilde{z}_\delta$ (Tihonov 1963a, 1963b). Le us restrict our investigation in the following to the former special case: that of the integral equation (2). If a regularizer $\mathbf{R}_\alpha$ has been found, it can be proved simply that for an appropriate choice of $\alpha=\alpha(\delta)$ $(\alpha(\delta)>0, \alpha(\delta)\to0$ if $\delta\to0)$, if $\tilde{u}_\delta$ is a "measured" value of $u(x)$ and $\mathbf{R}_\alpha\tilde{u}_\delta=\tilde{z}_\delta^{(\alpha)}$, then for a fixed exact solution $\bar{z}(x)$, $\varrho(\bar{z},\tilde{z}_\delta^{(\alpha)})\leqq\varepsilon(\delta)$ if $\varrho(\bar{u},\tilde{u}_\delta)\leqq\delta$ where $\varepsilon(\delta)\downarrow0$ as $\delta\downarrow0$; that is, $\tilde{z}_\delta^{(\alpha)}$ gives the solution of (2) in the above sense. It can be proved that $\tilde{z}_\delta^{(\alpha)}$ *can also be determined by minimizing some functional and this functional is simply* $M_n^{(\alpha)}[z,\tilde{u}_\delta]$ *formed by the above* $M_n^{(\alpha)}$ *provided that* α *satisfies the condition* $\dfrac{\delta^2}{\varepsilon_0(\delta)}\leqq\alpha\leqq\alpha_0(\delta)$ where $\alpha_0(\delta)$, $\varepsilon_0(\delta)$ are certain decreasing functions of δ. $\tilde{z}_\delta^{(\alpha)}$ tends to the exact solution if $\delta\to0$ or $\alpha\to0$. Thus the determination of $\mathbf{R}_\alpha$ can be avoided in this case also. However, it is most essential to choose α in accordance with the preceding condition and this is an intricate problem. In the case of a general operator equation (1), the result is similar to the above; details will be omitted here.

Here it should be mentioned that V. N. Strahov has developed a further regularization method for convolution integral equations (Strahov 1968). It is worthwhile, however, to consider his results in a subsequent point.

Let us mention here how A. N. Tihonov's regularization method is applied to *numerical* work. We consider again the special case of the integral equation

$$\int_a^b K(x,s)z(s)\,ds=u(x).$$

Let $Z\equiv C^{(n+1)}[a,b]$ (i.e. the space of the functions of one variable smooth in the nth order) and $U=L_2[c,d]$; let $K(x,y)$ be continuous. Let us consider the following discrete form of the above functional $M_n^{(\alpha)}[z,u]$:

$$\hat{M}_n^{(\alpha)}[z,u] = \sum_{i=1}^M\left(\sum_{j=1}^N K_{ij}z_j h-u_i\right)^2 h_1+$$

$$\alpha\sum_{j=1}^N\left\{\sum_{v=0}^{n+1}\frac{K_v^{(j)}}{h^{2v-1}}\left(z_{j+v}-\binom{v}{1}z_{j+v-1}+\binom{v}{2}z_{j+v-2}-\ldots\pm z_j\right)^2\right\},$$

where

$$u_i = u(x_i), \quad x_i = c+ih_1-\frac{h_1}{2}, \quad h_1=\frac{c-d}{M},$$

$$z_j = z(s_j), \quad s_j = a+jh-\frac{h}{2}, \quad h=\frac{a-b}{N} \quad (j=1,\ldots,N; \quad i=1,\ldots,M).$$

K_{ij} is defined by the quadrature formula

$$\int_a^b K(x, s)\, z(s)\, ds = \sum_{j=1}^{N} K_{ij} z_j h + 0(h^\gamma) \qquad (\gamma > 0)$$

and $K_\nu^{(j)} > 0$ $(\nu = 0, 1, ..., n+1)$ are given constants; it is convenient to take $K_\nu^{(j)} = 1$. A. N. Tihonov has proved (Tihonov 1963*b*) that if the values of the "measured" data $\tilde{u}_\delta(x)$ are $\tilde{u}_i$ at the points x_i, then in case of $\alpha > 0$ a set of numbers $\{\tilde{z}_{\alpha, j}\} \equiv \{\tilde{z}_\delta^{(\alpha)}(s_j)\}$ exists whose elements minimize the functional $\hat{M}_n^{(\alpha)}[z, \tilde{u}_\delta]$, and to the constants ε $(\varepsilon > 0)$, γ_1, γ_2 $(0 < \gamma_1 \leq \gamma_2)$, $\delta_0 = \delta_0(\varepsilon, \gamma_1, \gamma_2, z)$ and $h_0 = h_0(\varepsilon, \gamma_1, \gamma_2, z)$ can be found such that if $\varrho(u, \tilde{u}_\delta) \leq \delta$, and α satisfies the relation $\dfrac{\delta^2}{\gamma_1} \leq \alpha \leq \dfrac{\delta^2}{\gamma_2}$, then the quantities $\tilde{z}_{\alpha, j}$ will fall into the ε-neighbourhood of $z(s)$ if $\delta < \delta_0$, $h < h_0$; i.e. if δ is decreased, the points of ordinates $\tilde{z}_{\alpha, j}$ will lie nearer and nearer to the graph of the exact solution of our integral equation. Consequently, the approximate solution of our integral equations can be given by numerical tools. Since the errors of the quadrature formula cannot be included in our calculations, it is generally *accepted* that, in a given case, the relation $\alpha = C\delta^2$ (C is a constant) is taken and also a $K_i(x)$ belonging to a simple type and, then, the approximate solution of the integral equation at several different C-values is determined. If the graphs of the approximate solutions are compared, one of them is chosen according to considerations depending on the investigated case and, by this, the "best" constant C; one chooses e.g. the "smoothest" graph. The determination of error bounds is, in general, neglected.

The above numerical procedure needs different methods of minimization (see e.g. the survey article of Levitin, Poljak (1966)). However, they will not be dealt with here.

With regard to the above numerical methods it is possible to consider, instead of the discretized smoothing functional, the Euler equation whose investigation may be substituted for that of the functional. After discretizing the Euler equation its solution is determined; this will be, principally, an approximation belonging to the same type as $\tilde{z}_{\alpha, j}$. In this case, not a minimization, but the resolution of a certain system of linear equations is needed (Tihonov, Glasko 1964); a methodological example was considered by these authors, consequently, the optimal value of $\alpha = \alpha(\delta)$ was determined by subsequent trials.

The steps of the latter method may be demonstrated according to an example of Glasko, Zaikin (1966).

Let us consider again the integral equation

$$\int_a^b K(x, \xi) z(\xi) \, d\xi = u(x) \quad (c \leqq x \leqq d)$$

if for $u(x) \in U$, $z(\xi)$ is a solution. Its approximate solution $z_\delta^\alpha(\xi)$, — if instead of $u(x)$ only $u_\delta(x)$, $(\varrho(u_\delta(\xi), u(\xi)) \leqq \delta)$ is known — will be the function minimizing the functional

$$M_n^{(\alpha)}[z, u_\delta] = \int_c^d \left\{ \int_a^b K(x, \xi) z(\xi) \, d\xi - u_\delta(x) \right\}^2 dx + \alpha \int_a^b \sum_{i=0}^{n+1} k_i(\xi) [z^{(i)}(\xi)]^2 \, d\xi$$

where $k_i(\xi)$ is some given non-negative function, and α is the regularization parameter, varied during the calculations. Let us take $n=2$. Then this minimization problem is equivalent to the resolution of the boundary problem for the Euler equation

$$\int_a^b \overline{K}(\eta, \xi) z(\xi) \, d\xi + \alpha \left\{ \frac{d^2}{d\eta^2} \left[k_2 \frac{d^2 z}{d\eta^2} \right] - \frac{d}{d\eta} \left[k_1 \frac{dz}{d\eta} \right] + k_0 z \right\} = b(\eta) \quad (a \leqq \eta \leqq b)$$

at the boundary conditions

$$k_2(\eta) z''(\eta) \bigg|_{\eta = a,b} = 0; \quad k_1 z' - \frac{d}{d\eta} \left(k_2 \frac{d^2 z}{d\eta^2} \right) \bigg|_{\eta = a,b} = 0,$$

where

$$\overline{K}(\eta, \xi) = \int_c^d K(x, \eta) \, K(x, \xi) \, dx; \quad b(\eta) = \int_c^d K(x, \eta) \, u_\delta(x) \, dx.$$

This was given by Tihonov, Glasko (1964). Now to arrive at a numerical method of obtaining $z(\xi)$, the latter problem is discretized by taking some quadrature formula and finite difference derivative expressions instead of the exact terms in the Euler equation, and starting from "measured" $\tilde{u}_\delta(x)$ instead of $u_\delta(x)$. One finally gets a system of algebraic equations (with $n=2$)

$$\sum_{j=1}^N a_{\alpha, j} \tilde{z}_{\alpha, j} + \alpha \sum_{s=i-k}^{i+k} \delta_{is} \tilde{z}_{\alpha, s} = b_i \quad (i = 1, 2, ..., M)$$

where $\alpha_{ij} = \overline{K}_{ij} \sigma_j$, $\overline{K}_{ij}$ are the discretized values of $\overline{K}(\eta, \xi)$, σ_j are the used quadrature formula coefficients, δ_{is} the coefficients in the approximate expressions of the derivatives, used in a given case and based on k function values and b_i are the discretized values of $b(\eta)$. The solution $\tilde{z}_{\alpha, j}$ of this system of equations, taking the (discretized) boundary conditions into account, will yield values which approximate $\tilde{z}_\delta^{(\alpha)}(\xi)$ and also the exact solution $z(\xi)$. In Glasko, Zaikin (1960) the relevant computer programme is also given in detail.

The formulae to be applied in the case of the convolution equations appearing in our investigations can be deduced automatically from the above.

Supplements and problems to Ch. V, § 1

1. Although the *solution of Fredholm or Volterra integral equations of the* Ist *kind* is an incorrect problem and its numerical performance is thus a delicate problem, a vast number of articles have been devoted earlier to the latter because it appeared again and again in physics or astronomy. Theoretically the exact solution can be given, in most cases (cf. Schmeidler 1950). The early numerical methods consisted of solving a system of equations by discretizing the integrals; or the solution was assumed to be a function series $\sum_{n=0}^{M} a_n \varphi_n(x)$ where $\varphi_n(x)$ is fixed, M is appropriately varied and the coefficients a_n are calculated from a suitable system of algebraic equations or by some least square technique or by the principle of moments. A major part of the work was devoted, exclusively, to integral equations of convolution type. Some methods already stressed in essence the ideas to be considered in the present chapter.

From the relevant literature titles are given that may be regarded as representative of the main methods as well as yielding valuable ideas concerning the instrumental solution: Runge (1914); van Cittert (1931); Ornstein, van Wyk (1932); Burger, van Cittert (1932, 1933); Schulz (1934); Gyllenberg (1936) (a graphical method); Crout (1940); van de Hulst (1941); Kremer (1941); Righini (1941); Hildebrand, Crout (1941); Reiz (1943); van de Hulst (1946a), (1946b); Stokes (1948); Kreisel (1949); Keating, Warren (1952); Elste (1953); Brunk (1953); Trumpler, Weaver (1953) Chapter 1.5; Young (1954); Bracewell, Roberts (1954); Bracewell (1955a), (1955b); Kahn (1955); Unsöld (1955) pp. 252—265; Carver, Lokan (1957); Childers (1959); Loeb (1960); Flynn, Seymour (1960); Thies (1961); Jones (1961); Medgyessy (1961a) pp. 80—93, 172—188; Larson, Kenneth (1967); Linz (1969); see also the statements on the solving of convolution integral equations with Gaussian kernel described in Ch. III, § 1, Subsection 1.1.1.1. Some of the methods used in the latter can be extended also to integral equations having a more general form.

2. It should be mentioned that there is a relation between the regularized family of approximate solutions $\{z^{(\alpha)}\}$ and the quasi-solutions described in Ch. V, § 1, Section 3. It has been shown by Dombrovskaja, Ivanov (1965) that the family $\{z^{(\alpha)}\}$ is identical with the family of quasi-solutions, defined on a certain special compact set $\mathfrak{M}$ ($\mathfrak{M} \subset Z$ (a metric space)). In the procedure described in connection with the regularization method, however, no compact set has to be fixed (although the solution is the element of some special set and this fact is involved in the particular form of the smoothing functional).

3. In connection with the regularization method a great number of papers have been devoted to the investigation of a *more precise relation between* α *and*

δ in special cases (cf. e.g. Tihonov, Glasko 1964, 1965; Ivanov 1966a, 1966b, 1967) because the methods depend, essentially, on the value of α used.

4. In the regularization method the *dependence on δ of the distance between the approximate solution $\tilde{z}_\delta$ and the exact solution $\bar{z}$* (that is $\varrho(\bar{z}, \tilde{z}_\varrho)$, the goodness of the approximation) is also very important. In the case of the general operator equation (1) in Ch. V, § 1, Section 4 an exact description can be given, as shown by the investigations of V. K. Ivanov, only in that case, if for the operator equation (1) *"uniform regularization"* holds, it is carried out by means of an operator $\hat{R}_\alpha$ which is different from the R_α used there. $\hat{R}_\alpha$ is affected, among other things, by the fact that in point 3 of the definition of R_α in Ch. V, § 1, Section 4 convergence is substituted by uniform convergence; for further details see Ivanov (1966a), (1966b).

5. If the exact solution of the initial operator equation is known in *analytical* form, in many cases a regularizer R_α can be constructed from this in the regularization method in Ch. V, § 1, Section 4; then, by means of R_α, an approximate solution can be constructed. In addition, if for a function calculated in some way by the "measured" data $\hat{u}_\delta$ it can be proved that for $\varrho(u, \hat{u}_\delta) \leq \delta$ where u is the exact solution, $\varrho(z, \hat{z}) \leq \varepsilon(\delta)$, where $\varepsilon(\delta) \downarrow 0$ as $\delta \downarrow 0$, then $\hat{z}$ may justifiably be regarded as the solution of our operator equation (in such a case the regularizing parameter is one of the figuring parameters). In this case neither the construction of R_α nor the minimization of the functional $M_n^{(\alpha)}[z, \tilde{u}_\delta]$ in Ch. V, § 1, Section 4 is needed. ($\hat{z}$ can be defined e.g. by some iteration procedure built on $\tilde{u}_\delta$ etc.)

6. Up to now a fairly large number of possibilities have been presented for the application of the regularization method. But from the viewpoint of the present book it is worth presenting here, separately, a more detailed and generalized investigation of a special type of operator equation of the Ist kind than we have already considered, namely the *Fredholm integral equation of the Ist kind of convolution type*. (We have already considered it from the viewpoint of other methods in 1 above.)

In solving this equation the ideas of the regularization method could also be applied automatically; however, a separate special method for this has been worked out by Strahov (1968). This is as follows.

Let us consider the integral equation

(1) $$\frac{1}{\sqrt{2\pi}} \int_{-\infty}^{\infty} K(x-\xi)\,\varphi(\xi)\,d\xi = f(x) \qquad (-\infty < x < \infty)$$

where $\varphi \in L_2(-\infty, \infty)$, $f \in L_2(-\infty, \infty)$ and, if $k(t)$ is defined by $k(t) =$

$$k(t) = \frac{1}{\sqrt{2\pi}} \int_{-\infty}^{\infty} K(x) e^{-itx} dx, \quad |k(t)| \leq 1, \quad k(t) \neq 0 \text{ for almost every } t.$$ We
suppose that for a given $\hat{f}(x)$ a (unique) solution $\phi(\xi)$ of (1) exists; evidently,
this can be given analytically. Solving (1) is an incorrect problem in the sense
of Hadamard. Nevertheless, an approximate solution can be defined, if the
following concepts are introduced, even independently of (1).

1. The linear operator $\mathbf{L}$ ($\|\mathbf{L}\| < \infty$, ($\|\ \|$ denotes norm) acting in the space L_2 is
called *numerically finite dimensional* if, in determining the value of $\mathbf{L}f$ ($f \in L_2$) at
a point x one makes use of a finite number of the values of $f(x)$, performing a
finite number of arithmetic operations on them.

2. The two-parameter linear operator $\mathbf{T}_{\alpha,n}$ ($0 \leq \alpha \leq \alpha_1$; $n_0 \leq n < \infty$; α_1, n_0 are
given) is said to be *uniformly regularizing* our integral equation on a set $\mathfrak{M}$ ($\mathfrak{M} \subset L_2$)
if $\|\mathbf{T}_{\alpha,n}\| < \infty$ ($n \neq \infty$, $\alpha \neq 0$) and if for every $\varphi \in \mathfrak{M}$ and for at least one sequence
$(\alpha, n) \|\varphi - \mathbf{T}_{\alpha,n} A\varphi\| \to 0$, uniformly, as $\alpha \to 0$, $n \to \infty$. (This is an extension, according
to the meaning of the above concepts, of the "regularizing" of A. N. Tihonov
and V. K. Ivanov, respectively.)

In practice, only some "measured" $\hat{f}_\delta(x)$-values are at our disposal instead of
$\hat{f}(x)$ in (1) where $\|\hat{f} - \hat{f}_\delta\| \leq \delta$ (the metric being given by the norm, as usual). It can
be proved that if we know the two-parameter linear operator $\mathbf{T}_{\alpha,n}$ that uniformly
regularizes our equation on $\mathfrak{M}$, then $\mathbf{T}_{\alpha,n} \hat{f}_\delta$ will be an approximate solution which
approximates $\hat{\phi}$, at an appropriate $\alpha = \alpha(\delta)$, arbitrarily, if δ decreases. Criteria
for $\mathbf{T}_{\alpha,n}$ to be uniformly regularizing can be given.

V. N. Strahov took the operator $\mathbf{S}_{\alpha,n}$, defined, for a suitable $h(x)$, by
$\mathbf{S}_{\alpha,n}\{h(x)\} = \sum_{k=-n}^{n} c_k h(x+k\alpha)$ where the values $c_k = c_k(\alpha, n)$ are constants to be
fixed later, to be a uniformly regularizing, two-parameter linear operator to our
integral equation of the convolution type. Thus he constructed here an operator
generating an approximate solution and not the approximate solution itself, e.g.
by means of minimizing some functional. In this case the author assumed the
approximate solution of our equation, in accordance with the application of the
regularizer described above, in the form

$$\mathbf{S}_{\alpha,n}\{\hat{f}_\delta(x)\} \equiv \hat{\phi}_{\delta,n}^{(\alpha)}(x) = \sum_{k=-n}^{n} c_k \hat{f}_\delta(x+k\alpha).$$

Then he proved that if

1.

$$c_k = \frac{\alpha}{2\pi} \int_{-\frac{\pi}{\alpha}}^{\frac{\pi}{\alpha}} \frac{e^{ik\alpha t}}{k(t)} dt$$

(this choice is natural from the point of view of the analytical structure);

2. $\mathfrak{M}_1 \subset L_2$ is the set of bounded functions $\hat{\phi}_\sigma(x)$ $(\|\hat{\phi}_\sigma\| \leqq N)$ whose Fourier transform $\Phi_\sigma(t)$ satisfies the condition $\Phi_\sigma(t) \equiv 0$ if $|t| > \sigma$;

3. for $\alpha > 0$

$$\sup_{|t| \leqq \frac{\pi}{\alpha}} \left| \frac{1}{k(t)} \right| < \infty \quad \text{and} \quad \left| 1 - \frac{k(t)}{k_p^{(\alpha)}(t)} \right| \leqq C, \quad |t| > \frac{\pi}{\alpha},$$

where C is a constant, $k_p^{(\alpha)}(t)$ is the periodic repetition with period $\dfrac{2\pi}{\alpha}$ of the values of $k(t)$ in the interval $|t| \leqq \dfrac{\pi}{\alpha}$;

4.

$$\frac{1}{k_p^{(\alpha)}(t)} = \sum_{k=-\infty}^{\infty} c_k e^{-ik\alpha t} \quad (-\infty < t < \infty)$$

is uniformly convergent if $\alpha > 0$;

then $S_{\alpha,n}$ is uniformly regularizing on the set $\mathfrak{M}_1$; and if the solution $\hat{\phi}(x)$ satisfies our equation (1) and $\hat{\phi} \in \mathfrak{M}_1$ and the "measured data" $\hat{f}_\delta$ satisfy $\|f - f_\delta\| \leqq \delta$, then for the deviation of the function $\hat{\phi}_{\delta,n}^{(\alpha)}(x) = \sum\limits_{k=-n}^{n} c_k \hat{f}_\delta(x+k\alpha)$ from $\hat{\phi}$, $\|\hat{\phi} - \hat{\phi}_{\alpha,n}\|$, $\|\hat{\phi} - \hat{\phi}_{\alpha,n}\| \leqq \varepsilon(\delta)$ holds, where $\varepsilon(\delta) \downarrow 0$ as $\delta \downarrow 0$, provided that α and n have been chosen so that

$$(2) \qquad \frac{\delta}{N} \sup_{|t| \leqq \sigma} \left| \sum_{k=-n}^{n} c_k e^{-ik\alpha t} \right| + \sup_{|t| \leqq \sigma} \left| \frac{1}{k_p^{(\alpha)}(t)} - \sum_{k=-n}^{n} c_k e^{-ik\alpha t} \right|$$

assumes its minimal value. In other words: $\hat{\phi}_{\delta,n}^{(\alpha)}$ can then be considered as an approximate solution of our integral equation.

The course of the numerical solution of our convolution integral equation (1) is more simple than in applying automatically the regularization method. This is immediately shown by the form

$$\hat{\phi}_{\delta,n}^{(\alpha)}(x) = \sum_{k=-n}^{n} c_k \hat{f}_\delta(x+k\alpha)$$

of $\hat{\phi}_{\delta,n}^{(\alpha)}$, for $\hat{f}_\delta$ are the "measured data" at our disposal and the functional (2) yielding the optimal values of α and n has not to be discretized; it can be investigated in its original form. It is an advantage that if we already have α and n, then we *do not need* a minimization or an investigation of a system of linear equations to determine the approximate solution, in contrast to several other methods we have described. On the other hand, it is true that several newer difficulties appear. It is noticed also that the values c_k depend on δ.

The preceding clearly shows that it is possible to choose c_k in another way; if the linear operator defined by the sum enjoys certain properties, it is only essential that the operator we obtain should remain uniformly regularizing.

Furthermore, for a given problem there will be an optimal ensemble $\{c_k\}$; to find it is, however, an unsolved **problem**.

It was a weak point of the procedure that the Fourier transform $\Phi_\sigma(t)$ must be zero outside a given interval; in many instances this would not be fulfilled, but one conjectured that this introduces no great error. Indeed, it has been proved that, in the case of a function $\phi(x)$ whose Fourier transform $\Phi_\sigma(t)$ is rapidly (e.g. in $O(e^{-At^2})$) decreasing and, consequently, is approximately zero outside some interval, the method we have described can also be applied (Fiala 1975). Thus its practical importance may be ascribed to this.

7. The present method of V. N. Strahov is, essentially, the exact elaboration and investigation of the resolution of the integral equation

$$\int_{-\infty}^{\infty} K(x-\xi)\varphi(\xi)\,d\xi = f(x) \qquad (\varphi(\xi)\in L_2(-\infty,\infty))$$

given by him in an earlier paper (Strahov 1963). The basic idea of that is similar to the idea appearing in his method of computing numerically the convolution transform to be considered in *Supplements and problems to* Ch. V, § 4, Section 3. Namely, he wanted to approximate the exact solution of the present equation by $\bar{\varphi}(x) = \sum_{j=-m}^{m} b_j f(x+jh)$ (h is given). Putting $\mathbf{F}[K(x);t]=k(t)$, $\mathbf{F}[\varphi(x);t]=\Phi(t)$, he obtained for the error δ of this approximation, by the Parseval formula — after existence problems had been decided — that

$$\delta^2 = \int_{-\infty}^{\infty} |\varphi(x)-\bar{\varphi}(x)|^2\,dx = \frac{1}{2\pi}\int_{-\infty}^{\infty} |\Phi(t)|^2\left|\frac{1}{k(t)} - \sum_{j=-m}^{m} b_j e^{ijht}\right|^2\,dt.$$

Then he *assumed* that $|\Phi(t)|^2 \leq |H(t)|^2$, where $H(t)$ is *already known*, and determined the b_j from the condition that they minimize the integral

$$\frac{1}{2\pi}\int_{-\infty}^{\infty} |H(t)|^2\left|\frac{1}{k(t)} - \sum_{j=-m}^{m} b_j e^{ijht}\right|^2\,dt$$

which majorizes δ^2. With these b_j and "measured" values $\hat{f}(x)$ of $f(x)$ he took

$$\hat{\varphi}(x) = \sum_{j=-m}^{m} b_j \hat{f}(x+jh)$$

as an approximation of $\varphi(x)$. Errors and the question of correctness were not examined. The above exact result, as we have seen, is more restricted and modest than the latter, but the correctness has been ensured.

Further, V. N. Strahov obtained good approximations to the solutions of several special integral equations of convolution type. He gave values of b_j for

16   Medgyessy: Decomposition ...

these special cases. For instance he solved numerically the equation

$$f(x) = \frac{1}{2\lambda} \int\limits_{-\infty}^{\infty} \varphi(\xi) e^{-\frac{|x-\xi|}{\lambda}} d\xi;$$

here $k(t) = \dfrac{1}{1+\lambda^2 t^2}$ and $H(t) = e^{-H|t|}$ were taken and $m = 5$, $\dfrac{H}{h} = 2$, $\dfrac{\lambda}{h} = 1$.

These yielded a good approximate solution with the corresponding b_j-values

j	0	1	2	3	4	5
$b_j = b_{-j}$	2.2473	-2.1640	0.5817	-0.2345	0.0957	-0.0279

8. In the main text, we considered only the solution by the regularization method of incorrect problems connected with operator equations and Fredholm integral equations of the Ist kind. However, *several incorrect problems* in other fields are also treatable by this method and by its modifications; some of them have been used in the present work, too. From the extensive literature the following works are mentioned, according to the various spheres of problems: *Numerical differentiation:* Dolgopolova, Ivanov (1966); Tihonov (1967); Ramm (1968); Liskovec (1968); Vasin (1969). *Solution of systems of linear equations:* Tihonov (1965c), (1965e). *Partial differential equations:* Morozov, Ivaniščev (1966); Liskovec (1966); Gavurin (1967); Čudov (1967); Vasin (1968); Arcangeli (1968). *Ordinary differential equations:* Gorbunov (1967). *Linear programming:* Tihonov (1965d). *Theory of optimal processes:* Tihonov (1965b); Tihonov, Galkin, Zaikin (1967). — See also 9 below. — We do not need them, however, in the present book; hence the reader is referred to the preceding relevant literature.

9. Recently, incorrect problems related to operator equations of the Ist kind have been considered from *newer aspects*. Certain of these (without classifying them according to fields, are to be found in: Oettli, Prager (1964); Miller (1964); Bellman, Kalaba, Lockett (1965a), (1965b), (1966); Lattès, Lions (1967) (chiefly the subject of Ch. 4 and 8, i.e. the investigation of elliptic partial differential equations recalls the regularization method of A. N. Tihonov and his school applied to similar equations); Bellman (1969); Stallman (1970); Vemuri, Chen (1974). An idea similar to that of the regularization method can be found in the works of Grabar (1967a), (1967b) although the main objective here was to approximate the solution of integral equations and the results of a numerical differentiation, by means of Čebyšev polynomials orthogonal on equidistant points.

A treatment with a background of *probability theory* also occurs: see Sudakov, Halfin (1964); Lavrent'ev, Vasil'ev (1966); Petrov (1967); Turčin (1967), (1968); Strand, Westwater (1968a), (1968b); Franklin (1970). The relevant results are still under development and will not be considered here.

§ 2. A SPECIAL METHOD FOR THE NUMERICAL SOLUTION OF CERTAIN INTEGRAL EQUATIONS OF CONVOLUTION TYPE

For the numerical solution of the convolution integral equation

$$(1) \quad \int_{-\infty}^{\infty} K(y-x) f(x) \, dx = g(y) \qquad (f(x) \in L_2(-\infty, \infty), \; g(y) \in L_2(-\infty, \infty))$$

Medgyessy, Varga (1968) gave a special procedure similar to the regularization method and taking into account principally that $g(y)$ can be measured only with some error: some "noise" is superimposed on its values to be measured. The procedure, whose basic idea can be found without detailed working-out in Hamming's work (1962 pp. 321—322), preassumes that the conditions for the existence of the exact solution

$$(2) \quad f(x) = \frac{1}{2\pi} \int_{-\infty}^{\infty} e^{-ixt} \frac{\mathbf{F}[g(y); t]}{\mathbf{F}[K(y); t]} dt$$

are satisfied; further that the graph of $g(y)$ is known in such a long interval that the measured ordinate values at the interval ends are approximately zero and, in addition, equidistant (approximate) $g(y)$-values to be disposed of lie sufficiently densely.

The method makes use of (2); we want to compute it approximately. First the transform $\mathbf{F}[g(y); t]$ is formed numerically. Under our conditions it is best to use the trapezoidal formula. Thus the "spectrum" of the "measured" values of $g(y)$ is obtained for equidistant values of t. This "spectrum" is, theoretically, approximately zero outside some interval; however, in its numerical approximation, the same feature cannot be expected because of the "noise"; in fact we carry out the procedure also on the "measuring" errors of $g(y)$: on the "noise"; thus the spectrum approximation will show, after an initial diminishing character, an *oscillation without damping,* between approximately constant bounds outside some t-interval. The latter section originates, roughly, only from the "noise".

Now *this* interval is "cut off" (symmetrically) then multiplied by $\dfrac{1}{\mathbf{F}[K(x); t]}$ and finally, by means of the trapezoidal formula, the approximation of the inverse Fourier transform is formed on the product. The result is an approximation of $f(x)$ at equidistant points, but under some *"filtering"* of the effect of the "noise"; however, it also blurs the influence of the function $g(y)$ a little. The correctness of the filtering and, at the same time, of the approximation, depends on the place of the "cut off"; the disadvantages of the blurring could be significantly counterbalanced by the filtering, if the place T of the "cut off" has been chosen appropriately. This T has the same role as the regularization parameter in the

16*

regularization method. Neither is this known exactly; thus the "cut off" has to be made "by eye", at that point where the graph of the "spectrum" approximation is the narrowest. We work with what remains carrying out the described numerical procedure repeatedly, giving always another value to the "cut off" place, each being near to the previous one. Then the separate results are compared and, as usual, the "smoothest" is accepted as the graph of the best approximation to $f(x)$. Hence it follows, too, that the "cut off" can be carried out on the graph of the approximation of $\dfrac{\mathbf{F}[g(x); t]}{\mathbf{F}[K(x); t]}$, the rest being as in the above description — since an experimentation with the place of the "cut off" is used.

It can be shown that if $f(x)$ is unimodal then the "cut off" which means, theoretically, the multiplication of $\mathbf{F}[g(y); t]$ by a "filtering" function $D(t) = \begin{cases} 1 & (|t| < T) \\ 0 & \text{otherwise} \end{cases}$ ("cut off filter"), annihilates this unimodality from the beginning onwards. This situation arises *a priori* in the numerical case. The same feature can be observed in other numerical methods used in solving the integral equation (1). Details of the influence of the choice of the place of "cut off" on the result as well as some error analysis will be omitted here; so will the choice of the parameters in the used trapezoidal formula — e.g. the distance of the equidistant data points; see Medgyessy, Varga (1968). Clearly, "measured" ordinate values at equidistant points of small separation will, in principle, yield a better approximation to $f(x)$ if "optimality" is aimed at.

Sometimes the "cut-off" is carried out *after* the multiplication of the "spectrum" by $\dfrac{1}{\mathbf{F}[K(x); t]}$ because of certain advantages of this in practice. The effectiveness of the numerical procedure is not influenced by this, because one has, just the same, to choose from the results obtained by altering the parameters, the "optimal" one (according to some certain condition — smoothness etc.), after having considered each of them.

Supplements and problems to Ch. V, § 2

1. As to the role of the trapezoidal formula in calculating "spectra" and convolution integrals, see Jones (1961); de Balbine, Franklin (1966).

2. In the case of a (0) symmetrical "cut off" the whole procedure represents a regularization method for solving (1) in Ch. V, § 2, and the place of the "cut off" gives the value of the regularization parameter (cf. Marton, Varga 1971).

3. Naturally the "filter" $D(t)$ can have not only the described type; the simple "cut off" can be replaced by a strong smoothing. The different possibilities have different analytical advantages.

If e.g. $D(t)=e^{-\varepsilon t^2}$ ($\varepsilon>0$), i.e. it is (0) symmetrical, then for a certain type of $g(y)$

$$f(x) = \frac{1}{2\pi} \int_{-\infty}^{\infty} e^{-ixt} \frac{\mathbf{F}[g(y); t]}{\mathbf{F}[K(x); t]} e^{-\varepsilon t^2} dt$$

will be, theoretically, unimodal. This formula occurred in Ch. III, § 1, Subsection 2.2 in the construction of the test function for the decomposition of certain superpositions (e.g. of ch density functions) by means of a convolution transformation. Thus the procedure used there has also a "noise diminishing" effect if it is carried out numerically (cf. the results of Medgyessy 1967a), which are also the source of the preceding statements).

In applying the above filter $D(t)$,

$$f(x) = \frac{1}{2\pi} \int_{-\infty}^{\infty} e^{-ixt} \frac{\mathbf{F}[g(y); t]}{\mathbf{F}[K(x); t]} D(t) dt.$$

If

$$\frac{D(t)}{\mathbf{F}[K(x); t]} = \mathbf{F}[M(x); t],$$

i.e. the Fourier transform of a (non-density) function $M(x)$, then

$$f(x) = \int_{-\infty}^{\infty} M(x-y) g(y) dy;$$

thus $f(x)$ can be represented by a convolution transform which is a *correct problem*. $M(x)$ can be tabulated, in practice, for several types of $D(t)$ and $K(x)$; the computation of $f(x)$ is, however, practically the same in the end as in computing the "spectrum", product, and inverse Fourier transform in Ch. V, § 2. These statements appeared in Ch. III, § 1, Subsection 2.2 and were used in constructing a test function. However, in the application of the present convolution transform the "by eye" "cut off" of the "spectrum" of $g(y)$, obtained numerically, cannot be interpolated; the first result is to be calculated "unseen" with some sequence of filters of different parameters and, then, they are to be compared.

In practice $M(x)$ can also be tabulated for different types of $D(t)$ and $K(x)$. Calculating $f(x)$ in this way is, essentially, the same in practice as its described calculation by means of the construction of the "spectrum", the product and the inverse Fourier transform.

This supplement can essentially be found in Medgyessy (1961a p. 181).

4. The numerical application of the method described in the present § 2 is similar, in several respects, to the method suitable for the solution of convolution integral equations, applying the numerical or instrumental Fourier analysis and Fourier synthesis as described in Ch. III, § 1, Subsection 1.1.1.1. In the latter a "cut off" of the "spectrum" did not appear.

5. Clearly the present method can be applied also to the *numerical calculation of a convolution transform*. Namely, if

$$f^*(x) = \int_{-\infty}^{\infty} K^*(x-y)\, g^*(y)\, dy$$

is to be calculated (if it exists) by means of "measured" values of $g^*(y)$ then, under suitable conditions, we have

$$f^*(x) = \frac{1}{2\pi} \int_{-\infty}^{\infty} e^{-ixt} \mathbf{F}[g^*(y); t]\ \mathbf{F}[K^*(y); t]\, dt$$

and the computation procedure is *formally* the same as above in case of $f(x)$ in Point 3; we have merely to give the role of $\dfrac{1}{\mathbf{F}[K(x); t]}$ to $\mathbf{F}[K^*(y); t]$.

Evidently the "cut off" is not as significant here as in case of the integral equation because it is substituted by the fact that $\mathbf{F}[K^*(y); t]$ decreases, in general, rapidly if $t \to \pm\infty$.

This method often gives a better result than those described later, in Ch. V, § 4, e.g. if $K^*(x)$ varies "rapidly" or oscillates frequently but $\mathbf{F}[K^*(x); t]$ and $\mathbf{F}[g^*(y); t]$ do not do so.

§ 3. ANOTHER METHOD FOR THE NUMERICAL SOLUTION OF CERTAIN INTEGRAL EQUATIONS OF CONVOLUTION TYPE

In Ch. V, § 1 we have read that a) if instead of the right-hand side of an operator equation $\mathbf{A}x = y$ (defined in a convenient normed function space where for the "distance" between two functions the norm of their difference is taken) only an approximation, a "measured" value y_δ of y is known, and for its distance from y we have $\|y - y_\delta\| < \delta$; b) if, by analytical or numerical-analytical considerations, we can give the approximation to the solution x in the form $\mathbf{S}y_\delta = x_\delta$ where $\mathbf{S}$ is a convenient linear operator acting also on y and built on some analytical approximation to the solution of the equation; c) if in the relations (where "distances" are expressed by the norms)

$$\|x - x_\delta\| = \|x - \mathbf{S}y_\delta\| \leqq \|x - \mathbf{S}y\| + \|\mathbf{S}y - \mathbf{S}y_\delta\| \leqq$$
$$\|x - \mathbf{S}y\| + \|\mathbf{S}\| \cdot \|y - y_\delta\| \leqq \|x - \mathbf{S}y\| + \delta \cdot \|\mathbf{S}\|$$

(being valid here) we have, for a suitable choice of S, $\|x-S_y\| \leq \varepsilon(\delta)$ (where $\varepsilon(\delta) \downarrow 0$ as $\delta \downarrow 0$ and $\|S\|$ is bounded, see Ch. V, § 1), then the deviation between the exact solution and the approximate one, $\|x-x_\delta\|$, can be made arbitrarily small provided S is chosen conveniently and the "measurement" error δ is sufficiently small. That is x_δ can be considered as an approximate solution of $Ax=y$, obtained by the application of the linear operator S.

This sequence of ideas is the basis in most cases for finding the numerical solution of an operator equation of the Ist kind and for estimating the error of the approximate solution. In broad lines this could be seen also in the preceding point, e.g. in the regularization method.

It is on this that a method of solving approximately certain convolution integral equations different from the regularization method is based. Here the operator means the constructing of a linear expression of the function values measured at mesh points around a fixed point. This will be dealt with in the present paragraph.

The starting point is the following

Theorem 1. Let

$$(1) \qquad g(y) = \int_{-\infty}^{\infty} K(y-x)h(x)\,dx \qquad (-\infty < x < \infty; \quad -\infty < y < \infty)$$

be a Fredholm integral equation of the Ist integral convolution type where $K(x)$, $h(x)$, $g(y)$ are continuous and for which it holds that 1. *there exists a (unique) solution;* 2. $F[g(y); t] = \gamma(t)$, $F[K(x); t] = \varkappa(t)$, $F[h(x); t] = \chi(t)$ *exist and are absolutely integrable in* $(-\infty, \infty)$; 3. $\dfrac{1}{\varkappa(t)}$ *is an even integral function and*

$$\frac{1}{\varkappa(t)} = \sum_{n=0}^{\infty} a_{2n} t^{2n};$$ 4. $g^{(2r)}(y)$ $(r = 1, 2, ...)$ *exist;* 5. *the sum of the function*

series $\sum_{n=0}^{\infty} |a_{2n}\gamma(t)t^{2n}|$ *is integrable in* $(-\infty, \infty)$; *then*

$$(2) \qquad\qquad h(x) = \sum_{n=0}^{\infty} (-1)^n a_{2n} g^{(2n)}(x).$$

If, moreover, we have 6. $\dfrac{1}{\varkappa(t)} = \varkappa(it)$; 7. $g(z)$ *is an integral function for a*

complex variable z; and 8. *the sum of the function series* $\sum_{n=0}^{\infty} \left| g^{(2n)}(x) K(y) \dfrac{y^{2n}}{(2n)!} \right|$ *is*

integrable with respect to y in $(-\infty, \infty)$, *then*

$$(3) \qquad\qquad h(x) = \int_{-\infty}^{\infty} K(y) g(x+iy)\,dy$$

(John 1955b, Medgyessy 1966a).

Proof. By our assumptions

$$\chi(t) = \frac{\gamma(t)}{\varkappa(t)} = \sum_{n=0}^{\infty} a_{2n} \gamma(t) t^n$$

and

$$\frac{1}{2\pi} \int_{-\infty}^{\infty} \chi(t) e^{-iyt} dt = h(y) = \frac{1}{2\pi} \int_{-\infty}^{\infty} \left[\sum_{n=0}^{\infty} a_{2n} \gamma(t) t^{2n} \right] e^{-iyt} dt =$$

$$\sum_{n=0}^{\infty} a_{2n} \frac{1}{2\pi} \int_{-\infty}^{\infty} \gamma(t) t^{2n} e^{-iyt} dt = \sum_{n=0}^{\infty} (-1)^n a_{2n} g^{(2n)}(y).$$

If also 6—8 hold, $\varkappa(t) = \sum_{n=0}^{\infty} (-1)^n a_{2n} t^{2n}$ is also an even integral function. Thus $K(y)$ is even, the moments $\int_{-\infty}^{\infty} y^{2n} K(y) \, dy$ exist and $(-1)^n a_{2n} = \dfrac{\int_{-\infty}^{\infty} y^{2n} K(y) \, dy}{2n!}$, i.e.

$$h(x) = \sum_{n=0}^{\infty} g^{(2n)}(x) \int_{-\infty}^{\infty} \frac{y^{2n} K(y)}{(2n)!} dy =$$

$$\int_{-\infty}^{\infty} K(y) \left(\sum_{n=0}^{\infty} g^{(2n)}(x) \frac{y^{2n}}{(2n)!} \right) dy = \int_{-\infty}^{\infty} K(y) g(x+iy) \, dy.$$

The following part gives, *in the case of the validity of* (3), a method for the *numerical solution of* (1), based on the "measured" values $\hat{g}(x)$ of $g(x)$.

We shall define an approximate solution to (1), built on the values $\hat{g}(x)$. We shall show that this converges to the exact solution if the measurement error diminishes and if we make use of more and more $\hat{g}(x)$ values. Thus the solution of the integral equation (1) will be an approximate solution in the sense of A. N. Tihonov. Unfortunately the error in the numerical method may be rather large.

We ask for the approximate solution $\hat{h}(x)$ of (1) in the following form:

$$(4) \qquad\qquad \hat{h}(x) = \sum_{j=-m}^{m} w_j^{(m)} \hat{g}(x+jh)$$

(h is given), where $w_j^{(m)}$ is defined as follows. Being analytic, $g(x+iy)$ can be written by means of a complex interpolation formula built on the points $x+jh$. After some simplifications, we have

$$g(x+iy) = \sum_{j=-m}^{m} g(x+jh) \frac{(-1)^{m-j}}{h^{2m}(m+j)!(m-j)!} \frac{\prod_{\varrho=-m}^{m} (iy-\varrho h)}{(iy-jh)} + R_m(x,y)$$

where the remainder term is

$$R_m(x, y) = \frac{1}{2\pi i} \int_{\mathscr{C}} \frac{g(w)}{w - x - iy} \prod_{\varrho=-m}^{m} \left(\frac{iy - \varrho h}{w - x - \varrho h} \right) dw$$

where $\mathscr{C}$ is a simple closed curve containing the points $x+iy$, $x+jh$ ($j=0, \pm 1, \ldots, \pm m$) in its interior and $g(w)$ is analytic in the same domain. Inserting this into (3) we get

$$(5) \qquad h(x) = \sum_{j=-m}^{m} w_j^{(m)} g(x+jh) + \int_{-\infty}^{\infty} K(y) R_m(x, y) \, dy,$$

where

$$(6) \qquad w_j^{(m)} = \frac{(-1)^{m-j}}{h^{2m}(m+j)!(m-j)!} \int_{-\infty}^{\infty} K(y) \frac{\prod\limits_{\varrho=-m}^{m}(iy - \varrho h)}{(iy - jh)} \, dy.$$

Now for the approximation $\hat{h}(x)$ of $h(x)$ built on the "measured" values $\hat{g}(x)$ of $g(x)$ we take the first term $\sum\limits_{j=-m}^{m} w_j^{(m)} g(x+jh)$ on the right-hand side of (5), writing $\hat{g}(x)$ instead of $g(x)$. It is in this way that we obtain (4) and define $w_j^{(m)}$.

Let us apply the considerations of the introduction to the present paragraph. The deviation of $\hat{h}(x)$ from the exact solution $h(x)$ is

$$h(x) - \hat{h}(x) = \int_{-\infty}^{\infty} K(y) R_m(x, y) \, dy + \sum_{j=-m}^{m} w_j^{(m)} [g(x+jh) - \hat{g}(x+jh)].$$

The first term in the right-hand side is the difference of $h(x)$ and of the main part — as an approximation — of the expression of $h(x)$ given by the interpolation formula. The second term contains the "measurement error" of $g(x+jh)$. If the upper bound of the absolute value of the latter is δ, that is $|g(x) - \hat{g}(x)| < \delta$ where δ is known, then

$$|h(x) - \hat{h}(x)| \leq \max_x |h(x) - \hat{h}(x)| \leq$$

$$\max_x \left| \int_{-\infty}^{\infty} K(y) R_m(x, y) \, dy \right| + \delta \sum_{j=-m}^{m} |w_j^{(m)}| = E_1 + \delta E_2.$$

If we can prove that if $m \to \infty$, $\delta \to 0$ then $E_1 + \delta E_2 \to 0$, then the above expression for $\hat{h}(x)$, constructed by means of the linear operator S expressed by (4) and acting on $\hat{g}(x)$, can be considered the approximate solution of $h(x)$, in full accord with what has been said at the beginning of the present paragraph supposing that the norm of a function has been defined by the maximum of its absolute value.

The proof of all this is possible only after the concretization of $K(y)$. Anyway, the calculation of $w_j^{(m)}$ needs it minimally.

We argue as follows. In the integrand in (6) the product can be rewritten as

$$(-1)^{m-1} \frac{\prod\limits_{\varrho=1}^{m} (y^2 + \varrho^2 h^2)}{y^2 + j^2 h^2} (-y^2 + ijhy).$$

Introducing the new variable hu for y and taking into account that $K(y)$ is even, we finally obtain

$$w_j^{(m)} = \frac{(-1)^j h}{(m+j)!\,(m-j)!} \int\limits_{-\infty}^{\infty} K(hu) \frac{u^2(u^2+1^2)\dots(u^2+m^2)}{u^2+j^2}\,du.$$

It is easy to see that $w_{-j}^{(m)} = w_j^{(m)}$, $w_0^{(0)} = \int\limits_{-\infty}^{\infty} K(u)\,du$ and, *per definitionem*, $w_j(l) = 0$ if $l < j$ or $j < 0$; further *if* $K(x) \geqq 0$, sgn $w_j^{(m)} = (-i)^j$.

Clearly $w_j^{(m)}$ could be calculated from the even moments of $K(u)$ after the expansion of the product in the integrand into a polynomial. Indeed, let us put

(7)
$$\sum_{k=1}^{m} B_{j,k}^{(m)} u^{2k} = \frac{(-1)^j}{(m+j)!\,(m-j)!} \frac{u^2(u^2+1^2)\dots(u^2+m^2)}{u^2+j^2}.$$

The $B_{j,k}^{(m)}$ are absolute constants.

$$w_j^{(m)} = \int\limits_{-\infty}^{\infty} hK(hu)\left(\sum_{k=1}^{m} B_{j,k}^{(m)} u^{2k} \right) du = \sum_{k=1}^{m} B_{j,k}^{(m)} \frac{m_{2k}}{h^{2k}}$$

where $m_{2k} = \int\limits_{-\infty}^{\infty} u^{2k} K(u)\,du$, the $2k$th moment of $K(u)$. $B_{j,k}^{(m)}$ can be computed recursively from the relations

$$B_{j,k}^{(m)} = (-1)^{m+j} \binom{2m}{m+j} B_{m,k}^{(m)} + B_{j,k}^{(m-1)}$$

$$(m = 1, 2, \dots; \; j = 0, 1, \dots, m; \; k = 1, \dots, m),$$

$$B_{0,0}^{(m)} = 1 \quad (m = 0, 1, \dots), \quad B_{j,0}^{(m)} = 0 \quad (m = 1, 2, \dots; \; j = 1, \dots, m),$$

$$B_{j,m}^{(m)} = \frac{(-1)^j}{(m+j)!\,(m-j)!} \quad (m = 1, 2, \dots; \; j = 0, 1, \dots, m),$$

$$B_{m,k}^{(m)} = -\frac{(m-1)^2}{2m(2m-1)} B_{m-1,k}^{(m-1)} - \frac{1}{2m(2m-1)} B_{m-1,k-1}^{(m-1)}$$

$$(m = 1, 2, \dots; \; k = 1, 2, \dots, m),$$

$$B_{m,k}^{(m)} = 0 \text{ if } k < m \text{ or } k = 0 \text{ or } m = 0.$$

The first can be obtained, by comparing coefficients, from the transformation

$$\sum_{k=0}^{m}(-1)^{j}(m+j)!(m-j)!B_{j,k}^{(m)}u^{2k} = \frac{u^2(u^2+1^2)\dots(u^2+m^2)}{u^2+j^2} =$$

$$u^2(u^2+1^2)\dots[u^2+(m-1)^2] + (m^2-j^2)\frac{u^2(u^2+1^2)\dots[u^2+(m-1)^2]}{u^2+j^2} =$$

$$\sum_{k=0}^{m}(-1)^{m}(2m)!B_{m,k}^{(m)}u^{2k} + \sum_{k=0}^{m}(m^2-j^2)(-1)^{j}(m-1+j)!(m-1-j)!B_{j,k}^{(m-1)}u^{2k}.$$

The fourth relation is deduced, by comparing coefficients, from

$$\sum_{k=0}^{m}(-1)^{m}(2m)!B_{m,k}^{(m)}u^{2k} = u^2(u^2+1^2)\dots[u^2+(m-1)^2] =$$

$$u^2 \cdot u^2(u^2+1^2)\dots[u^2+(m-2)^2] + (m-1)^2 u^2(u^2+1^2)\dots[u^2+(m-2)^2] =$$

$$\sum_{k=0}^{m-1}(-1)^{m-1}(2m-2)!B_{m-1,k}^{(m-1)}u^{2k+2} + \sum_{k=0}^{m-1}(-1)^{m-1}(m-1)^2(2m-2)!B_{m-1,k}^{(m-1)}u^{2k}.$$

The others are almost trivial.

Putting $u=i$ ($i^2=-1$) in (7) we have the useful expression for checking the calculations of $B_{j,k}^{(m)}$:

$$\sum_{k=0}^{m}(-1)^{k}B_{j,k}^{(m)} = \begin{cases} 0 & (j = 0, 2, \dots, m), \\ 1/2 & (j = 1). \end{cases}$$

Further relations valid for $B_{j,k}^{(m)}$ can be written down, too.

Thus knowing $B_{j,k}^{(m)}$ we can calculate $w_j^{(m)}$ provided the moments m_{2k} of $K(u)$ are known since

(8) $$w_j^{(m)} = \sum_{k=0}^{m}B_{j,k}^{(m)}\frac{m_{2k}}{h^{2k}}.$$

A table of the constants $B_{j,k}^{(m)}$ for $m=0, 1, \dots, 5$ is given.

Table of the constants $B_{j,k}^{(m)}$

$m = 0$

k j	0
0	1

$m = 1$

k j	0	1
0	1	1
1	0	$-\dfrac{1}{2}$

$m = 2$

k j	0	1	2
0	1	$\dfrac{5}{4}$	$\dfrac{1}{4}$
1	0	$-\dfrac{2}{3}$	$-\dfrac{1}{6}$
2	0	$\dfrac{1}{24}$	$\dfrac{1}{24}$

$m = 3$

k j	0	1	2	3
0	1	$\dfrac{49}{36}$	$\dfrac{7}{18}$	$\dfrac{1}{36}$
1	0	$-\dfrac{3}{4}$	$-\dfrac{13}{48}$	$-\dfrac{1}{48}$
2	0	$\dfrac{3}{40}$	$\dfrac{1}{12}$	$\dfrac{1}{120}$
3	0	$\dfrac{1}{180}$	$\dfrac{1}{144}$	$\dfrac{1}{720}$

$m = 4$

k j	0	1	2	3	4
0	1	$\dfrac{205}{144}$	$\dfrac{91}{192}$	$\dfrac{5}{96}$	$\dfrac{1}{576}$
1	0	$-\dfrac{4}{5}$	$-\dfrac{61}{180}$	$-\dfrac{29}{720}$	$-\dfrac{1}{720}$
2	0	$\dfrac{1}{10}$	$\dfrac{169}{1440}$	$\dfrac{13}{720}$	$\dfrac{1}{1440}$
3	0	$-\dfrac{4}{315}$	$-\dfrac{1}{60}$	$-\dfrac{1}{240}$	$-\dfrac{1}{5040}$
4	0	$\dfrac{1}{1120}$	$\dfrac{7}{5760}$	$\dfrac{1}{2880}$	$\dfrac{1}{40320}$

$m = 5$

$j \diagdown k$	0	1	2	3	4	5
0	1	$\dfrac{5269}{3600}$	$\dfrac{1529}{2880}$	$\dfrac{341}{4800}$	$\dfrac{11}{2880}$	$\dfrac{1}{14400}$
1	0	$-\dfrac{5}{6}$	$-\dfrac{1669}{4320}$	$-\dfrac{323}{5760}$	$-\dfrac{1}{320}$	$-\dfrac{1}{17280}$
2	0	$\dfrac{5}{42}$	$\dfrac{4369}{90720}$	$\dfrac{13}{480}$	$\dfrac{17}{10080}$	$\dfrac{1}{30240}$
3	0	$-\dfrac{5}{252}$	$-\dfrac{541}{20160}$	$-\dfrac{29}{3840}$	$-\dfrac{23}{40320}$	$-\dfrac{1}{80640}$
4	0	$\dfrac{5}{2016}$	$\dfrac{1261}{362880}$	$\dfrac{19}{17280}$	$\dfrac{13}{120960}$	$\dfrac{1}{362880}$
5	0	$-\dfrac{1}{6300}$	$\dfrac{41}{181440}$	$\dfrac{13}{172800}$	$\dfrac{1}{121160}$	$-\dfrac{1}{3628800}$

E_1 and E_2 can be investigated only if the analytical form of $K(y)$ is explicitly known. Thus it is more useful to deal concretely with that special case which is the most interesting for us; i.e.

$$K(y) = \frac{e^{-\frac{y^2}{4\lambda}}}{\sqrt{4\pi\lambda}} \qquad (0 < \lambda < \Lambda, \quad \Lambda \text{ is given}).$$

For this function, $\varkappa(t)=e^{-\lambda t^2}$. Let us suppose that $g(x)$ is such that all conditions of Theorem 1 are satisfied, and further, that (3) also holds. $w_j^{(m)}$ will be a function of λ, thus let us write it in the form $c_j^{(m)}(\lambda)$. Now $m_{2k}=\dfrac{(2k)!}{k!}\,\lambda^k$ in this case and, by (8):

(9) $$c_j^{(m)}(\lambda) = \sum_{k=0}^{m} \beta_{j,k}^{(m)} \left(\frac{\lambda}{h^2}\right)^{k} \qquad (j = 0, 1, \ldots, m),$$

where

$$\beta_{j,k}^{(m)} = \frac{(2k)!}{k!}\, B_{j,k}^{(m)}.$$

Because of the importance of $K(y)$, a table of the constants $\beta_{j,k}^{(m)}$ is also given (Medgyessy 1966a).

Table of the constants $\beta_{j,k}^{(m)}$

$m = 0$

$_j\backslash^k$	0
0	1.00000

$m = 1$

$_j\backslash^k$	0	1
0	1.00000	2.00000
1	0	-1.00000

$m = 2$

$_j\backslash^k$	0	1	2
0	1.00000	2.50000	3.00000
1	0	-1.33333	-2.00000
2	0	0.08333	0.50000

$m = 3$

$_j\backslash^k$	0	1	2	3
0	1.00000	2.72222	4.66666	3.33333
1	0	-1.50000	-3.25000	-2.50000
2	0	0.15000	1.00000	1.00000
3	0	-0.01111	-0.08333	-0.16666

$m = 4$

$_j\backslash^k$	0	1	2	3	4
0	1.00000	2.84722	5.68750	6.24999	2.91666
1	0	-1.60000	-4.06666	-4.83333	-2.33333
2	0	0.20000	1.40833	2.16666	1.16666
3	0	-0.02539	-0.20000	-0.50000	-0.33333
4	0	0.00178	0.01458	0.04166	0.04166

$m = 5$

$_j\backslash^k$	0	1	2	3	4	5
0	1.00000	2.92722	6.37083	8.52500	6.41666	2.10000
1	0	-1.66666	-4.63611	-6.72916	-5.25000	-1.75000
2	0	0.23809	1.73373	3.25000	2.83333	1.00000
3	0	-0.03968	-0.32202	-0.90625	-0.95833	-0.37500
4	0	0.00496	0.04169	0.13194	0.18055	0.08333
5	0	-0.00031	-0.00271	-0.00902	-0.01388	-0.00833

For our special case i.e. for $K(x) = \dfrac{e^{-\frac{x^2}{4\lambda}}}{\sqrt{4\pi\lambda}}$ the estimation of E_1 and E_2 can be carried out. The result is as follows (Medgyessy 1966a, making use of John 1955b):

$$(10) \qquad E_1 \leqq \frac{2\mu}{\eta_0 - y_0}\left[\frac{\eta_0}{\sqrt{1 - \lambda/\Lambda}} + \frac{1}{mh}\sqrt{\frac{\lambda}{\pi}}\,\frac{1}{1 - \lambda/\Lambda}\right] e^{\frac{1}{\sigma m}}\, e^{m\,[P(y_0,\,\lambda) - P(\eta_0,\,\Lambda)]}$$

if we stipulate, in addition, that also

$$(11) \qquad mh^2 \leqq \pi\Lambda \qquad (m \geqq 1).$$

Here y_0 is the positive root of the equation

$$\pi - 2 \arctan y - \frac{mh^2 y}{2\lambda} = 0$$

and η_0 is the positive root of the equation

$$\pi - 2 \arctan \eta - \frac{mh^2 \eta}{2\Lambda} = 0,$$

$\eta_0 \geq 1$, $\eta_0 > y_0$ and $\eta_0 - y_0$ increases with m; μ is the upper bound of $|g(x)|$ ($|g(x)| \leq \mu$) and, finally

$$P(y_0, \lambda) - P(\eta_0, \Lambda) =$$

$$\log(1 + y_0^2) + y_0 \arctan\left(\frac{1}{y_0}\right) - \log(1 + \eta_0^2) - \eta_0 \arctan\left(\frac{1}{\eta_0}\right).$$

The right-hand side of (10) tends to 0 if $m \to \infty$ (John 1955b), i.e.

$$E_1 \to 0 \text{ if } m \to \infty.$$

Further, utilizing the equality $\sum\limits_{j=-m}^{m} |w_j^{(m)}| = \sum\limits_{j=-m}^{m} (-1)^j w_j^{(m)}$, we have

(12) $$E_2 \leq \delta \cdot 2e^{\frac{1}{6m}} e^{mA(y_0, \lambda)}$$

where

$$A(y_0, \lambda) = \frac{h^2 m y_0^2}{4\lambda} + \frac{m + 1/2}{m} \log(1 + y_0^2) - \frac{y_0^2}{m(1 + y_0^2)};$$

here y_0 is the root of the equation

$$\pi - 2 \arctan y_0 + \frac{y_0}{m(1 + y_0)^2} - \frac{h^2 m y_0}{2\lambda} = 0.$$

If $m \to \infty$,

(13) $$E_2 \leq \delta \cdot 2e^{\frac{\pi^2 \lambda}{h^2}}$$

which proves the boundedness of the norm of the operator occurring in (4).

Thus $\hat{h}(x)$, introduced by (4), can be considered as an approximate solution of (1).

It is seen that the values of h and m are to be chosen, among others, on the basis of (11) and (13) so that $\dfrac{\pi^2 \lambda}{h^2}$ should be as small as possible. It is advisable to experiment with several values of h and m examining, eventually, the correctness of the approximate solution obtained in a problem with given solution.

The approximation of $h(x)$ has, principally, *a local* character: we need only values of $\hat{g}(x)$ at x and at certain neighbouring points for its calculation and we do not need values taken from a long interval as in most numerical methods related to our work. If we have only few values of $\hat{g}(x)$ at our disposal e.g. from some section of the graph of $g(x)$, this fact can be advantageous. However, this advantage will be diminished in practice by the fact that, in the case of a small m, the error of the approximation may be great.

Supplements and problems to Ch. V, § 3

1. The conditions of Theorem 1 also include that some Fourier transform $\varkappa(t)$ must satisfy the functional equation $\dfrac{1}{\varkappa(t)} = \varkappa(it)$. The determination of all Fourier transforms, for which this is valid, is an unsolved **problem**. In the case of *characteristic* functions, hints at the solution of the problem are given in Lukács (1968).

2. The final result of Theorem 1 can eventually be obtained also under fewer conditions. For instance (Hirschman, Widder 1955 pp. 179—182; Medgyessy 1966a) one has the following:

If

$$K(y) = \frac{e^{-\frac{y^2}{4\lambda}}}{\sqrt{4\pi\lambda}} \qquad (0 < \lambda < \Lambda, \ \ \Lambda \ \ given),$$

and $g(y)$ is a polynomial or a non-negative, bounded, continuous function, then 1. $g^{(r)}(x)$ $(r = 1, 2, \ldots)$ *exists;* 2. *if*

$$\lim_{t \to 1} \sum_{r=0}^{\infty} \frac{(-\lambda)^r}{r!} g^{(2r)}(x) \, t^r = \sum_{r=0}^{\infty} \frac{(-\lambda)^r}{r!} g^{(2r)}(x),$$

then

$$h(x) = \sum_{r=0}^{\infty} \frac{(-\lambda)^r}{r!} g^{(2r)}(x)$$

(in accord with the general series solution); further 3. *$g(z)$ exists for any complex z and is analytic and*

$$h(x) = \int_{-\infty}^{\infty} \frac{e^{-\frac{y^2}{4\lambda}}}{\sqrt{4\pi\lambda}} g(x+iy)\, dy$$

holds.

(This was already given in Medgyessy 1954c.)

For newer results in this line see e.g. Cannon (1966), Varah (1973).

3. If, in (1)

$$K(y) = \frac{1}{2\lambda \operatorname{ch}(\pi x/2\lambda)} \qquad (\lambda > 0)$$

i.e. a ch density function, and

$$g(y) = \frac{1}{2\beta \operatorname{ch}(\pi x/2\beta)} \qquad (0 < \lambda < \beta),$$

then conditions 1—5 of Theorem 1 are fulfilled and the solution of the integral equation

$$\frac{1}{2\beta \operatorname{ch}(\pi y/2\beta)} = \int\limits_{-\infty}^{\infty} \frac{1}{2\lambda \operatorname{ch}[\pi(y-x)/2\lambda]} \; h(x)\,dx$$

will be given by

$$h(x) = \sum_{n=0}^{\infty} \frac{\lambda^{2n}}{(2n)!} \, g^{(2n)}(x)$$

as $\varkappa(t) = \dfrac{1}{\operatorname{ch}\lambda t}$ and $\dfrac{1}{\varkappa(t)} = \operatorname{ch}\lambda t = \sum\limits_{n=0}^{\infty} \dfrac{\lambda^{2n}}{(2n)!} t^{2n}.$

However, a representation in the form of an integral belonging to the type (3) cannot be given here for $h(x)$ because Condition 6 of Theorem 1 is *not* fulfilled.

4. Also under favourable circumstances the error estimates E_1 and E_2 are not much used in practice, and they may also be rather crude. It is commonly accepted that the approximate solutions with several m and h-values are calculated and the results compared to see if they show characteristics that can be evaluated or not.

The optimal choice of h and m may be adopted from the results of certain methodological examples.

5. We emphasize the very *important* fact that *the constants $w_j^{(m)}$ appearing in the numerical method depend on the* **moments** *of the kernel $K(x)$*, because the moments can be computed *by means of the Fourier transform of $K(x)$*, and in several cases this transform is all that is available, e.g. if $K(x)$ itself cannot be presented in a closed form.

§ 4. NUMERICAL COMPUTATION OF CONVOLUTION TRANSFORMS

For the numerical computation of convolution transforms a method has already been given in Point 5 of *Supplements and problems to* Ch. V, § 2. Here two other procedures will be presented.

17 Medgyessy: Decomposition ...

1. A method based on the trapezoidal formula

Let us consider the numerical computation, used frequently in our work, of the convolution transform

$$g(y) = \int\limits_{-\infty}^{\infty} K(x)f(y-x)\,dx = \int\limits_{-\infty}^{\infty} K(y-x)f(x)\,dx,$$

(if it exists) where $K(x)$ is a given kernel, the analytical form of $f(x)$ is known and $K(x)$, $f(x)$ are twice differentiable, by means of the "measured" values $\hat{f}(y)$ of $f(y)$ where $|f(y)-\hat{f}(y)| \leq \delta$. We remark that such a transformation represents a *correct* problem.

Adopting certain results of numerical analysis, we apply the trapezoidal formula (Jones 1961; Ralston 1965 pp. 94—97, 118—121). At a given "step-length" h, $g(y)$ is rewritten in the form

$$g(y) = \int\limits_{-Nh}^{Nh} K(x)f(y-x)\,dx + \int\limits_{|x|>Nh} K(x)f(y-x)\,dx$$

and — influenced, of course, by the given circumstances — N will be chosen so large that the value of the second integral (the error) be as small as possible (this is possible if the analytical form of $f(x)$ is known); i.e. let

$$\int\limits_{|x|>Nh} K(x)f(y-x)\,dx = I(y, Nh) \leq \varepsilon(Nh).$$

Now considering y as a parameter in the integral $\int\limits_{-Nh}^{Nh} K(x)f(y-x)\,dx$, let us apply the trapezoidal formula to it with increment h; supposing that we have any $\hat{f}(y)$-value at our disposal we write the trapezoidal formula with the values of $\hat{f}(y)$, with an appropriate error term.

In this case we have (Ralston 1965 p. 118):

$$\int\limits_{-Nh}^{Nh} K(x)f(y-x)\,dx = h\sum_{v=-N}^{N} w_v K(vh)f(y-vh) -$$

$$\frac{Nh^3}{\sigma}\left(\frac{d^2}{dx^2}[K(x)f(y-x)]\right)_{x=\eta} \qquad (-Nh \leq \eta \leq Nh)$$

where

$$w_v = \begin{cases} 1/2 & \text{if } v = \pm N \\ 1 & \text{otherwise} \end{cases},$$

supposing that the derivative exists.

Finally, introducing the values of $\hat{f}(x)$, we have

$$\int_{-\infty}^{\infty} K(x)f(y-x)\,dx = h \sum_{v=-N}^{N} w_v K(vh)\,\hat{f}(y-vh) + E,$$

where the error term E is

$$E = I(y, Nh) + h \sum_{v=-N}^{N} w_v K(vh)\,[f(y-vh) - \tilde{f}(y-vh)] -$$

$$\frac{Nh^3}{\sigma}\left(\frac{d^2}{dx^2}\,[K(x)\,f(y-x)]\right)_{x=\eta}.$$

An estimate of the absolute value of the error is given by

$$|E| \leqq |\varepsilon(Nh)| + \delta h \sum_{v=-N}^{N} w_v |K(vh)| + \frac{Nh^3}{\sigma}\left|\left(\frac{d^2}{dx^2}\,[K(x)\,f(y-x)]\right)_{x=\eta}\right|;$$

$\varepsilon(Nh)$ and the last term are to be estimated in each case separately. In concrete cases a finer error estimate can also be obtained. It can be conjectured that in the case of a slowly increasing $K(x)$ and a small Nh the error is, in general, great; since in a given case Nh cannot be increased in many instances because we dispose of $\hat{f}(x)$ only in a short interval — and its values may not lie sufficiently dense, either — the described procedure may involve, in fact, significant errors.

2. A method based on approximation by interpolatory polynomials

This procedure makes use of the basic idea of the method described in Ch. V, § 3 (Medgyessy 1971b, 1972b). Let us consider again the integral transform

$$g(y) = \int_{-\infty}^{\infty} K(x)f(y-x)\,dx.$$

Let $K(x)$ be a function *whose moments all exist*. An approximation $\hat{g}(y)$ to $g(y)$ based on the "measured" values $\hat{f}(x)$ of $f(x)$ will be constructed as follows.

$f(y-x)$ is written by means of an interpolation formula:

$$f(y-x) = \sum_{j=-m}^{m} f(y-jh)\frac{(-1)^{m-j}}{h^{2m}(m+j)!\,(m-j)!}\,\frac{\prod\limits_{\varrho=-m}^{m}(x-\varrho h)}{(x-jh)} + R_m(x, y)$$

where

$$R_m(x, y) = -\frac{\prod\limits_{\varrho=-m}^{m}(x-\varrho h)}{(2m+1)!}\,f^{(2m+1)}(\xi)$$

$(y-mh<\xi<y+mh)$ or, if $f(z)$ is analytic for complex z-values,

$$R_m(x, y) = -\frac{1}{2\pi i} \int_{\mathscr{C}} \frac{f(w)}{(w-y+x)} \prod_{\varrho=-m}^{m} \left(\frac{x+\varrho h}{w-y+\varrho h}\right)$$

where the definition of $\mathscr{C}$, taken from the general theory of interpolation, is the same as in Ch. V, § 3.

Making use of this,

$$g(y) = \int_{-\infty}^{\infty} K(x)f(y-x)\,dx = \sum_{j=-m}^{m} W_j^{(m)} f(y-jh) + \int_{-\infty}^{\infty} K(x)R(x, y)\,dx$$

where

$$W_j^{(m)} = \frac{(-1)^{m-j}}{h^{2m}(m+j)!\,(m-j)!} \int_{-\infty}^{\infty} K(x) \frac{\prod\limits_{\varrho=-m}^{m} (x-\varrho h)}{x-jh}\,dx.$$

By our assumptions the integral exists.

Now we take, as the *approximation to $g(x)$, constructed on the "measured" values $\hat{f}(x)$ of $f(x)$* the first term on the right-hand side of the equality for $g(y)$, writing $\hat{f}(x)$ instead of $f(x)$, i.e.

$$\hat{g}(y) = \sum_{j=-m}^{m} W_j^{(m)} \hat{f}(x+jh).$$

The deviation of this from the solution $g(y)$ is

$$g(y)-\hat{g}(y) = \int_{-\infty}^{\infty} K(x)R_m(x, y)\,dy + \sum_{j=-m}^{m} W_j^{(m)}[f(x+jh) - \hat{f}(x+jh)].$$

The first term is the error originating from the substitution of $g(y)$ by the main part of its own interpolation formula. The second term contains the "measurement error" of $f(x)$; we suppose that $|f(x)-\hat{f}(x)| < \delta$ with δ given. $\hat{g}(y)$ can be considered as an approximate solution if, for any y, $|g(y)-\hat{g}(y)| \to 0$ as $m \to \infty$, $\delta \to 0$. The approximate solution will be the expression of $\hat{g}(x)$ constructed by means of the above linear operator acting on $\hat{f}(x)$. Investigations in this respect can, however, be carried out only after the concretization of $K(x)$.

It is the calculation of $W_j^{(m)}$ which needs the least the knowledge of $K(x)$, if $K(x)$ is an *even* function. Then,

$$W_j^{(m)} = \frac{(-1)^{m+j}}{(m+j)!\,(m-j)!} \int_{-\infty}^{\infty} h\,K(hu) \frac{u^2(u^2-1)\ldots(u^2-m^2)}{u^2-j^2}\,du,$$

$$W_0^{(0)} = \int_{-\infty}^{\infty} K(u)\,du, \quad W_j^{(m)} = W_j^{(m)}, \quad W_j^{(m)} = 0 \quad \text{if } j > m, \quad \text{or} \quad j < 0.$$

Writing the polynomial in the integrand in the form

$$\frac{(-1)^{m+j}}{(m+j)!\,(m-j)!}\;\frac{u^2(u^2-1^2)\dots(u^2-m^2)}{u^2-j^2}=\sum_{k=1}^{m}D_{j,k}^{(m)}\,u^{2k},$$

we have

$$W_j^{(m)}=\int_{-\infty}^{\infty}hK(hu)\left(\sum_{k=1}^{m}D_{j,k}^{(m)}\,u^{2k}\right)du$$

i.e.

$$W_j^{(m)}=\sum_{k=1}^{m}D_{j,k}^{(m)}\,\frac{m_{2k}}{h^{2k}},$$

where $m_{2k}=\int_{-\infty}^{\infty}u^{2k}K(u)\,du$ — i.e. the $2k$th moment of $K(x)$.

However, *the constants $D_{j,k}^{(m)}$ can be reduced to the constants $B_{j,k}^{(m)}$ in Ch. V, § 3.* Namely, recalling that

$$\frac{(-1)^j}{(m+j)!\,(m-j)!}\;\frac{u^2(u^2+1^2)\dots(u^2+m^2)}{u^2+j^2}=\sum_{k=0}^{m}B_{j,k}^{(m)}\,u^{2k}$$

and putting iu $(i^2=-1)$ instead of u, we get

$$\sum_{k=1}^{m}B_{j,k}^{(m)}(-1)^k u^{2k}=\frac{(-1)^{m+j}}{(m+j)!\,(m-j)!}\;\frac{u^2(u^2-1)\dots(u^2-m^2)}{u^2-j^2}=\sum_{k=1}^{m}D_{j,k}^{(m)}\,u^{2k},$$

whence

$$D_{j,k}^{(m)}=(-1)^k B_{j,k}^{(m)}.$$

That is

$$W_j^{(m)}=\sum_{k=0}^{m}B_{j,k}^{(m)}(-1)^k\,\frac{m_{2k}}{h^{2k}}.$$

Thus, the moments of $m_{2k}\,K(u)$ being known, $W_j^{(m)}$ can be computed by using the table of the $B_{j,k}^{(m)}$, i.e. without any new table.

Neither special cases of $K(u)$ nor the correctness of the relevant approximate solutions will be treated here.

Supplements and problems to Ch. V, § 4

1. The error estimation in Ch. V, § 4, Section 1 can sometimes be improved by the method of Hämmerlin (1963).

2. It is well known that the advantage of the trapezoidal formula in integrating an empirical function lies in the fact that it accumulates the error of the "measured" function values relatively little in the final result (Ralston 1965 p. 119). This is

why it was chosen here. Of course, another quadrature formula could be applied if necessary.

3. In the foregoing, the absolute value of the deviation between the exact solution and the approximate one was taken as the error of approximation. If we took, instead of this, the integral of the square of the deviation, we should obtain, by minimizing it appropriately, an *approximation of a different character for the convolution transform.*

A method based on this remark is published in Strahov (1964) for the approximate computation of the integral transform

$$g(y) = \int\limits_{-\infty}^{\infty} K(x)f(y-x)\,dx.$$

(if it exists). In this the transform is approximated by the sum $\bar{g}(x) = \sum\limits_{j=-m}^{m} a_j f(y+jh)$, where the a_j minimize the integral $\delta^2 = \int\limits_{-\infty}^{\infty} |g(y)-\bar{g}(y)|^2\,dy$. If $\mathbf{F}[K(x);t]=\varkappa(t)$, $\mathbf{F}[f(x);t]=\varphi(t)$ and the Parseval formula can be applied, then

$$\int\limits_{-\infty}^{\infty} |g(y)-\bar{g}(y)|^2\,dy = \frac{1}{2\pi} \int\limits_{-\infty}^{\infty} |\varphi(t)|^2 \left|\varkappa(t) - \sum\limits_{j=-m}^{m} a_j e^{ijht}\right|\,dt.$$

Strahov (1964) assumed that a *known* estimate can be found for $|\varphi(t)|$, say $|\varphi(t)| < H(t)$. Then the author determined the coefficients a_j from the minimization of the integral

$$\int\limits_{-\infty}^{\infty} |H(t)|^2 \left|\varkappa(t) - \sum\limits_{j=-m}^{m} a_j e^{ijht}\right|^2\,dt$$

majorizing δ^2 (a_j is the Fourier coefficient of $\varkappa(t)$ with the weight function $H(t)$). Then he considered the following sum formed with the "measured" values $\hat{f}(x)$ of $f(x)$ as the approximation of $g(x)$:

$$\hat{g}(x) = \sum\limits_{j=-m}^{m} a_j \hat{f}(x+jh).$$

Error estimates were not investigated.

In the paper cited V. N. Strahov gave tables of a_j values for special types of the functions $\varkappa(t)$ and $H(t)$. For instance if $H(t)=e^{-H|t|}$ is a convenient majorization of $|\varphi(t)|$, then for $\varkappa(t)=e^{-\lambda|t|}$, i. e. for $K(x) = \dfrac{\lambda}{\pi(x^2+\lambda^2)}$, and $m=5$, $h=\lambda$, $H=2h$ he obtained the following table:

k	0	1	2	3	4	5
$a_k = a_{-k}$	0.1511	0.1691	0.0578	0.0397	0.0064	0.0330

With them, he reached a good approximation, for certain λ-values, of the so-called *Poisson transform*

$$g(y) = \frac{\lambda}{\pi} \int\limits_{-\infty}^{\infty} \frac{f(x)}{(y-x)^2 + \lambda^2} dx$$

occurring in this case.

4. Unfortunately, little has been written on the *numerical computation of convolution transforms*. They can be computed, of course, also by means of special mathematical instruments, e.g. product integrators. Such a one is described e.g. in Macnee (1953) or — assuming the use of a square planimeter — in Medgyessy (1954*d*).

5. In the case of the procedure given in Ch. V, § 4, Section 2, it would not be of much use to grapple with error estimates because in practice we always compute a series of approximations at different parameter values and other characteristics and then compare the results, choosing the most informative on the basis of "smoothness" and similar factors. Formulae for error estimates would seem to be insignificant.

6. We emphasize the very *important* fact that in the approximate solution in Ch. V, § 4, Section 2 *the constants $W_j^{(m)}$ can be expressed by means of the* **moments** *of the kernel $K(x)$*. The moments can be computed *by means of the Fourier transform of $K(x)$*, while in several cases $K(x)$ itself cannot be presented in a closed form.

POSTSCRIPT

The following aspects are summarized here:

1. **Relevant problems not considered in the present book; 2. the important unsolved problems in the field under consideration; 3. important tasks in future investigations.**

Ad 1.

Decomposition of superpositions, e.g. of density functions belonging to the type

$$k(x) = \sum_{k=1}^{N} p_k f(x, \alpha_k, \beta_k)$$

in which

(a) *several α_k are equal while all β_k are different;*

(b) $f(x, \alpha_k, \beta_k)$ *is multimodal, e.g.* $f(x, \alpha_k, \beta_k) = \dfrac{\beta_k}{\pi} \left[\dfrac{\sin \beta_k (x - \alpha_k)}{\beta_k (x - \alpha_k)} \right]^2,$

has not been considered at all here except for the superposition of density functions mentioned in *Supplements and problems to* Ch. III, § 1, 33 (cf. Medgyessy 1961*a* p. 164, 1961*c*). Similar examples can be found easily in the sphere of superpositions of discrete distributions.

Ad 2.

(a) There are superpositions that could not be decomposed by using any of the methods described. For instance the *decomposition of a superposition of translated Gamma density functions of the same order* is not possible using our methods; here

$$k(x) = \begin{cases} \displaystyle\sum_{k=1}^{N} p_k \dfrac{(x - \alpha_k)^{\gamma - 1} e^{-\frac{x - \alpha_k}{\beta_k}}}{\beta_k^{\gamma} \Gamma(\gamma)} & (x > \alpha_1) \\[4mm] 0 & (x \leqq \alpha_1) \end{cases}$$

$$(0 < \alpha_1 < \ldots < \alpha_N; \ 0 < \beta_1 < \ldots < \beta_N < \Lambda_2),$$

(Medgyessy 1961*a* p. 164, 1961*c*).

It is not possible to give the *decomposition of a superposition of uniform density functions* or the *decomposition of a superposition of "triangle" density functions* either (besides, these are not superpositions of infinitely divisible density functions). Also the *decomposition of a superposition of hypergeometrical distributions* or the *decomposition of a superposition of translated binomial distributions* cannot be carried out by any of the described methods.

(b) In the case of superpositions of density functions

$$k(x) = \sum_{k=1}^{N} p_k f(x - \alpha_k, \beta_k),$$

no exact method can be given for the *determination of β_k* (or it is impossible) but in practice this is basically needed.

(c) In the case of superpositions of density functions the numerical decomposition methods produce, by their essence, *"additional peaks"* in the graph of the test function. When surveying a series of such graphs it is often difficult (or, even impossible) to decide whether *a component hides behind a certain peak or not*. The experimental background may eventually provide hints for making such a decision. However, a purely mathematical test would be most necessary. Considering the test function as a realization of some stochastic process would give, maybe, some test to filter out peaks caused by the measurement (random) errors.

Ad 3.

The errors caused by the use of numerical methods in decomposing superpositions could be eliminated and, simultaneously, the *best type of test function* could be obtained if in the case of a superposition $k(x)$ (or $\{k_n\}$) it were possible to construct a test function belonging to the type

(1) $$b(y) = \sum_{j=A}^{B} a_j k(\alpha_j y + \beta_j)$$

or

(2) $$b(y) = \sum_{j=A}^{B} a_j(y) \, k(\beta_j)$$

or

(3) $$b(y) = \sum_{j=A}^{B} b_j(y) k_j$$

(or, to establish functional equations of similar type for $b(y)$); our numerical methods have provided *approximations* to test functions in the form

$$b(y) = \sum_{j=A}^{B} a_j k(\alpha_j y + \beta_j).$$

It is noticed that only values of $k(x)$ (or k_j) from the *shortest* possible interval for x (or j) should be used in this procedure; this would give a *local* character to the test function obtained in this way.

The construction of such a procedure would require that the above sums have unimodal components if $k(x)$ (or $\{k_j\}$) were so. Thus to find such a procedure is difficult and is an unsolved **problem**, from the viewpoint of investigations on unimodality also.

In the Example in Ch. III, § 2, the test function belonging to a superposition of exponential density functions showed the above type (1), and so did the test function belonging to a superposition of Laplace density functions, in *Supplements and problems to* Ch. III, § 3, 15. This proves the principal possibility of finding a test function of the above-mentioned type. Also in the case of superpositions of discrete distributions, a test function of type (3) above can be constructed in some instances (cf. Ch. IV, § 2, Section 1).

A general treatment of this problem and the former ones can, however, only be expected after further investigations.

REFERENCES

Aissen, M.—Schoenberg, I. J.—Whitney, A. M.
(1952)   On the generating functions of totally positive sequences. I. *J. Analyse Math.* **2** (1952)
          93—103.

Aitchison, J.—Brown, J. A. C.
(1957)   *The lognormal distribution with special reference to its uses in economics.* Cambridge
          University Press, Cambridge, 1957.

Aleksandrov, L.
(1970)   Reguljarizovannyĭ vyčislitel'nyĭ process dlja analiza eksponencial'nyh zavisimosteĭ.
          Ž. Vyčisl. Mat. i Mat. Fiz. **10** (1970) 1285—1287.

Arcangeli, R.
(1968)   Un problème de résolution rétrograde de l'équation de la chaleur. *Rev. Française
          Informat. Recherche Opérationelle* **2** (1968) No 13, 61—78.

Armstrong, B. H.
(1967)   Spectrum line profiles: The Voigt function. *J. Quant. Spectrosc. Radioactive Transfer*
          7 (1967) 61—88.

Arsenin, V. Ja.—Ivanov, V. V.
(1968)   O rešenii nekotoryh integral'nyh uravneniĭ I roda tipa svertki metodom reguljarizacii.
          Ž. Vyčisl. Mat. i Mat. Fiz. **8** (1968) 310—321.

Askey, R.
(1973)   Some characteristic functions of unimodal distributions. *J. Math. Anal. Appl.* **50**
          (1975) 465—469.

de Balbine, G.—Franklin, J. N.
(1966)   The calculation of Fourier integrals. *Math. Comp.* **20** (1966) 570—589.

Barndorff-Nielsen, O.
(1965)   Identifiability of mixtures of exponential families. *J. Math. Anal. Appl.* **12** (1965)
          115—121.

Békési, Gábor
(1967)   A Gauss-analízis. (The Gauss analysis.) Manuscript. Budapest, 1966—1967.

Bellman, R.
(1960)   On the separation of exponentials. *Boll. Un. Mat. Ital.* (3) **15** (1960) 38—39.
(1969)   A new method for the identification of systems. *Math. Biosci.* **5** (1969)   201—204.

Bellman, R. E.—Kagiwada, H. H.—Kalaba, R. E.
(1965)   On the identification of systems and the unscrambling of data, I. Hidden periodicities.
          Proc. Nat. Acad. Sci. U.S.A. **53** (1965) 907—910.

Bellman, R. E.—Kalaba, R. E.
(1965)   *Quasilinearization and nonlinear boundary-value problems.* American Elsevier Publishing
          Co., Inc., New York, 1965.

Bellman, R.—Kalaba, R. E.—Lockett, J.
(1965a) Dynamic programming and ill-conditioned linear systems. *J. Math. Anal. Appl.* **10** (1965) 206—215.
(1965b) Dynamic programming and ill-conditioned linear systems — II. *J. Math. Anal. Appl.* **12** (1965) 393—400.
(1966) Dynamic programming and ill-conditioned systems. *Numerical inversion of the Laplace-transform: Applications to biology, economics, engineering and physics.* American Elsevier Publ. Co., Inc., New York, 1966; pp. 135—173.

Berencz, Ferenc
(1955a) Megjegyzések az abszorpciós görbék analíziséhez. (Remarks to the analysis of the absorption curves.) *Magyar Fizikai Folyóirat* (Budapest) **3** (1955) 271—278.
(1955b) Bemerkungen zur Analyse der Absorptionskurven. *Acta Phys. Acad. Sci. Hungar.* **4** (1955) 317—326.

Bergström, H.
(1952) On some expansions of stable distribution functions. *Ark. Mat.* **2** (1952) 375—378.
(1953) Eine Theorie der stabilen Verteilungsfunktionen. *Arch. Math.* **4** (1953) 380—391.

Bernstein, F.
(1932) Verallgemeinertes Galtonbrett zur Durchführung von Funktionaltransformationen. *Z. Physik* **77** (1932) 104—113.

Bhattacharya, C. G.
(1967) A simple method of resolution of a distribution into Gaussian components. *Biometrics* **23** (1967) 115—135.

Birnbaum, Z. W.
(1948) On random variables with comparable peakedness. *Ann. Math. Statist.* **19** (1948) 76—81.

Blischke, W. R.
(1962) Moment estimations for the parameters of a mixture of two binomial distributions. *Ann. Math. Statist.* **33** (1962) 444—454.
(1964) Estimating the parameters of mixtures of binomial distributions. *J. Amer. Statist. Assoc.* **59** (1964) 510—528.
(1965) Mixtures of discrete distributions. *Classical and Contagious Discrete Distributions.* Proceedings of the International Symposium held at McGill University, Montreal, Que., Canada, August 15—August 20, 1963. (Ed. by Ganapati P. Patil.) Statistical Publishing Society, Calcutta, 1965; pp. 351—372.

Blomer, R. J.
(1974) Decomposition of compartment transition functions by means of the Pade-approximation. *Math. Biosci.* **19** (1974) 163—173.

Bracewell, R. N.
(1955a) Correcting for Gaussian aerial smoothing. *Austral. J. Phys.* **8** (1955) 54—60.
(1955b) A method of correcting the broadening of X-ray line profiles. *Austral. J. Phys.* **8** (1955) 61—67.

Bracewell, R. N.—Roberts, J. A.
(1954) Aerial smoothing in radio astronomy. *Austral. J. Phys.* **7** (1954) 615—640.

Brown, G. M.
(1933) On sampling from compound populations. *Ann. Math. Statist.* **4** (1933) 288—342.

Brownell, G. L.—Callahan. A. B.
(1963) Transform methods for tracer data analysis. *Ann. New York Acad. Sci.* **108** (1963) Article 1: Multi-compartment analysis of tracer experiments; pp. 172—181.

Brunk, H. D.
(1953)   Approximate solution of an initial value problem by generalized cardinal series. *Quart. Appl. Math.* **11** (1953) 285—294.

Buchanan-Wollaston, H. G.—Hodgeson, W. C.
(1929)   A new method of treating frequency curves in fishery statistics, with some results. *J. Cons.* **4** (1929) 207—225.

Buckingham, R. A.
(1957)   *Numerical methods.* Pitman and Sons, Ltd., London, 1957.

Burger, H. C.—van Cittert, P. H.
(1932)   Wahre und scheinbare Intensitätsverteilung in Spektrallinien. *Z. Physik* **79** (1932) 722—730.
(1933)   Wahre und scheinbare Intensitätsverteilung in Spektrallinien. II. *Z. Physik* **81** (1933) 428—434.

Burrau, C.
(1934)   The half-invariants of two typical laws of errors, with an application to the problem of dissecting a frequency curve into components. *Skand. Aktuarietidskr.* **17** (1934) 1—6.

Cannon, J. R.
(1966)   Some numerical results for the solution of the heat equation backwards in time. *Numerical solutions of nonlinear differential equations.* Proceedings of an advanced symposium conducted by the Mathematics Research Center, United States Army, at the University of Wisconsin, Madison, May 9—11, 1966. (Edited by Donald Greenspan.) J. Wiley and Sons, Inc., New York, 1966; pp. 21—63.

Carnahan, C. L.
(1964)   A method for the analysis of complex peaks occurring in gamma ray pulse height distributions. *Nuclear Instrum. Methods* **30** (1964) 165—183.

Carver, J. H.—Lokan, K. H.
(1957)   Det ermination of photonuclear cross sections. *Austral. J. Phys.* **10** (1957) 312—319.

Cassie, R. M.
(1954)   So me uses of probability paper for the graphical analysis of polymodal frequency distri butions. *Austral. J. Mar. Freshw. Res.* **5** (1954) 513—522.

Charlier, C. V. L.
(1905)   Researches into the theory of probability. *Acta Univ. Lund.* Neue Folge. Abt. 2. **1** (1905) No. 5, 33—38.

Charlier, C. V. L.—Wicksell, S. D.
(1925)   On the dissection of frequency functions. *Ark. Mat. Astr. Fys.* **18** (1924/25) No. 6, 1—64.

Childers, H. M.
(1959)   Analysis of single crystal pulse-height distribution. *Rev. Sci. Instrum.* **30** (1959) 810—814.

Chung, K. L.
(1953)   Sur les distributions unimodales. *C.R. Acad. Sci. Paris* **236** (1953) 583—584.

van Cittert, P. H.
(1931)   Zum Einfluss der Spektralbreite auf die Intensitätsverteilung in Spektrallinien. *Z. Physik* **69** (1931) 298—308.

Cohen, A. C. Jr.
(1967)   Estimation in mixtures of two normal distributions. *Technometrics* **9** (1967) 15—28.

270   REFERENCES

Craig, L. C.
(1944)   Identification of small amounts of organic compounds by distribution studies. II.
         Separation by counter-current distribution. *J. Biol. Chem.* **155** (1944) 519—534.
(1951)   Automatic counter-current distribution equipment. *Analyt. Chem.* **23** (1951) 1236—1244.

Cramér, H.
(1925)   On some classes of series used in mathematical statistics. *Matematiker Kongressen i
         København* 31. *August*—4. *September* 1925 (Den sjette Skandinaviske Matematiker-
         kongres). København, 1925; pp. 399—426.
(1946)   *Mathematical methods of statistics.* Princeton University Press, Princeton, 1946; pp.
         226—227.

Crout, P. D,
(1940)   An application of polynomial approximation to the solution of integral equations
         arising in physical problems. *J. Math. and Phys.* **19** (1940) 34—92.

Crum, W. L.
(1923)   The use of the median in determining seasonal variation. *J. Amer. Statist. Assoc.*
         1923, March, 607—614.

Čudov, L. A.
(1967)   Raznostnye shemy i nekorrektnye zadači dlja uravneniĭ s častnimi proizvodnymi.
         Vyčislitel'nye metody i programmirovanie. VIII. Izd. Moskov. Univ., Moskva, 1967;
         pp. 34—62.

Daeves, K.
(1933)   *Praktische Grosszahl-Forschung.* VDI-Verlag GMBH, Berlin, 1933.
(1951)   *Vorausbestimmungen im Wirtschaftsleben.* W. Girardet, Essen, 1951.

Daeves, K.—Beckel, A.
(1948)   *Grosszahlforschung und Häufigkeitsanalyse.* Verlag Chemie, Weinheim/Bergstr., 1948.
(1958)   *Grosszahl-Methodik und Häufigkeitsanalyse.* 2. Aufl. Verlag Chemie, Weinheim/Bergstr.,
         1958.

Defares, J. G.—Sneddon, I. N.
(1960)   *An introduction to the mathematics of medicine and biology.* North-Holland Publishing
         Co., Amsterdam, 1960; pp. 582—586.

Dobozy, O. K.—Volly, T.
(1970)   The anomalous properties of cystine and their effect on keratin. (Lecture: Fourth
         International Wool Textile Research Conference, Berkeley, California, 18—27. August
         1970.) Manuscript. Budapest, 1970.

Doetsch, G.
(1926)   Probleme aus der Theorie der Wärmeleitung. III. Mitteilung. Der lineare Wärmeleiter
         mit beliebiger Anfangstemperatur. Die zeitliche Fortsetzung des Wärmezustandes.
         Math. Z. **25** (1926) 608—626.
(1928)   Die Elimination des Dopplereffekts bei spektroskopischen Feinstrukturen und exakte
         Bestimmung der Komponenten. *Z. Physik* **49** (1928) 705—730.
(1936)   Zerlegung einer Funktion in Gauss'sche Fehlerkurven und zeitliche Zurückverfolgung
         eines Temperaturumstandes. *Math. Z.* **41** (1936) 283—318.
(1950)   *Handbuch der Laplace-Transformation.* Band I. Theorie der Laplace-Transformation.
         Birkhäuser, Basel, 1950.

Dolgopolova, T. F.—Ivanov, V. K.
(1966)   O čislennom differencirovanii. *Ž. Vyčisl. Mat. i Mat. Fiz.* **6** (1966) 570—576.

Dombrovskaja, I. N.—Ivanov, V. K.
(1965) K teorii nekotoryh linejnyh uravneniĭ v abstraktnyh prostranstvah. *Sibirsk. Mat. Ž.* **6** (1965) 499—508.

Douglas, J. Jr.
(1960) Mathematical programming and integral equations. *Symposium on the Numerical Treatment of Ordinary Differential Equations, Integral and Integro-Differential Equations* (Rome, 1960). Birkhäuser, Basel, 1960; pp. 269—274.

Dugué, D.—Girault, M.
(1955) Fonctions convexes de Pólya. *Publ. Inst. Statist. Univ. Paris.* **4** (1955) 3—10.

Dyson, F.
(1926) A method for correcting series of parallax observations. *Monthly Notices Roy. Astronom. Soc.* **86** (1926) 686—706.

Eddington, A. S.
(1913) On a formula for correcting statistics for the effects of a known probable error of observations. *Monthly Notices Roy. Astronom. Soc.* **73** (1913) 359—360.

Edrei, A.
(1952) On the generating function of totally positive sequences. II. *J. Analyse Math.* **2** (1952) 104—109.
(1953a) Proof of a conjecture of Schoenberg on the generating function of a totally positive sequence. *Canad. J. Math.* **5** (1953) 86—94.
(1953b) On the generating function of a doubly infinite, totally positive sequence. *Trans. Amer. Math. Soc.* **74** (1953) 367—383.

Elste, G.
(1953) Die Entzerrung von Spektrallinien unter Verwendung von Voigtfunktionen. *Z. Astrophys.* **33** (1953—1954) Heft 1, 39—73.

Emslie, L. A. J.—King, G. W.
(1953) Spectroscopy from the point of view of the communication theory, II. Line-width. *J. Opt. Soc. Amer.* **43** (1953) 658—663.

Faddeeva, V. N.—Terent'ev, N. M.
(1954) *Tablicy značeniĭ funkcii* $w(z) = e^{-z^2}\left(1 + \dfrac{2i}{\sqrt{\pi}} \int\limits_0^z e^{t^2}\,dt\right)$ *ot kompleksnogo argumenta.*

Gostehizdat, Moskva, 1954.

Fejér, L.
(1914) Nombre des changements de signe d'une fonction dans un intervalle et ses moments. *C.R. Acad. Sci. Paris* **158** (1914) 1328—1331.

Fekete, M.—Pólya, G.
(1912) Über ein Problem von Laguerre. *Rend. Circ. Mat. Palermo* **34** (1912) 89—120.

Feldheim, E.
(1939) Sur quelques propriétés des lois de probabilité stables. *Rev. Math. Union Interbalkan.* **2** (1939) Fasc. III—IV, 9—30.

Feller, W.
(1957) *An introduction to probability theory and its applications.* Volume I. Second edition. J. Wiley and Sons, Inc., New York, 1957.
(1966) *An introduction to probability theory and its applications.* Volume II. J. Wiley and Sons, Inc., New York, 1966.

Fiala, Tibor
(1975)   Eloszlásfüggvények szuperpozícióinak komponensekre bontása. (Decomposition of
          superpositions of distribution functions.) Thesis. Budapest, 1975.
de Finetti, B.
(1930)   Le funzioni caratteristiche di legge instantanea. *Atti Acad. Naz. Lincei Rend. Cl. Sci.
          Fis. Mat. Natur.* Ser. 6. **12** (1930) 278—282.
Fisz, M.
(1962)   Infinitely divisible distributions. Recent results and applications. *Ann. Math. Statist.*
          33 (1962) 68—84.
Flynn, C. P.—Seymour, E. F. W.
(1960)   The correction of spectral line shapes for instrumental and other broadening. *Proc.
          London Phys. Soc.* **75** (1960) 337—344.
Franklin, J. N.
(1970)   Well-posed stochastic extensions of ill-posed linear problems. *J. Math. Anal. Appl.*
          31 (1970) 682—716.
Fréchet, M.
(1939)   Sur les formules de répartition des revenus. *Rev. Inst. Internat. Statist.* **7** (1939) 32—38.
(1945)   Nouveaux essais d'explication de la répartition des revenus. *Rev. Inst. Internat. Statist.*
          13 (1945) 16—32.
Fridrik, F.
(1967)   Ob odnom eksperimente po rešeniju integral'nyh uravnenii I. roda. *Metody Vyčisl.*
          Vyp. 4. Izd. Leningr. Univ., Leningr. 1967; pp. 102—109.
Gardner, D. G.
(1963)   Resolution of multi-component exponential decay curves using Fourier transforms.
          Ann. New York Acad. Sci. **108** (1963) Article 1: Multi-compartment analysis of tracer
          experiments; pp. 195—203.
Gardner, D. G.—Gardner, J. C.—Laush, G.—Meinke, W. W.
(1959)   Method for the analysis of multi-component exponential decay curves. *J. Chem. Phys.*
          31 (1959) 978—986.
Gauss, C. F.
(1866)   Methodus nova integralium valores per approximationem inveniendi. *Werke,* Band 3.
          Königliche Gesellschaft der Wissenschaften, Göttingen, 1866; pp. 165—196.
Gavurin, M. K.
(1967)   O metode A. N. Tihonova rešenija nekorrektnyh zadač. *Metody Vyčisl.* Vyp. 4. Izd.
          Leningr. Univ., Leningrad, 1967; pp. 21—25.
Giaccardi, F.
(1939)   Di un tentativo di rappresentazione analitica dello sviluppo in peso dell'organismo
          umano. *Atti Accad. Sci. Torino Cl. Sci. Fis. Mat. Natur.* **74** (1939) 459—469.
Girault, M.
(1955)   Les fonctions caractéristiques et leurs transformations. *Publ. Inst. Statist. Univ. Paris*
          4 (1955) 223—299.
Glasko, V. B.—Zaikin, P. N.
(1966)   O programme reguljarizirujuščego algoritma dlja uravnenija Fredgol'ma pervogo roda.
          Vyčislitel'naja matematika i programmirovanie. V. Izd. Moskov. Univ., Moskva, 1966;
          pp. 61—73.
Gnedenko, B. V.
(1954)   *Kurs teorii verojatnostei.* Izdanie vtoroe. Gostehizdat, Moskva, 1954.

Gnedenko, B. V.—Kolmogorov, A. N.
(1954)   *Limit distributions for sums of independent random variables.* Addison-Wesley Publishing Co., Inc., Cambridge, Mass., 1954.

Godwin, H. J.
(1964)   *Inequalities on distribution functions.* Griffin, London, 1964.

Gorbunov, A. D.
(1967)   O rešenii nelinejnyh kraevyh zadač dlja sistemy obyknovennyh differencial'nyh uravneniĭ. *Vyčislitel'nye metody i programmirovanie.* VIII. Izd. Moskov. Univ., Moskva, 1967; pp. 186—199.

Gottschalk, V. H.
(1948)   Symmetrical bimodal frequency curves. *J. Franklin Inst.* **245** (1948) 245—252.

Grabar, L. P.
(1967a)   Primenenie polinomov Čebyševa, ortonormirovannyh na sisteme ravnootstojaščih toček, dlja rešenija integral'nyh uravneniĭ pervogo roda. *Dokl. Akad. Nauk SSSR* **172** (1967) 767—770.

(1967b)   Primenenie polinomov Čebyševa, ortonormirovannyh na sisteme ravnootstojaščih toček, dlja čislennogo differencirovanija. *Ž. Vyčisl. Math. i Mat. Fiz.* **7** (1967) 1375—1379.

Graf, U.—Henning, H. J.
(1960)   *Statistische Methoden bei textilen Untersuchungen.* III. ber. Neudruck. Springer, Berlin, 1960.

Gregor, J.
(1969)   An algorithm for the decomposition of a distribution into Gaussian components. *Biometrics* **22** (1969) 79—93.

Gumbel, E. J.
(1939)   La dissection d'une répartition. *Ann. Univ. Lyon Sect. A.* **11** (1939) 39—51.

Gyllenberg, W.
(1936)   A graphical method of determining the absolute luminosity distribution of the stars *Medd. Lunds Astron. Obs. Ser.* I. Nr. **145** (1936) 1—6.

Hadamard, J.
(1902)   Sur les problèmes aux dérivées partielles et leur signification physique. *Bull. Univ. Princeton* **13** (1902) 49—52.

(1932)   *Le problème de Cauchy et les équations aux dérivées partielles linéaires hyperboliques.* Hermann, Paris, 1932.

Hald, A.
(1952)   *Statistical theory with engineering applications.* John Wiley and Sons, Inc., New York, 1952.

Hämmerlin, G.
(1963)   Über ableitungsfreie Schranken für Quadraturfehler. *Numer. Math.* **5** (1963) 226—233.

Hamming, R. W.
(1962)   *Numerical methods for scientists and engineers.* McGraw-Hill Book Co. Inc., New York, 1962.

Harding, J. F.
(1949)   The use of probability paper for the graphical analysis of polymodal frequency distributions. *J. Mar. Biol. Assoc. U.K.* **28** (1949) 141—153.

Hildebrand, F. B.
(1956)   *Introduction to numerical analysis.* McGraw-Hill Book Co., Inc., New York, 1956.

Hildebrand, F. B.—Crout, P. D.
(1941)   A least square procedure for solving integral equations by polynomial approximation. *J. Math. and Phys.* **20** (1941) 310—335.

Hinčin, A. Ja.
(1937)   Ob arifmetike zakonov raspredelenija. *Bjull. Mosk. Gos. Univ. Math. Meh.* **1** (1937) No. 1, 6—17.
(1938*a*)   *Predel'nye teoremy dlja summ nezavisimyh slučajnyh veličin.* GONTI, Moskva, 1938.
(1938*b*)   Ob unimodal'nyh raspredelenijah. *Izv. Nauč.-Issled. Inst. Mat. Meh. Tomsk. Gos. Univ.* **2** (1938) No. 2, 1—7.

Hinčin, A. Ja. (Khintchine, A. Ya.) —Lévy, P.
(1937)   Sur les lois stables. *C. R. Acad. Sci. Paris* **202** (1937) 374—376.

Hirschman, I. I.
(1950)   Proof of a conjecture of I. J. Schoenberg. *Proc. Amer. Math. Soc.* **1** (1950) 63—65.

Hirschman, I. I.—Widder, D. W.
(1949)   The inversion of a general class of convolution transforms. *Trans. Amer. Math. Soc.* **66** (1949) 135—201.
(1955)   *The convolution transform.* Princeton University Press, Princeton, 1955.

Holgate, P.
(1970)   The modality of some compound Poisson distributions. *Biometrika* **57** (1970) 666—667.

Holt, D. R.—Crow, E. L.
(1973)   Tables and graphs of the stable probability density functions. *J. Res. Nat. Bur. Standards Sect. B* **77 B** (1973) 143—198.

Householder, A. S.
(1950)   Analyzing exponential decay curves. *Proceedings of the Seminar on Scientific Computation, November,* 1949. International Business Machines Corp., New York, 1950; pp. 28—32.

van de Hulst, H. C.
(1941)   The determination of the true profile of a spectral line. *Bull. Astronom. Inst. Netherlands* **9** (1941) 225—228.
(1946*a*)   Generalization of some methods for solving an integral equation of the first kind. *Bull. Astronom. Inst. Netherlands* **10** (1946) 75—79.
(1946*b*)   Instrumental distortion of weak spectral lines. *Bull. Astronom. Inst. Netherlands* **10** (1946) 79.

Hunt, B. R.
(1970)   The inverse problem of radiography. *Math. Biosci.* **8** (1970) 161—179.

Husung, E.
(1938)   Untersuchung über die Titergleichmässigkeit von Zellwollen und über deren praktische, zahlenmässige und zeichnerische Auswertung. *Melliand Textilber.* **19** (1938) 886—889, 956—960.

Ibragimov, I. A.
(1956)   O kompozicii odnoveršinnyh raspredeleniǐ. *Teor. Verojatnost. i Primenen.* **1** (1956) 283—288.

Ibragimov, I. A.—Černin, K. E.
(1959)   Od odnoveršinnosti ustoǐčivyh zakonov. *Teor. Verojatnost. i Primenen.* **4** (1959) 453—456.

Ibragimov, I. A.—Linnik, Ju. V.
(1965)   *Nezavisimye i stacionarno svjazannye veličiny.* Nauka, Moskva, 1965.

Idu, S. M.—Cucu, N. B.
(1959) Zur Auswertung von Elektrophorese-Diagrammen. *Hoppe-Seyler's Zeitschrift für physiologische Chemie* **314** (1959) 284—288.
Inouye, T.
(1964) The super resolution of gamma-ray spectrum. *Nuclear Instrum. Methods* **30** (1964) 224—228.
Isii, K.
(1958) Note on a characterization of unimodal distributions. *Ann. Inst. Statist. Math.* **9** (1958) 173—184.
Ito, K.
(1960) *Verojatnostnye processy.* Vypusk I. Izd. Inostrannoĭ Literatury, Moskva, 1960.
Ivanov, V. K.
(1962*a*) Integral'nye uravnenija pervogo roda i približennoe rešenie obratnoĭ zadači potenciala. *Dokl. Akad. Nauk SSSR* **142** (1962) 998—1000.
(1962*b*) O lineĭnyh nekorrektnyh zadačah. *Dokl. Akad. Nauk SSSR* **145** (1962) 270—272.
(1963) O nekorrektno postavlennyh zadačah. *Mat. Sb.* **61** (1963) 211—222.
(1966*a*) O ravnomernoĭ reguljarizacii neustoĭčivyh zadač. *Sibirsk. Mat. Ž.* **7** (1966) 546—558.
(1966*b*) O približennom rešenii operatornyh uravneniĭ pervogo roda. *Ž. Vyčisl. Mat. i Mat. Fiz.* **6** (1966) 1089—1094.
(1967) Ob integral'nyh uravnenijah Fredgol'ma pervogo roda. *Differencial'nye Uravnenija* **3** (1967) 410—421.
Jantzen, E.
(1932) *Das fraktionierte Destillieren und das fraktionierte Verteilen als Methoden zur Trennung von Stoffgemischen.* Verl. Chemie, Weinheim, 1932.
John, F.
(1955*a*) Numerical solution of the equation of heat conduction for preceding times. *Ann. Mat. Pura Appl. Ser. 4.* **40** (1955) 129—142.
(1955*b*) *Differential equations with approximate and improper data.* Lecture notes. New York, 1955.
Jones, J. G.
(1961) On the numerical solution of convolution integral equations. *Math. Comp.* **15** (1961) 131—142.
Jordan, Ch. (Károly)
(1927) *Statistique mathématique.* Gauthier-Villars, Paris, 1927.
Kahn, F. D.
(1955) The correction of observational data for instrumental band width. *Proc. Cambridge Philos. Soc.* **51** (1955) 519—525.
Kaplan, B. G.—Gurvič, K. E.
(1963) Kompleksnoe primenenie matematičeskih metodov k èlektroforetičeskomu issledovaniju belkovogo sostava krovi v norme i patologii. *Primenenie matematičeskih metodov v biologii.* II. Izd. Leningr. Univ., Leningrad, 1963; pp. 183—190.
Karlin, S.
(1968) *Total positivity.* Volume I. Stanford University Press, Stanford, California, 1968.
Keating, D. T.—Warren, B. E.
(1952) The effect of a low absorption coefficient on X-ray spectrometer measurements. *Rev. Sci. Instrum.* **23** (1952) 519—522.
Keilson, J.—Gerber, H.
(1971) Some results for discrete unimodality. *J. Amer. Statist. Assoc.* **66** (1971) 386—389.

Kindler, E.
(1969)   Simple use of pattern recognition in experiment analysis. *Kybernetika* (Prague)
          5 (1969) No. 3, 201—211.
(1973)   A heuristical algorithm for simple exponential analysis. *Apl. Mat.* **18** (1973) 391—398.
Kirillova, L. S.—Piontkovskiĭ, A. A.
(1968)   Nekorrektnye zadači v teorii optimal'nogo upravlenija (obzor). *Avtomat. i Telemeh.*
          1968, 10, 5—17.
Kiss, A.—Sándorfy, C.
(1948)   Sur les méthodes d'analyse des courbes d'absorption. *Acta Universitatis Szegediensis.*
          Acta Chemica et Physica **2** (1948) Fasc. 3, 71—76.
Knoll, F.
(1942)   Zur Grosszahlforschung: Über die Zerspaltung einer Mischverteilung in Normal-
          verteilungen. *Arch. Math. Wirtschafts- und Sozialforsch.* **8** (1942) Heft 1, 36—49.
Kovács, László Béla
(1963)   Kvázi-konkáv programozási feladat megoldása gradiens vetítési módszerrel. (Solution
          of a problem of quasi-concave programming by means of the gradient projection
          method.) *Magyar Tud. Akad. Mat. Fiz. Oszt. Közl.* **13** (1963) 157—178. (Shortened
          version: L. B. Kovács: Gradient projection method for quasi-concave program-
          ming. *Colloquium on applications of mathematics to economics,* Budapest, 1963. Aka-
          démiai Kiadó, Budapest, 1965; pp. 213—225.)
Kreisel, G.
(1949)   Some remarks on integral equations with kernels $L(\xi_1 - x_1, \ldots, \xi_n - x_n; \alpha)$. *Proc.*
          Roy. Soc. London Ser. A **197** (1949) 160—183.
Kremer, P.
(1941)   An apparatus for facilitating the elimination of the instrumental curve from observed
          Fraunhofer line profiles. *Bull. Astronom. Inst. Netherlands* **9** (1941) 229.
Kubik, L.
(1966)   On the class of infinitely divisible distributions and on its subclasses. *Studia Math.*
          27 (1966) 203—211.
Kurth, R.
(1965)   On Eddington's solution of the convolution integral equation. *Rend. Circ. Mat. Palermo*
          (2) **14** (1965) 76—84.
Kühnen, F.
(1909)   Methode zur Aufsuchung periodischer Erscheinungen in Reihen equidistanter Be-
          obachtungen. *Astronom. Nachr.* **182** (1909) 1—11.
Labhart, H.
(1947)   Ein Auswertegerät für Elektrophoresediagramme. *Experientia* **3** (1947) 36—37.
Laha R. G.
(1961)   On a class of unimodal distributions. *Proc. Amer. Math. Soc.* **12** (1961) 181—184.
Lánczos, C.
(1957)   *Applied analysis.* Pitman and Sons, Ltd., London, 1957.
Landahl, H. D.
(1963)   Some mathematical aspects of multi-compartment analysis of tracer experiments.
          Ann. New York Acad. Sci. **108** (1963) Article 1: Multi-compartment analysis of tracer
          experiments; pp. 331—335.
Lapin, A. I.
(1947)   O nekotoryh svoĭstvah ustoĭčivyh zakonov raspredelenija. Diplomnaja rabota. Manu-
          script. Moskva, 1947.

Larson, H. P.—Kenneth, L. A.
(1967)   A least squares deconvolution technique for the photoelectric Fabry—Perot spectrometer. *Appl. Opt.* **6** (1967) 1701—1705.
Lattès, R.—Lions, J. L.
(1967)   *Méthode de quasi-réversibilité et applications.* Dunod, Paris, 1967.
Lavrent'ev, M. M.
(1962)   *O nekotoryh nekorrektnyh zadačah matematičeskoǐ fiziki.* Izd. Sibirsk. Otdel. AN SSSR, Novosibirsk, 1962.
(1966)   O postanovke nekotoryh nekorrektnyh zadač matematičeskoǐ fiziki. *Nekotorye voprosy vyčislitel'noǐ i prikladnoǐ matematiki, Novosibirsk.* Nauka, Moskva, 1966; pp. 258—276.
Lavrent'ev, M. M.—Vasil'ev, V. G.
(1966)   O postanovke nekotoryh nekorrektnyh zadač matematičeskoǐ fiziki. *Sibirsk. Mat. Ž.* **7** (1966) 559—576.
Lehotai, L.
(1960)   Zerlegung der Absorptionskurven in Teilbanden mittels der Methode der Streuungsverminderung. *Acta Chim. Hungar.* **25** (1960) 25—32.
Levitin, E. S.—Poljak, B. T.
(1966)   Metody minimizacii pri naličii ograničeniǐ. *Ž. Vyčisl. Mat. i Mat. Fiz.* **6** (1966) 787—823.
Lévy, P.
(1924)   Théorie des erreurs. La loi de Gauss et les lois exceptionnelles. *Bull. Soc. Math. France* **52** (1924) 49—85.
(1925)   *Calcul des probabilités.* Gauthier-Villars, Paris, 1925.
(1937)   *Théorie de l'addition des variables aléatoires.* Gauthier-Villars, Paris, 1937.
Linnik, Ju. V.
(1953)   Lineǐnye formy i statističeskie kriterii. II. *Ukrain. Mat. Ž.* **5** (1953) 247—290.
(1954)   Ob ustoǐčivyh verojatnostnyh zakonah s pokazatelem men'šim edinicy. *Dokl. Akad. Nauk SSSR* **94** (1954) 619—621.
(1960)   *Razloženija verojatnostnyh zakonov.* Izd. Leningrad. Univ., Leningrad, 1960.
Linz, P.
(1969)   Numerical methods for Volterra integral equations of the first kind. *Comput. J.* **12** (1969) 393—397.
Lipka, István (St.)
(1938)   A Descartes-féle jelszabály kiterjesztéseiről. (Über die Erweiterung der Descartes-schen Zeichenregel.) *Matematikai és Fizikai Lapok* **45** (1938) 78—93.
(1942)   Über die Abzählung der reellen Wurzeln von algebraischen Gleichungen. *Math. Z.* **47** (1942) 343—351.
Liskovec, O. A.
(1966)   Regularizacija nekorrektnyh zadač dlja uravneniǐ matematičeskoǐ fiziki. *Differencial'nye Uravnenija* **2** (1966) 1128—1131.
(1968)   Čislennoe rešenie nekotoryh nekorrektnyh zadač metodom kvazi-reseniǐ. *Differencial'nye Uravnenija* **4** (1968) 735—742.
Loeb, J.
(1960)   La "déconvolution" en calcul numérique. *Ann. Télécommun.* **15** (1960) 84—91.
Lonn, E.
(1932)   Beweis der Eindeutigkeit der Zerlegung einer Intensitätskurve in ihre Komponenten. *Z. Physik* **75** (1932) 348—349.
Lukács, E.
(1957)   Remarks concerning characteristic functions. *Ann. Math. Statist.* **28** (1957) 717—723.

(1961)   Recent developments in the theory of characteristic functions. *Proceedings of the Fourth Berkeley Symposium on Mathematical Statistics and Probability*. Volume II. University of California Press, Berkeley and Los Angeles, 1961; pp. 307—335.

(1964)   *Fonctions caractéristiques*. Dunod, Paris, 1964.

(1968)   Contributions to a problem of D. van Dantzig. *Teor. Verojatnost. i Primenen.* **13** (1968) 114—125.

Macnee, A. B.

(1953)   A high speed product integrator. *Rev. Sci. Instrum.* **24** (1953) 207—211.

Magos, László—Medgyessy, Pál

(1954)   Elektroforetikus diagrammok matematikai kiértékelése. (Mathematical evaluation of electrophoretical diagrams.) *Kísérletes Orvostudomány* (Budapest) **6** (1954) 367—369.

Maĭorov, L. V.

(1965)   O nekorrektnosti odnoĭ obratnoĭ zadači. *Ž. Vyčisl. Mat. i Mat. Fiz.* **5** (1965) 363—365.

Marton, K.—Varga, L.

(1971)   Regularization of certain operator equations by filters. *Studia Sci. Math. Hungar.* **6** (1971) 457—465.

McPhee, W. N.

(1963)   *Formal theories of mass behaviour*. The Free Press, Glencoe, Ill., 1963.

Medgyessy, Pál

(1953)   Valószínűség-eloszlásfüggvények keverékének felbontása összetevőire. (Détermination des components d'un mélange des fonctions de distribution.) *Magyar Tud. Akad. Alkalm. Mat. Int. Közl.* **2** (1953) 165—177.

(1954a) Diszkrét valószínűség-eloszlások keverékének felbontása összetevőire. (Decomposition of discrete compound probability distributions.) *Magyar Tud. Akad. Alkalm. Mat. Int. Közl.* **3** (1954) 139—153.

(1954b) Újabb eredmények valószínűség-eloszlásfüggvények keverékének összetevőire bontásával kapcsolatban. (Some recent results concerning the decomposition of compound probability distributions.) *Magyar Tud. Akad. Alkalm. Mat. Int. Közl.* **3** (1954) 155—169.

(1954c) Valószínűség-eloszlásfüggvények keverékének felbontása összetevőire. (Decomposition into components of a superposition of distribution functions.) Dissertation. Budapest, 1954.

(1954d) Szorzatintegrálás, Fourier-szintézis és hasonló feladatok elvégzése kvadrátplaniméter és egy új készülék kombinációjának segítségével. (Product integration, Fourier-synthesis and similar operations carried out by means of a square planimeter and a new apparatus.) *Magyar Tud. Akad. Alkalm. Mat. Int. Közl.* **3** (1954) 129—137.

(1955a) Közelítő eljárás Cauchy-sűrűségfüggvények keverékének összetevőkre bontására. (An approximate method for the decomposition of a compound of Cauchy frequency functions.) *Magyar Tud. Akad. Alkalm. Mat. Int. Közl.* **3** (1955) 321—329.

(1955b) Kiegészítés az „Újabb eredmények valószínűség-eloszlásfüggvények keverékének összetevőire bontásával kapcsolatban" című dolgozathoz. (A supplement to the paper "Some recent results concerning the decomposition of compound probability distributions".) *Magyar Tud. Akad. Alkalm. Mat. Int. Közl.* **3** (1955) 331—341.

(1956)   Stabilis valószínűség-sűrűségfüggvényekre fennálló parciális differenciálegyenletek és alkalmazásaik. (Partial differential equations for stable density functions and their applications.) *Magyar Tud. Akad. Mat. Kutató Int. Közl.* **1** (1956) 489—518.

(1957a) Anwendungsmöglichkeiten der Analyse der Wahrscheinlichkeitsdichtefunktionen bei der Auswertung von Messungsergebnissen. *Z. Angew. Math. Mech.* **37** (1957) 128—139.

(1957b) A mechanical functional synthesizer. *Magyar Tud. Akad. Mat. Kutató Int. Közl.*
2 (1957) 33—42.

(1958) Partial integro-differential equations for stable density functions and their applications.
Publ. Math. Debrecen **5** (1958) 288—293.

(1961a) *Decomposition of superpositions of distribution functions.* Akadémiai Kiadó. Publishing
House of the Hungarian Academy of Sciences, Budapest, 1961.

(1961b) Diszkrét valószínűség-eloszlások áttranszformálása sűrűségfüggvénnyé. (Transforming
discrete probability distributions into density functions.) (Lecture: Seminar of the
Department of Calculus of Probability of the Mathematical Institute of the Hungarian
Academy of Sciences, 9. February 1961.) Abstract: *Magyar Tud. Akad. Mat. Kutató
Int. Közl.* **6** (1961) 528.

(1961c) On some unsolved problems in the theory of the decomposition of superpositions of
distribution functions. IIe *Congrès Mathématique Hongrois* Budapest, 24.—31. August
1960. Vol. II. Akadémiai Kiadó, Budapest, 1961; Section IV, pp. 16—18.

(1962) Karakterisztikus függvények előállítása keverék formájában. (On the representation
of characteristic functions in the form of mixtures.) Manuscript. Budapest, 1962.

(1963) On the interconnection between the representation theorems of characteristic functions
of unimodal distribution functions and of convex characteristic functions. *Magyar
Tud. Akad. Mat. Kutató Int. Közl.* **8** (1963) 425—430.

(1964) Eloszlás- és sűrűségfüggvény-grafikonok alakjának jellemzéséről, I. (On the characterization of the shape of graphs of distribution and density functions, I.) *Magyar Tud.
Akad. Mat. Fiz. Oszt. Közl.* **14** (1964) 279—292.

(1966a) Egy konvoluciós típusú integrálegyenlet numerikus megoldása és ennek felhasználása
Gauss-függvény szuperpozíciók felbontására. (Numerical solution of an integral equation of convolution type and its applications to the decomposition of superpositions
of Gauss' functions.) *Magyar Tud. Akad. Mat. Fiz. Oszt. Közl.* **16** (1966) 47—64.

(1966b) Remarks on the paper "On the separation of exponentials" of *R. Bellman.* Manuscript. Budapest, 1966.

(1966c) Valószínűség-eloszlások általánosítása és ezzel kapcsolatos faktorizációs problémák.
(A generalization of discrete distributions and problems of factorization connected
with it.) (Lecture: Colloquium on Mathematical Statistics, Debrecen (Hungary), 13—15.
October 1966. Sponsored by the Bolyai János Mathematical Society and the Mathematical Institute of the Kossuth Lajos University, Debrecen.) Abstract: *Matematikai
Statisztikai Kollokvium. Debrecen,* 1966. *október* 13—15. *Előadás kivonatok.* Debrecen,
1966; pp. 3—4.

(1967a) Sűrűségfüggvény szuperpozíciók felbontásának egy lényegileg új módszeréről. (On an
essentially new method of decomposing superpositions of density functions.) *Magyar
Tud. Akad. Mat. Fiz. Oszt. Közl.* **17** (1967) 383—390. (English translation: An essentially
new method for the decomposition of density functions. *Selected translations in mathematical statistics and probability.* Volume 10. American Mathematical Society, Providence, 1972; pp. 170—178.)

(1967b) Szilárd anyagok keverék-voltának megállapítása matematikai módszerekkel. (Determination by the aid of mathematical methods of whether a certain material is a mixture
or not.) *Tíz példa a matematika gyakorlati alkalmazására.* (Szerkesztette Vincze István.)
(Ten examples of the practical application of mathematics.) (Edited by István Vincze.)
Gondolat, Budapest, 1967; pp. 203—223.

(1967c) Eloszlás- és sűrűségfüggvény-grafikonok alakjának jellemzéséről, II. (On the characterization of the shape of graphs of distribution and density functions, II.) *Magyar Tud. Akad. Mat. Fiz. Oszt. Közl.* **17** (1967) 101—108.

(1967d) On a new class of unimodal infinitely divisible distribution functions and related topics. *Studia Sci. Math. Hungar.* **2** (1967) 441—446.

(1968) Valószínűség-eloszlásfüggvények és diszkrét valószínűségeloszlások egycsúcsúságával kapcsolatos problémákról. (On problems connected with the unimodality of distribution functions and discrete distributions.) Manuscript. Budapest, 1967—1968.

(1971a) Inkorrekt matematikai problémák vizsgálatának jelen állásáról, különös tekintettel I. fajú operátoregyenletek megoldására. (Áttekintés.) (On the present status of the investigations concerning incorrect problems in mathematics, with special regard to the resolution of operator equations of the Ist kind. (Survey.)) *Magyar Tud. Akad. Mat. Fiz. Oszt. Közl.* **20** (1971) 97—131.

(1971b) Sűrűségfüggvények és diszkrét eloszlások szuperpozícióinak felbontása. (Decomposition of superpositions of density functions and discrete distributions.) Dissertation. Budapest, 1971.

(1972a) Sűrűségfüggvények és diszkrét eloszlások szuperpozícióinak felbontása. (Decomposition of superpositions of density functions and discrete distributions.) *Magyar Tud. Akad. Mat. Fiz. Oszt. Közl.* **21** (1972) 129—200.

(1972b) Sűrűségfüggvények és diszkrét eloszlások szuperpozícióinak felbontása. II. (Decomposition of superpositions of density functions and discrete distributions. II.) *Magyar Tud. Akad. Mat. Fiz. Oszt. Közl.* **21** (1972) 261—382.

(1972c) On the unimodality of discrete distributions. *Periodica Mathematica Hungarica* **2** (1—4) (1972) 245—257.

(1973) Új módszerek szimmetrikus sűrűségfüggvények szuperpozicióinak felbontására. I. (New methods for the decomposition of superpositions of symmetrical density functions. I.) *Magyar Tud. Akad. Mat. Fiz. Oszt. Közl.* **22** (1973); to appear.

(1974a) On two methods of obtaining criteria of unimodality for density functions. *Publ. Math. Debrecen* **21** (1974) 305—308.

(1974b) Új módszerek szimmetrikus sűrűségfüggvények szuperpozicióinak felbontására. II. (New methods for the decomposition of superpositions of symmetrical density functions. II.) Manuscript. Budapest, 1974.

Medgyessy, Pál—Rényi, Alfréd—Tettamanti, Károly—Vincze, István
(1954) A kémiai frakcionáló megosztás matematikai tárgyalása nem teljes diffúzió esetében. (Mathematical investigation of chemical countercurrent distribution, in case of noncomplete diffusion.) *Magyar Tud. Akad. Alkalm. Mat. Int. Közl.* **3** (1954) 81—97.

Medgyessy, Pál—Varga, László
(1968) Gauss-függvény keverékek numerikus felbontására szolgáló egyik eljárás javításáról. (On the improving of a method for the numerical decomposition of mixtures of Gaussian functions.) *Magyar Tud. Akad. Mat. Fiz. Oszt. Közl.* **18** (1968) 31—39.

Meszéna, György
(1968) Valószínűségeloszlások és idősorok felbontása. (The break-down of probability distributions and time series.) *Szigma* **1** (1968) 60—75.

Meszéna, György—Scherf, Emil
(1960) Matematikai-statisztikai vizsgálatok a természetes vizek uránban való feldúsulásának fizikai feltételeiről. (Mathematical-statistical studies of the physical conditions for uranium enrichment of natural waters.) *ATOMKI Közl.* **2** (1960) 109—143.

Miller, G. F.
(1974) Fredholm equations of the first kind. *Numerical solution of integral equations.* (Ed. L. M. Delves—J. Walsh.) Clarendon Press, Oxford, 1974; pp. 175—188.

Miller, K.
(1964) Three circle theorems in partial differential equations and applications to improperly posed problems. *Arch. Rational Mech. Anal.* **16** (1964) 126—154.

Molodenkova, I. D.—Kovalev, I. F.
(1972) K voprosu primenenija odnogo iteracionnogo processa dlja razdelenija perekryvajuščihsja konturov spektral'nyh liniĭ. *Differencial'nye uravnenija i vyčislitel'naja matematika.* Vypusk perviĭ. Izdatel'stvo Saratovskogo universiteta, Saratov. 1972; pp. 54—64.

Morison Smith, D.—Bartlet, J. C.
(1961) Calculation of the areas of isolated or overlapping normal probability curves. *Nature* **191** (1961) 688—689.

Morozov, V. A.
(1973) Lineĭnye i nelineĭnye nekorrektnye zadači. *Itogi nauki i tehniki. Matematičeskiĭ analiz.* Tom 11. VINITI, Moskva, 1973; pp. 129—178.

Morozov, V. A.—Ivanišcev, V. F.
(1966) Primenenie metoda reguljarizacii k rasčetu aročnyh plotin. *Vyčislitel'nye metody i programmirovanie.* V. Izd. Moskov. Univ., Moskva, 1966; pp. 171—186.

Morozova, I. D.
(1970) Opredelenie istinnyh značeniĭ parametrov izolirovannyh liniĭ kombinacionnogo rassejanija po nabljudaemym. *Vyčislitel'nye metody i programmirovanie.* Vyp. 4. Saratov. un-t, Saratov, 1970; pp. 14—22.

Morozova, I. D.—Balabanova, N. A.
(1970) Razdelenie konturov perekryvajuščihsja spektral'nyh liniĭ. Primenenie funkcii proizvedenija funkciĭ Koši i Gaussa. *Vyčislitel'nye metody i programmirovanie.* Vyp. 4. Saratov. un-t, Saratov, 1970; pp. 23—29.

Nedelkov, I. P.
(1972) Improper problems in computational physics. *Comput. Phys. Comm.* **4** (1972) 157—164.

Noble, W.—Hayes, J. E. Jr.—Eden, M.
(1959) Repetitive analog computer for analysis of sums of distribution functions. *Proc. IRE* **47** (1959) 1952—1956.

Oettli, W.—Prager, W.
(1964) Compatibility of approximate solution of linear equations with given error bounds for coefficients and right-hand sides. *Numer. Math.* **6** (1964) 405—409.

Oka, M.
(1954) Ecological studies on the kidai by the statistical method II. On the growth of kidai (*Taius tumifrons*). *Bull. Fac. Fish. Nagasaki* **2** (1954) 8—25.

Olshen, R. A.—Savage, L. J.
(1970) A generalized unimodality. *J. Appl. Prob.* **7** (1970) 21—34.

Ornstein, L. S.—van Wyk, W. R.
(1932) Optische Untersuchung des Akkomodationskoeffizienten der Molekulartranslation und dessen Verteilungsfunktion in einem verdünnten Gase. *Z. Physik* **78** (1932) 734—743.

Ovseevič, I. A.—Jaglom, A. M.
(1954) Monotonnye perehodnye processy v odnorodnyh dlinnyh linijah. *Izv. Akad. Nauk SSSR Otdel. Tehn. Nauk* **1954**, No. 7, 13—20.

Papoulis, A.
(1955)   Method of correction for the $\alpha_1\alpha_2$ Doublet in the X-ray diffraction lines. *Rev. Sci. Instrum.* **26** (1955) 423—426.
(1973)   A different approach to the analysis of tracer data. *SIAM J. Control* **11** (1973) 466—474.

Parsons, D. H.
(1968)   Biological problem involving sums of exponential functions of time: A mathematical analysis that reduces experimental time. *Math. Biosci.* **2** (1968) 123—128.
(1970)   Biological problems involving sums of exponential functions of time: An improved method of calculation. *Math. Biosci.* **9** (1970) 37—47.

Pearson, K.
(1894)   Contributions to the mathematical theory of evolution. *Philos. Trans. Roy. Soc. London Ser. A* **185** (1894) 71—110.
(1915a)   On the problem of sexing osteometric material. *Biometrika* **10** (1915) 479—487.
(1915b)   On certain types of compound frequency distributions in which the components can be individually described by binomial series. *Biometrika* **11** (1915) 139—144.

Pearson, K. —Lee, A.
(1909)   On the generalized probable error in multiple normal correlation. *Biometrika* **6** (1908—09) 59—68.

Peter, L.—Peter, W.
(1960)   Abkling- und Sättigungsfunktion bei Radionukliden. *Praxis Math.* **2** (1960) 57—61.

Petrov, A. P.
(1967)   Ocenki lineĭnyh funkcionalov dlja resenija nekotoryh obratnyh zadač. *Ž. Vyčisl. Mat i Mat. Fiz.* **7** (1967) 648—654.

Phillips, D. L.
(1962)   A technique for the numerical solution of certain integral equations of the first kind. *J. Assoc. Comput. Mach.* **9** (1962) 84—97.

Pollard, H. (S.)
(1934)   On the relative stability of the median and arithmetic mean, with particular reference to certain frequency distributions which can be dissected into normal distributions. *Ann. Math. Statist.* **5** (1934) 227—262.
(1946)   The representation of $e^{-x^{\lambda}}$ as a Laplace integral. *Bull. Amer. Math. Soc.* **52** (1946) 908—910.
(1953)   Distribution functions containing a Gaussian factor. *Proc. Amer. Math. Soc.* **4** (1953) 578—582.

Pólya, G.
(1915)   Algebraische Untersuchungen über ganze Funktionen vom Geschlechte Null und Eins. *J. Reine Angew. Math.* **145** (1915) 224—249.
(1949)   Remarks on characteristic functions. *Proceedings of the Berkeley Symposium on Mathematical Statistics and Probability.* University of California Press, Berkeley and Los Angeles, 1949; pp. 115—122.

Poulik, M. D.—Pinteric, L.
(1955)   An electronic computer for the evaluation of results of filterpaper electrophoresis. *Nature* **176** (1955) 1226—1227.

Preston, E. J.
(1953)   A graphical method for the analysis of statistical distributions into two normal components. *Biometrika* **40** (1953) 460—464.

de Prony, G. C. F. M. R.
(1795)   Essai expérimental et analytique ... *Journal de l'École Polytechnique* (Paris) **1** (2) (1795) 24—76.

Ralston, A.
(1965)   *A first course in numerical analysis.* McGraw-Hill Book Co. Inc., New York, 1965.

Ramm, A. G.
(1968)   O čislennom differencirovanii. *Izv. Vysš. Učebn. Zaved. Matematika* **11** (1968) 131—134.

Ramsthaler, K.
(1949)   Analyse der Faserfestigkeitsverteilung einer Zellwollmischung durch statistische Verfahren und Grosszahlforschung. *Textil-Praxis* **4** (1949) 358—360, 438—440.

Rao, C. R.
(1948)   The utilisation of multiple measurements in problems of biological classification. *J. Roy. Statist. Soc. Ser. B* **10** (1948) 159—193.
(1952)   *Advanced statistical methods in biometric research.* J. Wiley and Sons, Inc., New York, 1952.

Reiz, A.
(1943)   On the numerical solution of certain types of integral equations. *Ark. Mat. Astr. Fys.* Band **29 A**, No 29. 1943. 21 pp.

Rider, P. R.
(1961)   The method of moments applied to a mixture of two exponential distributions. *Ann. Math. Statist.* **32** (1961) 143—147.
(1962)   Estimating the parameters of mixed Poisson, binomial and Weibull distributions by the Method of Moments. *Bull. Inst. Internat. Statist.* **39** (1962) 2e Livraison, 225—232.

Righini, G.
(1941)   Integratore ottica per la determinazione de profilo vero delle righe spettrali. *Osserv. e. Mem. Arcetri* **60** (1941) 27.

Rosenstiehl, P.—Ghouila-Houri, A.
(1960)   *Les choix économiques, décisions séquentielles et simulation.* Dunod, Paris, 1960.

Ruff, I.
(1965)   A simple approximative method for calculation of the positions and extinction coefficients of overlapping absorption bands. *Acta Chim. Hungar.* **45** (1965) 13—21.

Runge, C.
(1914)   Über eine besondere Art von Integralgleichungen. *Math. Ann.* **75** (1914) 130—132.

Runge, C.—König, R.
(1924)   *Vorlesungen über numerisches Rechnen.* Springer, Berlin, 1924.

Rust, B. W.—Burrus, W. R.
(1972)   *Mathematical programming and the numerical solution of linear equations.* American Elsevier Publ. Co., Inc., New York, 1972; Chapter 1.

Schellenberg, O.
(1932)   Zur Analyse der ultravioletten Emissionen der Erdalkaliphosphore. *Ann. Physik* **5/13** (1932) 249—264.

Schilling, W.
(1947)   A frequency distribution represented as the sum of two Poisson distributions. *J. Amer. Statist. Assoc.* **42** (1947) 407—424.

Schmaedeke, W. W.
(1968)   Approximate solutions for Volterra integral equations of the first kind. *J. Math. Anal. Appl.* **23** (1968) 604—613.

Schmaedeke, W. W.
(1969)   A new approach to unstable problems using variational techniques. *J. Math. Anal. Appl.*
         25 (1969) 272—275.
Schmeidler, W.
(1950)   *Integralgleichungen mit Anwendungen in Physik und Technik* I. Lineare Integralgleichun-
         gen. Akademische Verlagsgesellschaft Geest & Portig K.-G., Leipzig, 1950.
Schoenberg, I. J.
(1947)   On totally positive functions, Laplace integrals and entire functions of the Laguerre—
         Pólya—Schur type. *Proc. Nat. Acad. Sci. U.S.A.* **33** (1947) 11—17.
(1948)   Some analytical aspects of the problem of smoothing. *Studies and essays presented to
         R. Courant.* Interscience Publishers, Inc., New York, 1948; pp. 351—370.
(1951)   On Pólya frequency functions. I. The totally positive functions and their Laplace
         transforms. *J. Analyse Math.* **1** (Deuxième Partie) (1951) 331—374.
(1953)   On smoothing operations and their generating functions. *Bull. Amer. Math. Soc.*
         59 (1953) 199—230.
(1954)   On multiply positive sequences and functions. *Bull. Amer. Math. Soc.* **60** (1954) 160–161.
(1955)   On the zeros of the generating functions of multiply positive sequences and functions.
         Ann. of Math. **62** (1955) 447—471.
Schulz, G.
(1934)   Umkehrung von Integraltransformationen. *Z. Angew. Math. Mech.* **14** (1934) 373—374.
Ščigolev, B.
(1924)   O razloženii asimmetričeskoĭ krivoĭ raspredelenija na dve krivye Gaussa. *Russkiĭ
         Astronom. Ž.* **1** (1924): 3—4, 76—89.
Sen, N.
(1922)   Über den Einfluss des Dopplereffekts auf spektroskopische Feinstrukturen und seine
         Elimination. *Physikalische Zeitschrift* **23** (1922) 397—399.
Serebrennikov, M. G.
(1948)   *Garmoničeskiĭ analiz.* Ogiz-Gostehizdat, Moskva—Leningrad, 1948.
Serebrennikov, M. G.—Pervozvanskiĭ, A. G.
(1965)   *Vyjavlenie skrytyh periodičnosteĭ.* Nauka, Moskva, 1965.
Skorohod, A. V.
(1954)   Asimptotičeskie formuly dlja ustoĭčivyh zakonov raspredelenija. *Dokl. Akad. Nauk
         SSSR* **98** (1954) 731—734.
Solodovnikov, V. V.
(1952)   *Vvedenie v statističeskoju dinamiku sistem avtomatičeskogo upravlenija.* Gostehizdat,
         Moskva, 1952.
Stallman, F. W.
(1970)   Numerical solution of integral equations. *Numer. Math.* **15** (1970) 297—305.
Stene, S.
(1945)   A contribution to the theory of systematic extraction and other related convection
         problems. *Ark. Kem. Miner. Geol.* **18**/A (1945) 1—121.
Sticker, B.
(1930a) Untersuchungen über Sternfarben. II. Analyse der Farbenhäufigkeitsfunktion. *Veröf-
         fentlichungen der Universitäts-Sternwarte Bonn* Nr. 23 (1930) 20—32.
(1930b) Über die Farbenhäufigkeitsfunktion in Sternhaufen. *Z. Astrophys.* **1** (1930) 174—191.
Stokes, A. R.
(1948)   A numerical Fourier-analysis method for the correction of widths and shapes of line
         on X-ray powder photographs. *Proc. London Phys. Soc.* **61** (1948) 382—391.

Stone, M. H.
(1927)   The normal probability function and general frequency functions. *Amer. J. Math.* **49** (1927) 543—552.
Strahov, V. N.
(1963)   Ob odnom čislennom metode rešenija linejnyh integral'nyh uravneniĭ tipa svertki. *Dokl. Akad. Nauk. SSSR* **153** (1963) 533—536.
(1964)   Ob odnom novom metode vyčislenija integralov tipa svertki. *Izv. Akad. Nauk SSSR Ser. Geofiz.* **1964.** No. 12, 1819—1822.
(1968)   O čislennom rešenii nekorrektnyh zadač, predstavljaemyh integral'nymi uravnenijami tipa svertki. *Dokl. Akad. Nauk SSSR* **178** (1968) 299—302.
Strand, O. N.—Westwater, (E.) R.
(1968a) Statistical estimation of the numerical solution of a Fredholm integral equation of the first kind. *J. Assoc. Comput. Mach.* **15** (1968) 100—114.
(1968b) Minimum-RMS estimation of the numerical solution of a Fredholm integral equation of the first kind. *SIAM J. Numer. Anal.* **5** (1968) 287—295.
Strömgren, B.
(1934)   Tables and diagrams for dissecting a frequency curve into components by the half-invariant method. *Skand. Aktuarietidskr.* **17** (1934) 7—54.
Sudakov, V. N.—Halfin, L. A.
(1964)   Statističeskiĭ podhod k korrektnosti zadač matematičeskoĭ fiziki. *Dokl. Akad. Nauk SSSR* **157** (1964) 1058—1060.
Svedberg, Th.—Pedersen, K. D.
(1940)   *Die Ultracentrifuge.* Th. Steinkopf, Dresden, 1940.
Sydow, A.—Dittmann, H.
(1963)   Statistische Analysen mittels elektronischer Analogrechner. *Messen, Steuern, Regeln* **6** (1963) 501—504.
Szigeti, György
(1947)   Lumineszkáló anyagok. (Luminescent materials.) *Elektrotechnika* (Budapest) **39** (1947) 70—73, 81—86.
Szőke, J.—Varga, L.—Nagypál, I.
(1967)   Experimental and computer analysis of the spectral fine structure. *Proceedings of the* XIV. *Colloquium Spectroscopicum Internationale.* Debrecen, Hungary, 1967; pp. 1205—1218.
Tallis, G. M.
(1969)   The identifiability of mixtures of distributions. *J. Appl. Probability* **6** (1969) 389—398.
Tanaka, S.
(1962)   A method of analysing of polymodal frequency distribution and its application to the length distribution of the Porgy, *Taius tumifrons* (J. and S.). *J. Fish. Res. Bd. Can.* **19** (1962) 1143—1159.
Teicher, H.
(1960)   On the mixture of distributions. *Ann. Math. Statist.* **31** (1960) 55—73.
(1961)   Identifiability of mixtures. *Ann. Math. Statist.* **32** (1961) 244—248.
(1963)   Identifiability of finite mixtures. *Ann. Math. Statist.* **34** (1963) 1265—1269.
Thies, H. H.
(1961)   Resolution of Bremsstrahlung experiments. *Austral. J. Phys.* **14** (1961) 174—187.
Thionet, P.
(1966)   Note sur les mélanges de certaines distributions de probabilités. *Publ. Inst. Statist. Univ. Paris* **15** (1966) 61—80.

Tihonov, A. N.

(1944) Ob ustoĭčivosti obratnyh zadač. *Dokl. Akad. Nauk SSSR* **39** (1944) 195—198.

(1963a) O rešenii nekorrektno postavlennyh zadač i metode reguljarizacii. *Dokl. Akad. Nauk SSSR* **151** (1963) 501—504.

(1963b) O reguljarizacii nekorrektno postavlennyh zadač. *Dokl. Akad. Nauk SSSR* **153** (1963) 49—52.

(1963c) O rešenii nekorrektno postavlennyh zadač i metode reguljarizacii. *Materialy k Sovmestnomu sovetsko-amerikanskomu simpoziumu po uravnenijam s častnymi proizvodnymi* (Novosibirsk, avg. 1963). Izd. Sibirsk. Otdel. AN SSSR, Novosibirsk, 1963; pp. 261—265.

(1965a) O nelinejnyh uravnenijah pervogo roda. *Dokl. Akad. Nauk SSSR* **161** (1965) 1023–1026.

(1965b) O metodah reguljarizacii zadač optimal'nogo upravlenija. *Dokl. Akad. Nauk SSSR* **162** (1965) 763—765.

(1965c) O nekorrektnyh zadačah lineĭnoĭ algebry i ustoĭčivyh metodah ih rešenija. *Dokl. Akad. Nauk SSSR* **163** (1965) 591—595.

(1965d) O nekorrektnyh zadačah optimal'nogo planirovanija i ustoĭčivyh metodah ih rešenija. *Dokl. Akad. Nauk SSSR* **164** (1965) 507—510.

(1965e) Ob ustoĭčivosti algoritmov dlja rešenija vyroždennyh sistem lineĭnyh algebraičeskih uravneniĭ. *Ž. Vyčisl. Mat. i Mat. Fiz.* **5** (1965) 718—722.

(1967) O nekorrektno postavlennyh zadačah. *Vyčisliteľnaja matematika i programmirovanie*. VIII. Izd. Moskov. Univ., Moskva, 1967; pp. 3—33.

Tihonov, A. N. — Arsenin, V. Ja.

(1974) *Metody rešenija nekorrektnyh zadač*. Izdateľstvo "Nauka", Moskva, 1974.

Tihonov, A. N.—Galkin, V. Ja.—Zaikin, P. N.

(1967) O prjamyh metodah rešenija zadač optimal'nogo upravlenija. *Ž. Vyčisl. Mat. i Mat. Fiz.* **7** (1967) 416—423.

Tihonov, A. N.—Glasko, V. B.

(1964) O približennom rešenii integral'nyh uravneniĭ Fredgol'ma pervogo roda. *Ž. Vyčisl. Mat. i Mat. Fiz.* **4** (1964) 564—571.

(1965) Primenenie metoda reguljarizacii v nelineĭnyh zadačah. *Ž. Vyčisl. Mat. i Mat. Fiz.* **5** (1965) 463—473.

Tricomi, F. (G.)

(1935) Sur la rappresentazione di una legge di probabilità mediante esponenziali di Gauss e la transformazione di Laplace. *Giorn. Ist. Ital. Attuari* **6** (1935) 135—140.

(1936) Ancora sulla rappresentazione di una legge di probabilità mediante esponenziali di Gauss. *Giorn. Ist. Ital. Attuari* **7** (1936) 42—44.

(1938) Les transformations de Fourier, Laplace, Gauss et leurs applications au calcul des probabilités et à la statistique. *Ann. Inst. H. Poincaré* **8** (1938) 111—149.

(1968) *Repertorium der Theorie der Differentialgleichungen*. Springer, Berlin, 1968; pp. 130–132.

Trumpler, R. J.

(1951) Correction of frequency functions for observational errors of the variables. *Proceedings of the Second Berkeley Symposium on Mathematical Statistics and Probability*. University of California Press, Berkeley and Los Angeles, 1951; pp. 437—440.

Trumpler, R. J.—Weaver, H. F.

(1953) *Statistical astronomy*. University of California Press, Berkeley and Los Angeles, 1953.

Turčin, V. F.

(1967) Rešenie uravnenija Fredgol'ma I. roda v statističeskom ansamble gladkih funkciĭ. *Ž. Vyčisl. Mat. i Mat. Fiz.* **7** (1967) 1270—1284.

Turčin, V. F.
(1968) Vybor ansamblja gladkih funkciĭ pri rešenii obratnoĭ zadači. *Ž. Vyčisl. Mat. i Mat. Fiz.*
        8 (1968) 230—238.
Twomey, S.
(1963) On the numerical solution of Fredholm integral equations of the first kind by the in-
        version of the linear system produced by quadrature. *J. Assoc. Comput. Mach.*
        10 (1963) 97—101.
Unsöld, A.
(1955) *Physik der Sternatmosphären.* II. Auflage. Springer, Berlin, 1955.
Varah, J. M.
(1973) On the numerical solution of ill-conditioned linear systems with applications to ill-posed
        problems. *SIAM J. Numer. Anal.* **10** (1973) 257—267.
Varga, László
(1966a) Gauss-függvények keverékének komponensekre bontásáról. (Decomposition of super-
        imposed Gaussian distribution functions.) *Magyar Tud. Akad. Közp. Fiz. Kutató Int.*
        Közl. **14** (1966) 383—389.
(1966b) Exponenciális bomlásgörbe paramétereinek egy egyszerű meghatározásáról. (Simple
        method for the evaluation of the parameters of exponential decay curves.) *Magyar Tud.*
        Akad. Közp. Fiz. Kutató Int. Közl. **14** (1966) 21—24.
(1968) Nemlineáris becslési feladatok numerikus módszerei. (Numerical methods for non-
        linear problems of estimation.) Dissertation. Budapest, 1968.
Vasin, V. V.
(1968) Reguljarizacija nelineĭnyh differencial'nyh uravneniĭ v častnyh proizvodnyh. *Dif-*
        ferencial'nye Uravnenija **4** (1968) 2268—2274.
(1969) Reguljarizacija zadači čislennogo differencirovanija. *Ural. Gos. Univ. Mat. Zap.*
        7 (1969) No. 2, 29—33.
Vemuri, V.—Chen, Fang-pai
(1974) An initial value method for solving Fredholm integral equation of the first kind.
        J. Franklin Inst. **297** (1974) 187—200.
Wallner, A.—Ulke, R.
(1952) Auswertung von Diagrammen aus Elektrophoreseversuchen. *Hoppe-Seyler's Zeit-*
        schrift für physiologische Chemie **290** (1952) Heft 3—6, 81—91.
Weichselberger, K.
(1961) Über ein graphisches Verfahren zur Trennung von Mischverteilungen und zur Identifika-
        tion kupierter Normalverteilungen bei grossem Stichprobenumfang. *Metrika* **4** (1961)
        178—229.
Whittaker, E.—Robinson, G.
(1949) *The calculus of observations.* A treatise on numerical mathematics. Fourth edition.
        Blackie and Sons Ltd., London and Glasgow, 1949.
Wiedemann, E.
(1947) Über die Auswertung von Elektrophorese-Diagrammen nach L. G. Longsworth und
        Philpot—Svensson. *Helv. Chim. Acta* **30**, Pars I (1947) 892—900.
Willers, F. A.
(1923) *Numerische Integration.* Walter de Gruyter und Co., Berlin, 1923.
(1950) *Methoden der praktischen Analysis.* 2. Auflage. Walter de Gruyter und Co., Berlin,
        1950.
Wintner, A.
(1936) On a class of Fourier transforms. *Amer. J. Math.* **58** (1936) 45—90.

(1941)   The singularities of Cauchy's distributions. *Duke Math. J.* **8** (1941) 678—681.
(1956)   Cauchy's stable distribution and an "explicit formula" of Mellin. *Amer. J. Math.* **78** (1956) 819—861.
Wolfe, S. J.
(1971)   On the unimodality of *L* functions. *Ann. Math. Statist.* **42** (1971) 912—918.
Wuhrmann, F.—Wunderly, Ch.
(1957)   *Die Bluteiweisskörper des Menschen.* Dritte, vollständig neu bearbeitete Auflage. B. Schwabe und Co. Verlag, Basel/Stuttgart, 1957.
Yakowitz, S. J.—Spragins, J. D.
(1968)   On the identifiability of finite mixtures. *Ann. Math. Statist.* **39** (1968) 209—214.
Young, A.
(1954)   The application of approximate product integration to the numerical solution of integral equations. *Proc. Roy. Soc. London Ser. A* **224** (1954) 561—573.
Židkov, N. P.—Ščedrin, B. M.—Rambidi, N. G.—Egorova, N. M.
(1968)   Primenenie metoda reguljarizacii dlja rešenija nekotoryh zadač gazovoĭ èlektronografii. *Vyčislitel'nye metody i programmirovanie.* X. Izd. Moskov. Univ., Moskva, 1968; pp. 215—222.
Zolotarev, V. M.
(1954)   Vyraženie plotnosti ustoĭčivogo raspredelenija s pokazatelem α, bol'šim edinicy, čerez plotnost' s pokazatelem $\frac{1}{\alpha}$. *Dokl. Akad. Nauk SSSR* **98** (1954) 734—738.
(1956)   Ob analitičeskih svoĭstvah ustoĭčivyh zakonov raspredeleniĭ. *Vestnik Leningrad. Univ.* **11** (1956) No. 1, 49—52.
van Zwet, W. R.
(1964a) *Convex transformations of random variables.* Mathematical Centre Tracts 7. Mathematisch Centrum, Amsterdam, 1964.
(1964b) Convex transformations. A new approach to skewness and kurtosis. *Statistica Neerlandica* **18** (1964) 433—441.

CHRONOLOGICAL BIBLIOGRAPHY
TO THE PROBLEMS OF DECOMPOSITION

1922

Sen, N.: Über den Einfluss des Dopplereffekts auf spektroskopische Feinstrukturen und seine Elimination. *Physikalische Zeitschrift* 23 (1922) 397—399.

1928

Doetsch, G.: Die Elimination des Dopplereffekts bei spektroskopischen Feinstrukturen und exakte Bestimmung der Komponenten. *Z. Physik* 49 (1928) 705—730.

1930

Sticker, B.: Untersuchungen über Sternfarben. II. Analyse der Farbenhäufigkeitsfunktion. *Veröffentlichungen der Universitäts-Sternwarte Bonn* Nr. 23 (1930) 20—32.
Sticker, B.: Über die Farbenhäufigkeitsfunktion in Sternhaufen. *Z. Astrophys.* 1 (1930) 174—191.

1932

Schellenberg, O.: Zur Analyse der ultravioletten Emissionen der Erdalkaliphosphore. *Ann. Physik* 5/13 (1932) 249—264.

1936

Doetsch, G.: Zerlegung einer Funktion in Gauss'sche Fehlerkurven und zeitliche Zurückverfolgung eines Temperaturzustandes. *Math. Z.* 41 (1936) 283—318.

1938

Tricomi, F.: Les transformations de Fourier, Laplace, Gauss et leurs applications au calcul des probabilités et à la statistique. *Ann. Inst. H. Poincaré* 8 (1938) 111—149.

1947

Labhart, H.: Ein Auswertegerät für Elektrophoresediagramme. *Experientia* 3 (1947) 36—37
Szigeti, György: Lumineszkáló anyagok. (Luminescent materials.) *Elektrotechnika* (Budapest 39 (1947) 70—73, 81—86.
Wiedemann, E.: Über die Auswertung von Elektrophorese-Diagrammen nach L. G. Longsworth und Philpot—Svensson. *Helv. Chim. Acta* 30, Pars I (1947) 892—900.

1948

Kiss, A.—Sándorfy, C.: Sur les méthodes d'analyse des courbes d'absorption. *Acta Universitatis Szegediensis. Acta Chemica et Physica* **2** (1948) Fasc. 3, 71—76.

1949

Harding, J. F.: The use of probability paper for the graphical analysis of polymodal frequency distributions. *J. Mar. Biol. Assoc. U.K.* **28** (1949) 141—153.

1952

Wallner, A.—Ulke, R.: Auswertung von Diagrammen aus Elektrophoreseversuchen. *Hoppe-Seyler's Zeitschrift für physiologische Chemie* **290** (1952) Heft 3—6, 81—91.

1953

Medgyessy, Pál: Valószínűség-eloszlásfüggvények keverékének felbontása összetevőire. (Détermination des components d'un mélange des fonctions de distribution.) *Magyar Tud. Akad. Alkalm. Mat. Int. Közl.* **2** (1953) 165—177.

1954

Cassie, R. M.: Some uses of probability paper for the graphical analysis of polymodal frequency distributions. *Austral. J. Mar. Freshw. Res.* **5** (1954) 513—522.

Medgyessy, Pál: Diszkrét valószínűség-eloszlások keverékének felbontása összetevőire. (Decomposition of discrete compound probability distributions.) *Magyar Tud. Akad. Alkalm. Mat. Int. Közl.* **3** (1954) 139—153.

Medgyessy, Pál.: Újabb eredmények valószínűség-eloszlásfüggvények keverékének összetevőire bontásával kapcsolatban. (Some recent results concerning the decomposition of compound probability distributions.) *Magyar Tud. Akad. Alkalm. Mat. Int. Közl.* **3** (1954) 155—169.

Medgyessy, Pál: Valószínűség-eloszlásfüggvények keverékének felbontása összetevőire. (Decomposition into components of a superposition of distribution functions.) Dissertation. Budapest, 1954.

1955

Berencz, Ferenc: Megjegyzések az abszorpciós görbék analíziséhez. (Remarks to the analysis of the absorption curves.) *Magyar Fizikai Folyóirat* (Budapest) **3** (1955) 271—278.

Berencz, F.: Bemerkungen zur Analyse der Absorptionskurven. *Acta Phys. Acad. Sci. Hungar.* **4** (1955) 317—326.

Medgyessy, Pál: Közelítő eljárás Cauchy-sűrűségfüggvények keverékének összetevőkre bontására. (An approximate method for the decomposition of a compound of Cauchy frequency functions.) *Magyar Tud. Akad. Alkalm. Mat. Int. Közl.* **3** (1955) 321—329.

Medgyessy, Pál: Kiegészítés az "Újabb eredmények valószínűség-eloszlásfüggvények keverékének összetevőire bontásával kapcsolatban" c. dolgozathoz. (A supplement to the paper "Some recent results concerning the decomposition of compound probability distributions.") *Magyar Tud. Akad. Alkalm. Mat. Int. Közl.* **3** (1955) 331—341.

Poulik, M. D.—Pinteric, L.: An electronic computer for the evaluation of results of filterpaper electrophoresis. *Nature* **176** (1955) 1226—1227.

1956

Medgyessy, Pál.: Stabilis valószínűség-sűrűségfüggvényekre fennálló parciális differenciálegyenletek és alkalmazásaik. (Partial differential equations for stable density functions and their applications.) *Magyar Tud. Akad. Mat. Kutató Int. Közl.* **1** (1956) 489—518.

1957

Medgyessy, Pál.: Anwendungsmöglichkeiten der Analyse der Wahrscheinlichkeitsdichtefunktionen bei der Auswertung von Messungsergebnissen. *Z. Angew. Math. Mech.* **37** (1957) 128—139.

1958

Daeves, K.—Beckel, A.: *Grosszahl-Methodik und Häufigkeitsanalyse.* 2. Aufl. Verlag Chemie, Weinheim/Bergstr., 1958.

Medgyessy, P.: Partial integro-differential equations for stable density functions and their applications. *Publ. Math. Debrecen* **5** (1958) 288—293.

1959

Gardner, D. G.—Gardner, J. C.—Laush, G.—Meinke, W. W.: Method for the analysis of multicomponent exponential decay curves. *J. Chem. Phys.* **31** (1959) 978—986.

Idu, S. M.—Cucu, N. B.: Zur Auswertung von Elektrophorese-Diagrammen. *Hoppe-Seyler's Zeitschrift für physiologische Chemie* **314** (1959) 284—288.

Noble, W.—Hayes, J. E. Jr.—Eden, M.: Repetitive analog computer for analysis of sums of distribution functions. *Proc. IRE* **47** (1959) 1952—1956.

1960

Bellman, R.: On the separation of exponentials. *Boll. Un. Mat. Ital.* (3) **15** (1960) 38—39.

Defares, J. G.—Sneddon, I. N.: *An introduction to the mathematics of medicine and biology.* North-Holland Publishing Co., Amsterdam, 1960; pp. 582—586.

Lehotai, L.: Zerlegung der Absorptionskurven in Teilbanden mittels der Methode der Sreuungsverminderung. *Acta. Chim. Hungar.* **25** (1960) 25—32.

1961

Medgyessy, Pál: *Decomposition of superpositions of distribution functions.* Akadémiai Kiadó. Publishing House of the Hungarian Academy of Sciences, Budapest, 1961.

Medgyessy, P.: On some unsolved problems in the theory of the decomposition of superpositions of distribution functions. IIe *Congrès Mathématique Hongrois* Budapest, 24—31. August 1960. Vol. II. Akadémiai Kiadó, Budapest, 1961; Section IV, pp. 16—18.

Medgyessy, Pál: Diszkrét valószínűség-eloszlások áttranszformálása sűrűségfüggvénnyé. (Transforming discrete probability distributions into density functions.) (Lecture: Seminar of the Department of Calculus of Probability of the Mathematical Institute of the Hungarian Academy of Sciences, 9 February 1961.) Abstract: *Magyar Tud. Akad. Mat. Kutató Int. Közl.* **6** (1961) 528.

Morison Smith, D.—Bartlet, J. C.: Calculation of the areas of isolated or overlapping normal probability curves. *Nature* **191** (1961) 688—689.

1963

Brownell, G. L.—Callahan, A. B.: Transform methods for tracer data analysis. *Ann. New York Acad. Sci.* **108** (1963) Article 1: Multi-compartment analysis of tracer experiments; pp. 172—181.

Gardner, D. G.: Resolution of multi-component exponential decay curves using Fourier transforms. *Ann. New York Acad. Sci.* **108** (1963) Article 1: Multi-compartment analysis of tracer experiments; pp. 195—203.

Landahl, H. D.: Some mathematical aspects of multi-compartment analysis of tracer experiments. *Ann. New York Acad. Sci.* **108** (1963) Article 1: Multi-compartment analysis of tracer experiments; pp. 331—335.

Kaplan, B. G.—Gurvič, V. E.: Kompleksnoe primenenie matematičeskih metodov k èlektroforetičeskomu issledovaniju belkovogo sostava krovi v norme i patologii. *Primenenie matematičeskih metodov v biologii.* II. Izd. Leningr. Univ., Leningrad, 1963; pp. 183—190.

Sydow, A.—Dittmann, H.: Statistische Analysen mittels elektronischer Analogrechner. *Messen, Steuern, Regeln* 6 (1963) 501—504.

1966

Medgyessy, Pál: Egy konvolúciós típusú integrálegyenlet numerikus megoldása és ennek felhasználása Gauss-függvény szuperpozíciók felbontására. (Numerical solution of an integral equation of convolution type and its application to the decomposition of superpositions of Gauss' functions.) *Magyar Tud. Akad. Mat. Fiz. Oszt. Közl.* **16** (1966) 47—64.

Medgyessy, P.: Remarks on the paper "On the separation of exponentials" of *R. Bellman.* Manuscript. Budapest, 1966.

Medgyessy, Pál: Valószínűség-eloszlások általánosítása és ezzel kapcsolatos faktorizációs problémák. (A generalization of discrete distributions and problems of factorization connected with it.) (Lecture: Colloquium on Mathematical Statistics, Debrecen (Hungary), 13—15. October 1966. Sponsored by the Bolyai János Mathematical Society and the Mathematical Institute of the Kossuth Lajos University, Debrecen.) Abstract: *Matematikai Statisztikai Kollokvium. Debrecen,* 1966. *október* 13—15. *Előadás kivonatok. Debrecen,* 1966; 3—4.

Thionet, P.: Note sur les mélanges de certaines distributions de probabilités. *Publ. Inst. Statist. Univ. Paris* **15** (1966) 61—80.

Varga, László: Gauss-függvények keverékének komponensekre bontásáról. (Decomposition of superimposed Gaussian distribution functions.) *Magyar Tud. Akad. Közp. Fiz. Kutató Int. Közl.* **14** (1966) 383—389.

1967

Bhattacharya, C. G.: A simple method of resolution of a distribution into Gaussian components. *Biometrics* **23** (1967) 115—135.

Medgyessy, Pál: Sűrűségfüggvény szuperpozíciók felbontásának egy lényegileg új módszeréről. (On an essentially new method of decomposing superpositions of density functions.) *Magyar. Tud. Akad. Mat. Fiz. Oszt. Közl.* **17** (1967) 383—390. (English translation: An essentially new method for the decomposition of density functions. *Selected translations in mathematical statistics and probability*. Volume 10. American Mathematical Society, Providence, 1972; pp. 170—178.)

Medgyessy, Pál: Szilárd anyagok keverék-voltának megállapítása matematikai módszerekkel. (Determination by the aid of mathematical methods of whether a certain material is a mixture or not.) *Tíz példa a matematika gyakorlati alkalmazására*. (Szerkesztette Vincze István.) (Ten examples of the practical application of mathematics.) (Edited by István Vincze.) Gondolat, Budapest, 1967; pp. 203—223.

Szőke, J.—Varga, L.—Nagypál, I.: Experimental and computer analysis of the spectral fine structure. *Proceedings of the XIV. Colloquium Spectroscopicum Internationale*. Debrecen, Hungary, 1967; pp. 1205—1218.

1968

Medgyessy, Pál—Varga, László: Gauss-függvény keverékek numerikus felbontására szolgáló egyik eljárás javításáról. (On the improving of a method for the numerical decomposition of mixtures of Gaussian functions.) *Magyar Tud. Akad. Mat. Fiz. Oszt. Közl.* **18** (1968) 31—39.

Parsons, D. H.: Biological problem involving sums of exponential functions of time: A mathematical analysis that reduces experimental time. *Math. Biosci.* **2** (1968) 123—128.

Tricomi, F. G.: *Repertorium der Theorie der Differentialgleichungen*. Springer, Berlin, 1968; pp. 130—132.

Varga, László: Nemlineáris becslési feladatok numerikus módszerei. (Numerical methods for nonlinear problems of estimation.) Dissertation. Budapest, 1968.

Židkov, N.P.— Ščedrin, B .M.—Rambidi, N. G.—Egorova, N. M.: Primenenie metoda reguljarizacii dlja rešenija nekotoryh zadač gazovoĭ èlektronografii. *Vyčislitel'nye metody i programmirovanie*, X. Izd. Moskov. Univ., Moskva, 1968; pp. 215—222.

1969

Gregor, J.: An algorithm for the decomposition of a distribution into Gaussian components. *Biometrics* **22** (1969) 79—93.

1970

Aleksandrov, L.: Reguljarizovannyĭ vyčislitel'nyĭ process dlja analiza eksponencial'nyh zavisi-mosteĭ. *Ž. Vyčisl. Mat. i. Mat. Fiz.* **10** (1970) 1285—1287.

Dobozy, O. K.—Volly, T.: The anomalous properties of cystine and their effect on keratin. (Lecture: Fourth International Wool Textile Research Conference, Berkeley, California, 18—27 August 1970.) Manuscript. Budapest, 1970.

Parsons, D. H.: Biological problems involving sums of exponential functions of time: An improved method of calculation. *Math. Biosci.* **9** (1970) 37—47.

1971

Medgyessy, Pál: Sűrűségfüggvények és diszkrét eloszlások szuperpozícióinak felbontása. (Decomposition of superpositions of density functions and discrete distributions.) Dissertation. Budapest, 1971.

1972

Medgyessy, Pál: Sűrűségfüggvények és diszkrét eloszlások szuperpozícióinak felbontása. (Decomposition of superpositions of density functions and discrete distributions.) *Magyar Tud. Akad. Mat. Fiz. Oszt. Közl.* **21** (1972) 129—200.

Medgyessy, Pál: Sűrűségfüggvények és diszkrét eloszlások szuperpozícióinak fe[bontása. II. (Decomposition of superpositions of density functions and discrete distributions. II.) *Magyar Tud. Akad. Mat. Fiz. Oszt. Közl.* **21** (1972) 261—382.

1973

Medgyessy, Pál: Új módszerek szimmetrikus sűrűségfüggvények szuperpozícióinak felbontására. I. (New methods for the decomposition of superpositions of symmetrical density functions. I.) *Magyar Tud. Akad. Mat. Fiz. Oszt. Közl.* **22** (1973); to appear.

Papoulis, A.: A different approach to the analysis of tracer data. *SIAM J. Control* **11** (1973) 466—474.

1974

Blomer, R. J.: Decomposition of compartment transition functions by means of the Pade approximation. *Math. Biosci.* **19** (1974) 163—173.

Medgyessy, Pál: Új módszerek szimmetrikus sűrűségfüggvények szuperpozícióinak felbontására. II. (New methods for the decomposition of superpositions of symmetrical density functions. II.) Manuscript. Budapest, 1974.

AUTHOR INDEX

SUBJECT INDEX

INDEX OF UNSOLVED PROBLEMS

INDEX OF NOTATIONS